# Concrete: Material Science to Application

## *A Tribute to Surendra P. Shah*

*Editors*
P. Balaguru
A. Naaman
W. Weiss

**international** ®

SP-206

DISCUSSION of individual papers in this symposium may be submitted in accordance with general requirements of the ACI Publication Policy to ACI headquarters at the address given below. Closing date for submission of discussion is October 2002. All discussion approved by the Technical Activities Committee along with closing remarks by the authors will be published in the January/February 2003 issue of either ACI Structural Journal or ACI Materials Journal depending on the subject emphasis of the individual paper.

The Institute is not responsible for the statements or opinions expressed in its publications. Institute publications are not able to, nor intended to, supplant individual training, responsibility, or judgment of the user, or the supplier, of the information presented.

The papers in this volume have been reviewed under Institute publication procedures by individuals expert in the subject areas of the papers.

Printed in the United States of America

Editorial production: Bonnie L. Gold

Library of Congress catalog card number: 2002103339
ISBN: 0-87031-075-5

# PREFACE

Surendra (Suru) Poonamchand Shah, the son of a Gujarati businessman, grew up in Mumbai (Bombay), India and attended Modern School and Elphinstone College where he became President of the Literary Society. While in Engineering College, he played the leading role in a drama that traveled to New Delhi and was performed in front of the then Prime Minister, Jawaharlal Nehru. For Suru, this tour of India was a memorable introduction to the delights of travel, an aspect of his academic career that he genuinely enjoys. Among Suru's other passions are exploring cuisine and wines in restaurants worldwide, investigating archaeological and architectural wonders of the world, analyzing theatre, opera, and film productions, and entertaining visitors to Chicago and Evanston, as well as working with students, post-docs, and colleagues.

After earning a B.E. at Sardar Vallabhbhai Vidyaputh Gujarat University, Suru came to the United States for graduate study at Lehigh University where he completed a Master of Science degree. For two years, he worked as a design engineer for Modjeski and Masters, a bridge design firm in Harrisburg, Pennsylvania, where he met Dorothie Crispell, whom he married before he enrolled in the Ph.D. program in civil engineering at Cornell University. There, Professor George Winter, Professor Richard White, and Professor Floyd Slate were his research mentors.

In 1965, Suru joined the faculty of the Materials Engineering Department at the University of Illinois in Chicago. In addition to teaching a variety of civil engineering classes, he developed a state-of-the-art research laboratory and built a graduate program. He was rapidly promoted to associate, then full, professor. In 1981, he joined the faculty at Northwestern University where he holds the Walter P. Murphy Chair in Civil Engineering.

Suru was key in establishing the pioneering National Science Foundation Science and Technology Center for Advanced Cement Based Materials and has served as director since its inception in 1989. The ACBM Center is a multi-disciplinary consortium comprised of five institutions: Northwestern University, University of Illinois at Urbana-Champaign, University of Michigan at Ann Arbor, Purdue University, and the National Institute of Science and Technology. The goal of the center is to improve knowledge of cement and concrete so that stronger, tougher, and more-durable materials are economically produced for infrastructure facilities and other structures. With ACBM Center colleagues, post-docs, and graduate students, Suru continues his tireless pursuit of fundamental understanding of cement-based materials and related fiber-reinforced cement composites. Currently, a consortium of world-class industrial partners provide strong support for the ACBM Center and assist the center in promoting improved teaching of the materials science of concrete.

Suru's accomplishments are too numerous to list here; suffice it to say that anyone involved in the materials science of concrete, fracture mechanics, high-performance concrete, or fiber-reinforced concrete has encountered and utilized his contributions. He has co-authored two textbooks, edited over a dozen books, and published more than 400 papers. Suru has received many major professional and technical awards, including the ACI Arthur R. Anderson Award, the RILEM Gold Medal, the Swedish Concrete Award, the Alexander von Humboldt Award, Engineering Newsletter Award, ASCE–CERF Charles Pankow Award, and ASTM Thompson Award. He has been the principal advisor of almost 100 graduate students and supervised the work of over 60 post-docs and visiting scholars. Contributors to this symposium in his honor include many of his former students. A list of former students and post-docs appear in the Appendix.

Suru and Dorothie Shah have two married sons, Byron and Daniel, and a grandson, Kian, born in 1999.

It is with great respect, admiration, and affection that members of the organizing committee of this symposium dedicate this volume in tribute to their colleague, mentor, and friend, Surendra P. Shah.

**The Organizing Committee**
April 2002

P. Balaguru
*Rutgers University*

D. Lange
*University of Illinois at Urbana-Champaign*

N. Banthia
*University of British Columbia*

A. Naaman
*University of Michigan*

J. Biernacki
*Tennessee Technical University*

W. Weiss
*Purdue University*

V. Gopalaratanam
*University of Missouri*

# TABLE OF CONTENTS

## LAMINATED AND FIBER REINFORCED CEMENT COMPOSITES

THE FUTURE OF RESEARCH AND EDUCATION IN CONCRETE

# ADVANCES IN FATIGUE
# AND
# FRACTURE

# Failure Mechanism of Reinforced Concrete Under Cyclic Loading

## by T. C. Hsu and M. Y. Mansour

**Synopsis:** The load-deformation response of R/C membrane elements (panels) subjected to reversed cyclic shear shows that the orientation of the steel bars with respect to the principal coordinate of the applied stresses has a strong effect on the "pinched shape" of the post-yield hysteretic loops. When the steel bars in a panel are oriented in the coordinate of the applied principal stresses, there is no "pinching effect," and the panel exhibits ductile behavior and high capacity of energy dissipation. Whereas, when the steel bars are oriented at an angle of 45° to the applied principal stresses, severe pinching effect is observed and the panel becomes more brittle.

This paper presents concisely a rational theory, called the *Cyclic Softened Membrane Model* (CSMM). This new rational theory is capable of predicting the entire history of the hysteretic loops (pre- and post-yield); can explain the mechanism behind the "pinching effect"; and can elucidate the failure mechanism that causes the deterioration of reinforced concrete structures under cyclic loading.

**Keywords:** cyclic loading; mechanism; pinching; reinforced concrete; shear; strain; steel; stress; softened truss models

1

ACI Fellow Thomas T. C. Hsu is Moores Professor in the Dept. of Civil and Envir. Engineering at the University of Houston. He was the recipient of ACI's Anderson Award for research in 1990, and Wason Medal for material research in 1965. He is a member of ACI committee 215, Fatigue of Concrete; and joint ACI-ASCE committee 343, concrete bridge design; and 445, shear and torsion.

M. Y. Mansour is Assistant Professor in the Dept. of Civil & Construction Engineering at Bradley University, Peoria, IL. He received his B.E. and M.E. from the American University of Beirut, Lebanon, and his Ph. D. from University of Houston in 2001. His research interests include design of concrete structures, analytical modeling, and experimental testing.

## INTRODUCTION

Under earthquake condition, the behavior of wall-type reinforced concrete structures (such as shear walls) can each be visualized as assemblies of membrane elements subjected to in-plane cyclic stresses. The key to rational analysis of these structures is to study and thoroughly understand the behavior of isolated reinforced concrete elements, separated from the complications of whole-structure behavior. By studying panel elements' behavior under cyclic loading, rational analytical models are developed to predict their cyclic behavior. Then by incorporating these models into a finite element program, one can predict the behavior of whole structures under earthquake loading.

Structures located in high earthquake regions are designed to withstand not only moderate seismic loading within the elastic range, but must also be able to absorb the energy of high seismic loading in the inelastic range. Thus, it is necessary to evaluate the inelastic response of structures under high earthquake loading. When the shear force governs the response, as in the case of low-rise shear walls, the effect of shear on the panel response was thought to be responsible for the "pinching effect" in the hysteretic loops. The "pinching effect" which results in the degradation of stiffness and the reduction of energy dissipation capacity, was also surmised to be caused by the bond slips between concrete and steel bars.

Since 1997 a total of 15 panels had been subjected to reversed cyclic loading at the University of Houston using the Universal Panel Tester (Hsu, Belarbi and Pang, 1995), (1), equipped with a servo-control system that was capable of conducting strain-control tests (Hsu, Zhang and Gomez, 1995). Using this strain-control feature, the panels were loaded beyond the yield point up to 10 times the yield strain. It was found that the primary factor affecting the pinched behavior of hysteretic loops was the orientation of the steel bars with respect to the applied principal stresses. When the steel bars were oriented in the principal coordinate of the applied stresses (i.e. at a fixed angle $\alpha_2$ of 90°), there was no pinching effect. When the steel bars orientation was $\alpha_2 = 45°$ or 68.2° to the principal coordinate, there was severe pinching effect.

Based on the testing of panels during the past 13 years, a series of theories were developed at UH to predict the monotonic behavior of cracked RC membrane elements. These theories include the rotating-angle softened truss models (RA-STM) (Hsu, 1993; Belarbi and Hsu, 1994, 1995; Pang and Hsu, 1995), (3,4,5,6), the fixed-angle softened truss model (FA-STM) (Pang and Hsu, 1996; Hsu and Zhang, 1997, Zhang and Hsu, 1998), (7,8,9), and the softened membrane model (SMM) (Zhu, 2000; Hsu & Zhu, 2001; Zhu and Hsu, 2002; Hsu and Zhu, 2002), (10,11,12,13). The SMM is an extension of the FA-STM with two improvements: First, SMM can predict the entire load-deformation history of panel behavior, including the post-peak descending branches, because the Hsu/Zhu ratios (or Poisson effect) are taken into account. Second, the complicated and empirical shear modulus of concrete in FA-STM is replaced by a simple and rational shear modulus in SMM. As a result, the solution algorithm of SMM is considerably simpler and the prediction more accurate than those of FA-STM.

In order to predict the behavior of membrane elements under cyclic loading, SMM for monotonic loading was extended for application to cyclic loading by adding the constitutive models of materials (concrete and embedded steel bars) in the unloading and reloading regions (Mansour, 2001; Mansour, Lee and Hsu, 2001; Mansour, Hsu and Lee, 2001), (14,15,16). This paper presents a new model, called the "Cyclic Softened Membrane Model (CSMM)," that is capable of predicting the entire cyclic history of load-deformation relationship, including the post-yield hysteretic loops and the pinching effect.

Applying the CSMM to predict the behavior of two panels CA3 ($\alpha_2 = 45°$) and CE3 ($\alpha_2 = 90°$), this paper discusses rationally the presence and absence of the pinching effect in the hysteretic loops. The comparison of the behavior of these two panels not only reveals the mechanism inherent in the pinching phenomenon, it also elucidates the failure mechanism of reinforced concrete composites under cyclic loading.

## TESTS OF R/C SHEAR PANELS CA3 AND CE3

The two test panels, CA3 and CE3, have a size of 1398 mm x 1398 mm x 178 mm, and the steel bars are placed at angles $\alpha_2$ of 45 and 90 degrees, respectively, as shown in Fig. 1, to form orthogonal steel grids. The reinforcing ratios of panels CA3 and CE3 are 1.7% and 1.2%, respectively, in each direction. The material properties for these two panels are summarized in Table 1.

The panels were subjected to reversed cyclic stresses in the horizontal and vertical directions using the Universal Panel Tester. When these two principal applied stresses were equal in magnitude and opposite in direction, a state of pure shear stress $\tau_{45°}$ was created at the $45°$ direction to the applied principal stresses. The testing facility was equipped with a servo-control system capable of switching from load-control mode to strain-control mode as the yielding load was approached. In the strain-control mode, the shear strain (i.e. the algebraic sum of the horizontal and vertical strains), which followed a specified strain history, was used as an input signal to control the horizontal principal stress. The horizontal principal stress was, in turn, used to control the vertical principal stress such that they were always equal in magnitude and opposite in direction.

The hysteretic loops of the two panels, CA3 and CE3, are shown in Fig. 2 (a) and (b), respectively. In these figures, the vertical and horizontal axes represent the shear stress $\tau_{45°}$ and the shear strain $\gamma_{45°}$ at 45° degrees to the principal coordinate of applied stresses. In Fig. 2 (a), the hysteretic loops of panel CA3 displayed a highly pinched shape that are generally associated with shear dominated behavior. The envelope curve of this panel also exhibited a distinct descending branch indicating a severe strength degradation of the panel with increasing shear strain magnitude. In contrast, no pinching effect was observed in Fig. 2 (b) for the hysteretic loops of panel CE3, with its steel grid parallel to the applied principal stresses. The envelope curve of panel CE3 did not have a descending branch and the strength deterioration was not noticeable.

## CYCLIC SOFTENED MEMBRANE MODEL (CSMM)

In the basic concept of CSMM, the cracks are smeared throughout the reinforced concrete elements, and the reinforcing bars are uniformly distributed in two orthogonal directions ($\ell$ and t). The behavior of the panels is, therefore, formulated in terms of smeared (average) stresses and smeared (average) strains, and the continuum mechanics can be applied. The equilibrium equations, compatibility equations, and the material constitutive models are summarized in this paper:

Equilibrium Equations:

$$\sigma_\ell = \sigma_V^c \cos^2 \alpha_2 + \sigma_H^c \sin^2 \alpha_2 + \tau_{VH}^c 2\sin\alpha_2 \cos\alpha_2 + \rho_\ell f_\ell \tag{1}$$

$$\sigma_t = \sigma_V^c \sin^2 \alpha_2 + \sigma_H^c \cos^2 \alpha_2 - \tau_{VH}^c 2\sin\alpha_2 \cos\alpha_2 + \rho_t f_t \tag{2}$$

$$\tau_{\ell t} = (-\sigma_V^c + \sigma_H^c)\sin\alpha_2 \cos\alpha_2 + \tau_{VH}^c (\cos^2 \alpha_2 - \sin^2 \alpha_2) \tag{3}$$

Compatibility Equations:

$$\varepsilon_\ell = \varepsilon_V \cos^2 \alpha_2 + \varepsilon_H \sin^2 \alpha_2 + \frac{\gamma_{VH}}{2} 2\sin\alpha_2 \cos\alpha_2 \tag{4}$$

$$\varepsilon_t = \varepsilon_V \sin^2 \alpha_2 + \varepsilon_H \cos^2 \alpha_2 - \frac{\gamma_{VH}}{2} 2\sin\alpha_2 \cos\alpha_2 \tag{5}$$

$$\frac{\gamma_{tt}}{2} = (-\varepsilon_V + \varepsilon_H)\sin\alpha_2\cos\alpha_2 + \frac{\gamma_{VH}}{2}(\cos^2\alpha_2 - \sin^2\alpha_2) \tag{6}$$

where the symbols are given in the list of Notations.

The constitutive laws of steel bars and concrete are summarized in the following sections.

## Constitutive Relationships of Steel Bars

The proposed cyclic stress-strain relationship of steel bar embedded in concrete is shown by the solid curves in Fig. 3. This solid curves can be divided into two groups: the backbone envelope curves and the unloading and reloading curves. The figure also gives the monotonic stress-strain curves of bare steel bars as shown by the dotted lines.

**Backbone Envelope Curves (Stage 1, Stage 2T and Stage 2C)** - The monotonic tensile stress-strain curve of embedded steel bars proposed by Belarbi and Hsu (8) can be used to approximate the backbone envelope curve of the cyclic tensile stress-strain curves of steel bars. This monotonic, bilinear stress-strain relationship, which was adopted for Stage 1 and Stage 2T, is expressed as:

(Stage 1)  $\quad f_s = E_s\varepsilon_s$  $\hfill (\varepsilon_s \leq \varepsilon_n) \qquad (7a)$

(Stage 2T)  $\quad f_s = f_y\left[(0.91 - 2B) + (0.02 + 0.25B\frac{\varepsilon_s}{\varepsilon_y})\right] \qquad (\varepsilon_s > \varepsilon_n) \qquad (7b)$

where $f_s$ and $\varepsilon_s$ are the average stress and strain of mild steel bars, respectively; $f_y$ and $\varepsilon_y$ are the yield stress and strain of bare mild steel bars, respectively; $E_s$ is the modulus of elasticity of steel bars; and $\varepsilon_n = \varepsilon_y(0.93 - 2B)$. The parameter B is given by $B = (f_{cr}/f_y)^{1.5}/\rho$, where $\rho$ is the reinforcement steel ratio and $\geq 0.5\%$. $f_{cr}$ is the cracking strength of concrete given by $f_{cr} = 0.31\sqrt{f'_c}(MPa)$.

If the steel stress progresses in the compression region, the average maximum stress $f_s$ is limited to the compressive yield stress $-f_y$, Eq. (7c), as indicated in Stage 2C.

(Stage 2C)  $\hspace{4em} f_s = -f_y$  $\hfill (f_s \leq -f_y) \qquad (7c)$

**Unloading and Reloading Curves (Stage 3 and Stage 4)** - The cyclic stress-strain relationship for steel bars subjected to reversed cyclic loading must include the unloading and the reloading branches. In this analysis, the unloading and reloading stress-strain relationships are expressed by the Ramberg-Osgood type of equations:

$$\text{(Stage 3 and Sage 4)} \quad \varepsilon_s - \varepsilon_i = \frac{f_s - f_i}{E_s}\left[1 + A^{-R}\left|\frac{f_s - f_i}{f_y}\right|^{R-1}\right] \quad \text{(7d)}$$

where $f_s$ and $\varepsilon_s$ are the smeared stress and smeared strain of an embedded steel bar; $f_i$ and $\varepsilon_i$ are the smeared stress and smeared strain of steel bars at the load reversal point.

The coefficients A and R in Eq. (7d) were determined from the reversed cyclic loading tests at the University of Houston (Mansour, Lee and Hsu, 2000), (15), to best fit the test results: $A = 1.9k_p^{-0.1}$, $R = 10k_p^{-0.2}$. The parameter in the coefficients A and R is the plastic strain ratio $k_p$ which is defined as the ratio $\varepsilon_p/\varepsilon_n = (\varepsilon_i - \varepsilon_n)/\varepsilon_n$. In this expression $\varepsilon_p$ is the plastic strain, and $\varepsilon_n$ is the initial yield strain.

## Constitutive Relationships of Concrete

The cyclic stress-strain curves of concrete in the CSMM are shown in Fig. 4. The curves are divided into three groups: the compressive backbone envelope curves, the tensile backbone envelope curves, and the unloading and reloading curves. In addition, a rational shear modulus of concrete is adopted in CSMM. This shear modulus is simply a function of the stress-strain relationships of concrete in compression and in tension.

**Compressive Backbone Envelope Curves (C1 and C2)** - As pointed out by Mansour, Lee and Hsu (2001), (15), the backbone envelope curves for the cyclic compressive stress-strain curves of concrete can be expressed by a small modification of the monotonic compressive stress-strain curve of concrete proposed by Belarbi and Hsu (1995), (5):

$$\text{(Stage C1)} \quad \sigma_c = D(\zeta f_c' - f_{cT4}')\left[2\left(\frac{\varepsilon_c}{\zeta\varepsilon_o}\right) - \left(\frac{\varepsilon_c}{\zeta\varepsilon_o}\right)^2\right] + f_{cT4}' \quad \varepsilon_o \le \varepsilon_c < 0 \quad \text{(8a)}$$

$$\text{(Stage C2)} \quad \sigma_c = D\zeta f_c'\left[1 - \left(\frac{\varepsilon_c/\varepsilon_o - 1}{4/\zeta - 1}\right)^2\right] \quad \varepsilon_c < \varepsilon_o \quad \text{(8b)}$$

where $\sigma_c$ is the smeared stress of concrete; $\varepsilon_c$ is the smeared strain of concrete; $f_c'$ and $\varepsilon_o$ are the maximum concrete compressive cylinder strength and the peak cylinder compressive strain, respectively ; $f_{cT4}'$ is the stress at point TD (Fig. 4) between Stage C1 and Stage T4.

The softening coefficient $\zeta$ in Eqs. (8a) and (8b) was given by Zhang and Hsu (1998), (9):

$$\zeta = \frac{5.8}{\sqrt{f_c'(MPa)}} \frac{1}{\sqrt{1+k\varepsilon_1}} \leq 0.9 \tag{8c}$$

where $\varepsilon_1$ is the tensile strains, either $\varepsilon_H$ or $\varepsilon_V$ in cyclic loading; $k$ is the loading coefficient taken as $400/\eta$ for proportional loading, where $\eta=1$ for the case of pure shear and equal volume of steel in the $\ell$ and t directions.

The damage coefficient, D, takes into account the effect of concrete damage in one direction on the concrete strength in the perpendicular direction:

$$D = 1 - 0.4\frac{\varepsilon_{mc}}{\varepsilon_o} \tag{8d}$$

where $\varepsilon_{mc}$ is the maximum compression strain occurring in the immediate opposite direction and perpendicular to the compression strain being considered. $\varepsilon_o$ is the strain corresponding to the peak cylinder strength. The damage coefficient, which is not considered in Mansour, Hsu and Lee (2001), (16), and Mansour, Lee and Hsu (2001), (15), does take care of the descending branches of the backbone envelope curves, as shown in Fig. 2 (a) for panel CA3.

**Tensile Backbone Envelope Curves (T1 and T2)** - The envelope curves for the cyclic tensile stress-strain curves of concrete can be expressed by the monotonic tensile stress-strain curve of concrete proposed by Belarbi and Hsu (1994), (4):

(Stage T1)             $\sigma_c = E_c\varepsilon_c$                    $0 \leq \varepsilon_c \leq \varepsilon_{cr}$             (8e)

(Stage T2)             $\sigma_c = f_{cr}\left(\dfrac{\varepsilon_{cr}}{\varepsilon_c}\right)^{0.4}$                    $\varepsilon_c > \varepsilon_{cr}$             (8f) where $E_c$ is the modulus of elasticity of concrete taken as $3875\sqrt{f_c'(MPa)}$; $f_{cr}$ is the cracking stress of concrete taken as $0.31\sqrt{f_c'(MPa)}$; and $\varepsilon_{cr}$ is the cracking strain of concrete taken as $0.00008$.

**Unloading and Reloading Curves** - A linear expression is proposed for the unloading and reloading curves as follows:

$$\sigma_c = \sigma_{ci} + E_{cc}(\varepsilon_{ci} - \varepsilon_c)$$  (8g)

where $\sigma_{ci}$ and $\varepsilon_{ci}$ are concrete stress and strain at the load reversal point "i" or at the point where the stages change; $E_{cc}$ is the slope of the linear expression and is taken to be:

$$E_{cc} = \frac{\sigma_{ci} - \sigma_{ci+1}}{\varepsilon_{ci} - \varepsilon_{ci+1}}$$  (8h)

where $\sigma_{ci+1}$ and $\varepsilon_{ci+1}$ are the concrete stress and strain at the end of the stage under consideration. The points for $\varepsilon_{ci}$, $\sigma_{ci}$, $\varepsilon_{ci+1}$ and $\sigma_{ci+1}$ are specified in Fig. 4. Notice that the concrete stress of point TD is $1.5f_{cr} + 0.8f_{cT2}'$, rather than $f_{cr} + 0.8f_{cT2}'$ given in Mansour, Hsu and Lee (2001), (16) and Mansour, Lee and Hsu (2001), (15). This slight modification is taken to improve the agreement of crack closing strains between the CSMM predictions and the experimental results for all the test panels.

**Constitutive Relationship of Concrete in Shear** - Zhu, Hsu and Lee (2001), (17), showed that a rational relationship exists between the shear stress and the shear strain of concrete. This rational relationship in the (H,V) system of axes is given by the following expression:

$$\tau_{VH}^c = \frac{\sigma_H^c - \sigma_V^c}{2(\varepsilon_H - \varepsilon_V)}\gamma_{VH}$$  (9)

**Poisson Effect** – the Poisson effect of cracked reinforced concrete subjected to monotonic loading is characterized by two Hsu/Zhu ratios based on the smeared crack concept (Zhu, 2000; Hsu and Zhu, 2001), (10,11). The Hsu/Zhu ratio $\upsilon_{CT}$ (compression strain caused by tensile strain) was found to be zero for the entire post-cracking range. The Hsu/Zhu ratio $\upsilon_{TC}$ (tensile strain caused by compression strain), however, was found to be a function of the maximum steel strain and varied from 0.2 to 1.9. Before the Hsu/Zhu ratios of cracked reinforced concrete subjected to cyclic loading can be determined, an average value of $v_{TC} = 1.0$ is assumed in this study.

## COMPARISON OF CSMM PREDICTIONS WITH EXPERIMENTS

### Pinching Effect

The CSMM-predicted hysteretic loops of the two panels CA3 and CD3 are plotted in Fig. 2 (a) and (b), together with the experimental curves. It can be seen that the CSMM is capable of predicting the pinched shape of the hysteretic loops of panels CA3 as well as the fully-rounded hysteretic loops of panel CE3.

**Panel CA3 ($\alpha_2 = 45°$)** - The first cycle of the hysteretic loops beyond yielding for panel CA3 is plotted in Fig. 5(a). Four points A, B, C, and D are chosen in Fig. 5(a) to illustrate the *presence* of the pinched shape. Point A is at the maximum shear strain of the first cycle beyond yielding. Point B is at the stage where the shear stress is zero after unloading. Point C with a very low shear stress is taken in the negative shear strain region at the end of the low-stress pinching zone just before the sudden increase in stiffness. Point D is at the maximum negative shear strain of the first negative cycle. The three segments of curves from point A to point D in Fig. 5(a) clearly define the pinched shape of the hysteretic loops.

**Panel CE3 ($\alpha_2 = 90°$)** - The first cycle of the hysteretic loops beyond yielding for panel CE3 is plotted in Fig. 5(b). Four points A, B, C, and D are chosen in Fig. 5(b) to illustrate the *absence* of the pinched shape. Points A, B and D correspond to the same three points in Fig. 5(a). However, point C with a high shear stress is taken in the negative shear strain region when the compression steel reaches yielding. The three segments of curves from point A to point D in Fig. 5(b) clearly show the absence of pinching.

### Stresses and Strains Represented by Mohr Circles

The smeared strains, the applied stresses, the smeared concrete stresses, and the smeared steel stresses at the four points (A, B, C and D) chosen in Fig. 5 (a) and (b) are represented by Mohr circles as shown in Figs. 6 and 7 for panels CA3 and CE3, respectively. Using the Mohr circles has two advantages: First, Mohr circles represent the entire stress or strain state in an element, i.e. stresses or strains in all directions. Second, Mohr circles are the best means to help illustrate the mechanism behind the pinching effect and the failure mechanism of reinforced concrete elements under cyclic loading. To better visualize what is happening in the two panels, CA3 and CE3, subjected to reversed cyclic loading, we will examine the stresses and the strains as the applied shear stresses change from point A to point D.

**Panel CA3 ($\alpha_2 = 45°$)** - At point A in the first post-yield cycle, Fig. 6, the maximum applied shear stress, $\tau_{45°}$, of 6.84 MPa produces a large shear strain, $\gamma_{45°}$, of 0.00828 (twice the number shown in the Mohr circle because the vertical axis represents $\gamma/2$). To resist this applied shear stress, the concrete is subjected to a maximum vertical compressive stress, $\sigma_V^C$, of 13.64 MPa, and a maximum smeared steel stresses, $\rho f_s$, of 6.80 MPa in both the longitudinal and transverse directions.

When the panel is unloaded from point A to point B, the applied shear stress of 6.84 MPa is reduced to virtually zero. From equilibrium, the compressive stress in the concrete and the tensile stress in the steel are also approaching zero. Correspondingly, the shear strain is reduced from 0.00828 at point A to a value of 0.00288 at point B. This unloading process produces an almost linear shear stiffness due to the normal relaxation of steel and concrete. These nearly proportional reductions of stresses in the concrete and steel are also related to the closing of cracks in the vertical direction. The horizontal strain, $\varepsilon_H$, decreases from 0.00762 at point A to 0.00278 at point B, but remains in tension with significant size of crack width. The compressive strain in the vertical direction, $\varepsilon_V$, decrease from –0.00067 at point A to -0.00011 at point B.

When the positive shear strain of 0.00288 at point B is reversed to become a negative shear strain of –0.00268 at point C, the vertical strain increases to a tensile strain of 0.00273, while the horizontal strain further decreases to a small tensile value of 0.000039 (not in compression). This large change of shear strain through the origin, however, is not accompanied by a corresponding change in the applied shear stress, meaning that the shear stiffness in the BC region is very small. This is because at point C the vertical cracks are not fully closed, and the concrete compressive stress cannot be developed. Without forming an effective set of concrete compressive struts, the stresses in the steel bars also cannot be developed. Hence, both the concrete stress and steel stress remain small at point C.

When the negative shear strain reaches –0.00862 at point D, the vertical cracks are fully closed ($\varepsilon_H = -0.00067$) and the concrete compressive struts are fully formed. The concrete struts can now resist a compressive stress of 13.50 MPa, and the smeared steel in both the longitudinal and transverse directions are resisting a stress of 6.59 MPa. Correspondingly, the element is resisting an applied shear stress of 6.91 MPa. In other words, in the reloading CD range the shear stiffness is restored to its normal magnitude.

The pinched shape of the hysteretic loop of panel CA3, Fig. 2 (a), is formed by a small shear stiffness in the BC region, sandwiched between two large shear stiffnesses in the AB and the CD regions.

smeared strain, however, continues to increase. At point D the shear strain is − 0.00386, and the vertical strain reaches a value of 0.00400.

## Physical Visualization

Panel CA3 ($\alpha_2 = 45°$) - The presence of the pinching mechanism in panel CA3 can be explained intuitively by examining a cracked element with $45°$ steel bars as shown in Fig. 8(a). A state of pure shear in the $45°$ direction of this element is equivalent to applying a horizontal compressive stress $\sigma_H$ and a vertical tensile stress $\sigma_V$ of equal magnitude. In the reverse loading stage from point B to point C (Fig. 5(a)), which defines the region where pinching occurs, both the vertical and the horizontal cracks are open. The concrete struts have not yet been formed, and the applied stresses $\sigma_H$ and $\sigma_V$ must be resisted by the two 45° steel bars. Separate the effect of $\sigma_H$ and $\sigma_V$ as shown in Fig. 8(b) and Fig. 8(c), respectively. The horizontal compressive stress $\sigma_H$ induces a compressive stress in the two $45°$ steel bars, Fig. 8(b), while the vertical tensile stress $\sigma_V$ induces a tensile stress of equal magnitude in the same two $45°$ bars, Fig. 8(c). These two stresses in the two $45°$ steel bars cancel out each other. As a result, the element offers no shear resistance to the applied shear stress $\tau_{45°}$ in the 45° direction, while the shear strain $\gamma_{45°}$ increases rapidly due to the opening of the horizontal cracks and the closing of the vertical cracks. The resulting near-zero shear stiffness in the BC regions creates the pinched shape in the hysteretic loops of the shear stress vs. shear strain curves.

Panel CE3 ($\alpha_2 = 90°$) - The absence of pinching mechanism in panel CE3 can be intuitively visualized by considering a cracked element with $90°$ steel bars as shown in Fig. 9(a). This element is also subjected to a pure shear state in the 45° direction, which is equivalent to applying a horizontal compressive stress $\sigma_H$ and vertical tensile stress $\sigma_V$ of equal magnitude. In the reverse loading stage from point B to point C (Fig. 5(b)) both the vertical and horizontal cracks are open. Consequently, the horizontal compressive stress $\sigma_H$ is resisted by the compressive stress in the horizontal steel bar (Fig. 9(b)), while the vertical tensile stress $\sigma_V$ is resisted by a tensile stress in the vertical bar (Fig. 9(c)). Both the compressive stress in the horizontal steel bar and the tensile stress in the vertical steel bar contribute to the shear stress $\tau_{45°}$ in the 45° direction. As a result, the shear stress increases proportionally to the shear strain, and the shear stiffness in the BC region becomes large. This large stiffness creates a smooth and robust hysteretic loop without the pinched shape.

## FAILURE MECHANISM UNDER CYCLIC LOADING

### Structures Subjected to Static Loading

When a reinforced concrete structure is subjected to static loading, the principal compression stresses in the structure can be resisted by concrete, but cracks will occur due to principal tensile stresses, because concrete is very week in tension. To resist additional loads, steel bars must be added to resist the principal tensile stresses. Thus, reinforced concrete can be visualized as a truss consisting of concrete struts in compression and steel ties in tension. This truss is capable of resisting high magnitude of applied static loads. In this strut-and-tie model, equilibrium condition is satisfied by maintaining force equilibrium at the joints (or nodes) among struts and ties.

As we know, this strut-and-tie concept can be used to design all reinforced concrete structures under static loads. The discussion below addresses the question whether this concept is applicable to structures subjected to reversed cyclic loading.

### Structures Subjected to Cyclic Loading or Earthquake Loading

Comparison of the cyclic behavior of panel CE3 ($\alpha_2 = 90°$) and panel CA3 ($\alpha_2 = 45°$) clearly shows that the strut-and-tie model is no longer valid for application to cyclic loading and earthquake. When reinforced concrete structures are resisting cyclic loading beyond the yielding of steel, crack widths increase in both directions with each cyclic of loading (elements are expanding). Crack widths will not close after unloading and the concrete struts cannot be formed to resist the reversed loading while the cracks are still open. Therefore, the steel bars must be designed to resist principal compression as well as principal tension. Panel CE3 ($\alpha_2 = 90°$), which is designed based on this principle, does perform very well. It does not have the pinching problem, and it exhibits ductile behavior and high capacity in energy dissipation.

When the steel bars are oriented at an angle of 45° to the principal stress coordinate as in panel CA3 ($\alpha_2 = 45°$), steel bars in both directions are always in tension. Then the cracks are forced to close in order to form the concrete struts required in establishing a truss of struts and ties. This forced closing of cracks and the subsequent reopening of cracks in each cycle of loading represent a very destructive failure mechanism. This mechanism leads to the weakening of concrete in compression and the rapid deterioration of bond between the concrete and the steel bars. This menacing failure mechanism is responsible for the pinching effect in the hysteretic loops, the early arrival of the descending branch, and the low capacity in energy dissipations.

## CONCLUSION

The analysis of the two panels CA3 ($\alpha_2 = 45°$) and CE3 ($\alpha_2 = 90°$) clearly reveal the working mechanism of cracked reinforced concrete elements under cyclic shear. It is found that the "pinching effect" in the post-yield hysteretic loops is not an inherent behavior attributed to shear per se. Also, it is not caused by the bond slips between concrete and steel bars as has often been mistaken as the main reason. This pinching phenomenon is actually caused by the orientation of the steel bars with respect to the principal coordinate of the applied stresses, and can be predicted by a rational theory.

Whereas the strut-and-tie concept can be used to design reinforced concrete structures subjected to static loading (i.e., the concrete bears compression and steel bars resist tension), it cannot be applied to concrete structures subjected to cyclic or earthquake loading. This is because cracked concrete cannot be relied upon to resist compression after the yielding of steel. Thus, the steel bars must be designed to take on the role of resisting principal compression, in addition to principal tension. This design principle for cyclic loading is applicable to shear as well as bending actions.

## REFERENCES

(1) Hsu, T. T. C., Belarbi, A., and Pang, X. B. "A Universal Panel Tester," *Journal of Testing and Evaluations*, ASTM, Vol. 23, No. 1, 1995, pp. 41-49.

(2) Hsu, T. T. C., Zhang, L. X. and Gomez, T. "A Servo-Control System for Universal Panel Tester," *Journal of Testing and Evaluations*, ASTM, Vol. 23, No. 6, 1995, pp. 424-430.

(3) Hsu, T. T. C., *Unified Theory of Reinforced Concrete*, CRC Press Inc., Boca Raton, FL, 1993, 329 pp.

(4) Belarbi, A. and Hsu, T. T. C. "Constitutive Laws of Concrete in Tension and Reinforcing Bars Stiffened by Concrete", *Structural Journal of the American Concrete Institute*, V. 91, No. 4, July-Aug., 1994, pp. 465-474.

(5) Belarbi, A. and Hsu, T. T. C. "Constitutive Laws of Softened Concrete in Biaxial Tension-Compression", *Structural Journal of the American Concrete Institute*, V. 92, No. 5, Sept.-Oct., 1995, pp. 562-573.

(6) Pang, X. B. and Hsu, T. T. C. "Behavior of Reinforced Concrete Membrane Elements in Shear," *Structural Journal of the American Concrete Institute*, Vol.92, No.6, Nov.-Dec., 1995, pp.665-679.

(7) Pang, X. B. and Hsu, T. T. C. "Fixed-Angle Softened-Truss Model for Reinforced Concrete," *Structural Journal of the American Concrete Institute*, Vol. 93, No. 2, Mar.-Apr., 1996, pp. 197-207.

(8) Hsu, T. T. C. and Zhang, L. X. "Nonlinear Analysis of Membrane Elements by Fixed-Angle Softened-Truss Model," *Structural Journal of the*

*American Concrete Institute*, Vol. 94, No. 5, Sept.-Oct., 1997, pp. 483-492.

(9) Zhang, L. X. and Hsu, T. T. C., "Behavior and Analysis of 100 MPa Concrete Membrane Elements," *Journal of Structural Engineering*, ASCE, Vol. 124, No. 1, Jan. 1998, pp. 24-34.

(10) Zhu, R. R. H. "Softened Membrane Model of Cracked Reinforced Concrete Considering Poisson effect," *Ph.D. Dissertation*, Department of Civil Engineering, University of Houston, Houston, TX, 2000.

(11) Hsu, T. T. C. and Zhu, R. R. H. "Post-yield Behavior of Reinforced Concrete Membrane Elements - the HSU/ZHU ratios," *ASCE Publication: Modeling of Inelastic Behavior of RC Structures Under Seismic Loads*, American Society of Civil Engineers, Reston, VA, April 2001, pp. 139-157.

(12) Zhu, R. R. H., and Hsu, T. T. C. "Poisson Effect in Reinforced Concrete Membrane Elements," *Structural Journal of the American Concrete Institute* (submitted for publication in 2002).

(13) Hsu, T. T. C. and Zhu, R. R. H. "Softened Membrane Model for Reinforced Concrete," *Structural Journal of the American Concrete Institute* (submitted for publication in 2002).

(14) Mansour, M. Y. "Behavior of Reinforced Concrete Elements Under Cyclic Shear: Experiments to Theory" *Ph.D. Dissertation*, Department of Civil Engineering, University of Houston, Houston, TX, 2001.

(15) Mansour, M. Y., Lee, J. Y., and Hsu, T. T. C. "Cyclic Stress-Strain Curve of Concrete and Steel Bars in Membrane Elements", *Journal of Structural Engineering*, ASCE, Vol. 127, No. 12, Dec. 2001, (scheduled).

(16) Mansour, M. Y., Hsu, T. T. C., and Lee J. Y. "Pinching effect in Hysteretic Loops of R/C Shear Elements", *ACI Special Publication*, American Concrete Institute (accepted for publication in 2001).

(17) Zhu, R. R. H., Hsu, T. T. C. and Lee, J. Y. "Rational Shear Modulus for Smeared Crack Analysis of Reinforced Concrete," *Structural Journal of the American Concrete Institute*, Vol. 98, No. 4, July-Aug. 2001, pp.443-450.

## NOTATIONS

A    = coefficient in the unloading and reloading equation of a steel bar, given as $1.9k_p^{-0.1}$.

B    = parameter equal to $(f_{cr}/f_y)^{1.5}/\rho$.

D    = damage coefficient for concrete in compression.

$E_c$    = modulus of elasticity of concrete.

$E_{cc}$    = slope of the linear unloading and reloading expression of concrete.

$E_s$    = modulus of elasticity of a bare steel bar.

$f_c'$    = maximum compressive strength of a concrete cylinder.

$f_{cr}$   = cracking stress of concrete.

$f_\ell$   = smeared stress of steel bars embedded in concrete in the longitudinal direction.

$f_s$   = smeared stress of steel bars embedded in concrete.

$f_t$   = smeared stress of steel bars embedded in concrete in the transverse direction.

$f_y$   = yield stress of bare steel bars.

k   = coefficient in the equation for softened coefficient $\zeta$, taken as 400 for proportional loading.

$k_p$   = plastic strain ratio given as $\varepsilon_p / \varepsilon_n$.

R   = coefficient in the unloading and reloading equation of a steel bar, given as $10k_p^{-0.2}$.

$\varepsilon_1$   = tensile strain in 1 – direction, could be $\varepsilon_H$ or $\varepsilon_V$ under cyclic loading.

$\varepsilon_c$   = concrete strain.

$\varepsilon_{comp}$ = maximum compression strain occurring in the immediate opposite direction and perpendicular to the compression strain being considered.

$\varepsilon_{cr}$   = cracking strain of concrete taken as 0.00008.

$\varepsilon_{ci}$   = concrete strain at the load reversal point "i".

$\varepsilon_{ci+1}$ = concrete strain at the end of a stage.

$\varepsilon_H$   = smeared strain in horizontal direction.

$\varepsilon_\ell$   = smeared strain in longitudinal direction.

$\varepsilon_n$   = yield strain of steel bars embedded in concrete.

$\varepsilon_p$   = plastic strain of steel bars embedded in concrete.

$\varepsilon_s$   = smeared strain of steel bars embedded in concrete.

$\varepsilon_t$   = smeared strain in transverse direction.

$\varepsilon_V$   = smeared strain in vertical direction.

$\varepsilon_y$   = yield strain of a bare steel bar.

$\varepsilon_0$   = concrete strain at maximum compressive cylinder strength.

$\zeta$   = stress or strain softening coefficient

$\rho$   = reinforcement steel ratio.

$\rho_\ell$   = reinforcement ratio in longitudinal direction.

$\rho_t$   = reinforcement ratio in transverse direction.

$\sigma_c$   = concrete stress.

$\sigma_{ci}$   = concrete stress at the load reversal point "i".

$\sigma_{ci+1}$ = concrete stress at the end of a stage.

$\sigma_\ell$    = longitudinal applied stress.

$\sigma_t$    = transverse applied stress.

$\sigma_H^c$    = concrete stress in horizontal direction.

$\sigma_V^c$    = concrete stress in vertical direction.

$\tau_{45°}$    = shear stress at $45°$ to the horizontal direction.

$\tau_{\ell t}$    = shear stress in ( $\ell - t$ ) coordinate system.

$\tau_{VH}^c$    = concrete shear stress in (H-V) coordinate system.

$\gamma_{45°}$    = shear strain at $45°$ to horizontal direction.

$\gamma_{\ell t}$    = shear strain in ( $\ell - t$ ) coordinate system.

$\gamma_{VH}^c$    = concrete shear strain in (H-V) coordinate system.

$\upsilon_{CT}$    = Hsu/Zhu ratio that takes into account the effect of tensile strain on compression strain.

$\upsilon_{TC}$    = Hsu/Zhu ratio that takes into account the effect of compression strain on tensile strain.

Table 1  Material properties of test specimens

| Panel | Concrete | | Steel in $\ell$ direction | | | Steel in t direction | | |
|---|---|---|---|---|---|---|---|---|
| | $f_c'$ (MPa) | $\varepsilon_o$ | Spacing of #6 bars | $\rho_\ell$ | $f_{\ell y}$ (MPa) | Spacing of #6 bars | $\rho_t$ | $f_{ty}$ (MPa) |
| CA3 | 45 | 0.0025 | 189 mm | 1.7% | 428 | 189 mm | 1.7% | 428 |
| CE3 | 50 | 0.0023 | 267 mm | 1.2% | 428 | 267 mm | 1.2% | 428 |

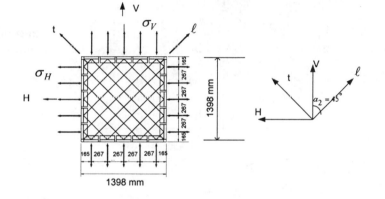

(a) Panel CA3 ($\alpha_2 = 45°$).

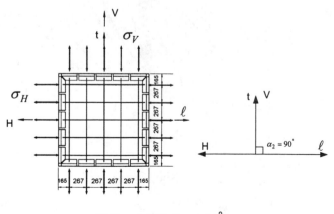

(b) Panel CE3 ($\alpha_2 = 90°$).

Fig. 1  Steel layout and dimensions of test specimens.

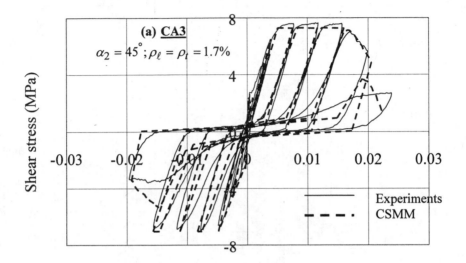

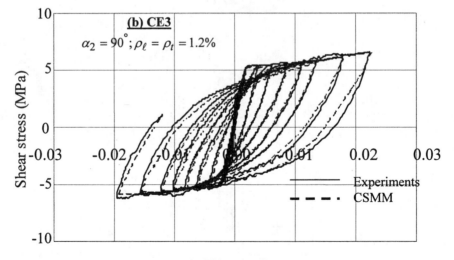

Fig. 2 Experimental and analytical shear stress-strain curves of
panels CA3 and CE3.

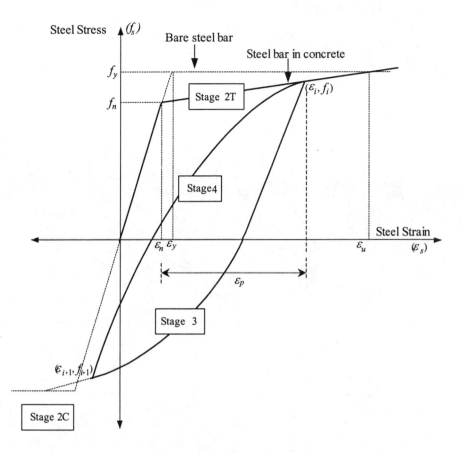

Fig. 3  Average cyclic stress-strain curves of steel bars in concrete.

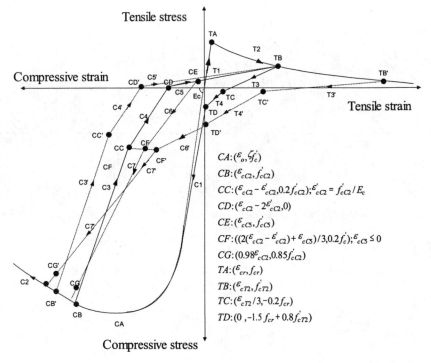

$CA: (\varepsilon_o, \zeta f_c^{'})$

$CB: (\varepsilon_{cC2}, f_{cC2}^{'})$

$CC: (\varepsilon_{cC2} - \varepsilon_{cC2}^{'}, 0.2 f_{cC2}^{'}); \varepsilon_{cC2}^{'} = f_{cC2}^{'}/E_c$

$CD: (\varepsilon_{cC2} - 2\varepsilon_{cC2}^{'}, 0)$

$CE: (\varepsilon_{cC5}, f_{cC5}^{'})$

$CF: ((2(\varepsilon_{cC2}^{'} - \varepsilon_{cC2}^{'}) + \varepsilon_{cC5})/3, 0.2 f_c^{'}); \varepsilon_{cC5}^{'} \leq 0$

$CG: (0.98\varepsilon_{cC2}, 0.85 f_{cC2}^{'})$

$TA: (\varepsilon_{cr}, f_{cr})$

$TB: (\varepsilon_{cT2}, f_{cT2}^{'})$

$TC: (\varepsilon_{cT2}/3, -0.2 f_{cr})$

$TD: (0, -1.5 f_{cr} + 0.8 f_{cT2}^{'})$

**Not to Scale**

Fig. 4 Proposed model for cyclic stress-strain curves of concrete.

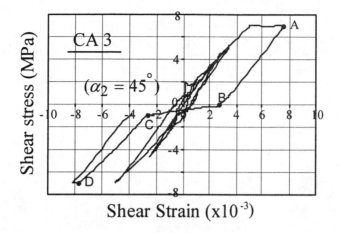

Fig. 5 (a) Pinching mechanism for panel CA 3.

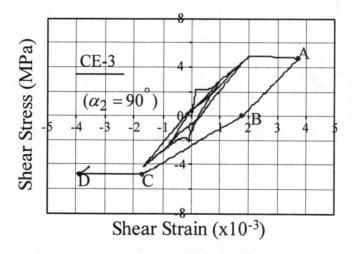

Fig. 5 (b) Absence of pinching mechanism for panel CE 3.

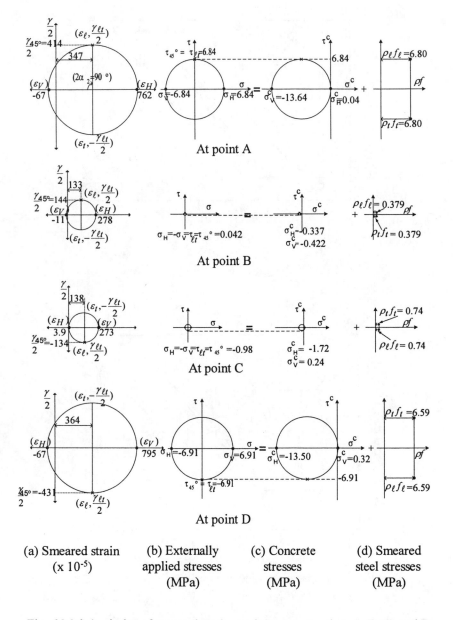

(a) Smeared strain
(x 10⁻⁵)

(b) Externally
applied stresses
(MPa)

(c) Concrete
stresses
(MPa)

(d) Smeared
steel stresses
(MPa)

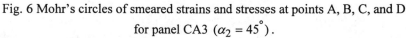

Fig. 6 Mohr's circles of smeared strains and stresses at points A, B, C, and D
for panel CA3 $(\alpha_2 = 45°)$.

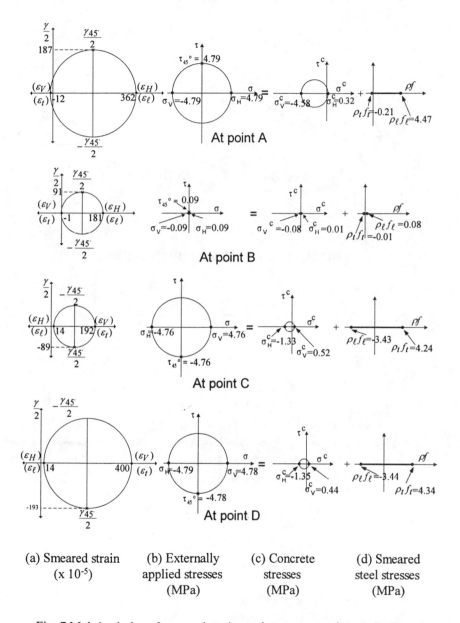

(a) Smeared strain
(x 10⁻⁵)

(b) Externally
applied stresses
(MPa)

(c) Concrete
stresses
(MPa)

(d) Smeared
steel stresses
(MPa)

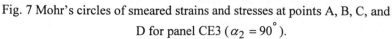

Fig. 7 Mohr's circles of smeared strains and stresses at points A, B, C, and D for panel CE3 ($\alpha_2 = 90°$).

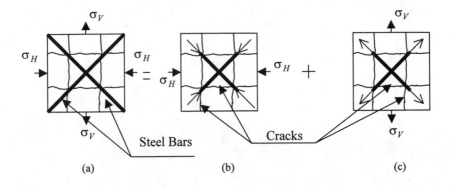

Fig. 8 Cracked R/C element with 45° steel bars.

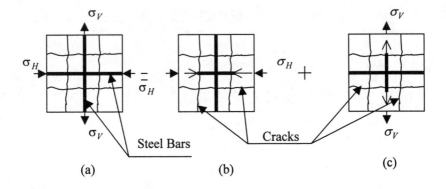

Fig. 9 Cracked R/C element with 90° steel bars.

# Fatigue Fracture and Crack Propagation in Concrete Subjected to Tensile Biaxial Stresses

## by K. V. Subramaniam, J. S. Popovics, and S. P. Shah

**Synopsis:** The objective of this paper is to characterize the quasi-static and fatigue response of concrete subjected to biaxial stresses in the t-C-T region, where the principal tensile stress is larger in magnitude than the principal compressive stress. An experimental investigation of material behavior is conducted. The failure of concrete in the stated biaxial region is shown to be a local phenomenon under both quasi-static and fatigue loading, wherein the specimen fails owing to a single crack. The crack propagation is studied using the principles of fracture mechanics. It is observed that crack growth in constant amplitude fatigue loading is a two-phase process: a deceleration phase followed by an acceleration stage. The quasi-static load envelope is shown to predict the crack length at fatigue failure. A fracture-based fatigue failure criterion is proposed, wherein the fatigue failure can be predicted using the critical mode I stress intensity factor obtained from the quasi-static response. A material model for the damage evolution during fatigue loading of concrete in terms of crack propagation is proposed. The model parameters obtained from uniaxial fatigue tests are shown to be sufficient for predicting the biaxial fatigue response.

Keywords:  biaxial; concrete; crack; damage; failure; fatigue; fracture; uniaxial

**Kolluru V Subramaniam** is an Assistant Professor in the Civil Engineering Department at the City College of the City University of New York. He obtained his PhD from Northwestern University. He is a member of ACI committees 215, Fatigue of Concrete. His research interests include material characterization, fatigue and fracture of concrete, and nondestructive testing.

**John S. Popovics** is an Assistant Professor in the Civil and Architectural Engineering at Drexel University. He received his PhD from Pennsylvania State University. He is a member of ACI committees 215, Fatigue of Concrete, and 228 nondestructive Testing of Concrete. His research interests include nondestructive testing of concrete and wave mechanics.

**Surendra P. Shah (FACI)** is the Walter P. Murphy Professor of Civil Engineering at Northwestern University, and the Director of the Center for Advanced Cement Based Materials. His research interests include constitutive relationships, failure and fracture of concrete, nondestructive testing, and impulse loading.

## INTRODUCTION

The Compression-Tension (C-T) region within the biaxial stress space is the area in which the principal stresses are of opposite signs. The biaxial C-T space can be further divided into two regions: (a) the tensile-Compression-Tension (t-C-T) region, where the magnitude of the principal tensile stress is greater than or equal to that of the principal compressive stress, and (b) the compressive-Compression-Tension (c-C-T) region, where the principal compressive stress has a larger magnitude. The t-C-T response of the material represents the material behavior subjected to combined shear and tension and is investigated in this paper.

Different methods have been tried to test concrete subjected to a biaxial state of stress. These include: subjecting cylindrical specimens to hydrostatic pressure in radial and axial directions (1); subjecting hollow cylinders to torsion and axial compression or to internal pressure and axial compression (2,3,4); and subjecting concrete plates to in-plane loading for different biaxial stress combinations (5,6). In these investigations the strength of concrete subjected to different ratios of principal stresses was determined and the failure surface of concrete in the biaxial stress space was established. However, no data are available in the published literature for the behavior of concrete subjected to biaxial loading after the peak stress. From the modes of failure reported in the literature (5), the damage localizes into a single crack in the t-C-T region. The predominant mode of failure seems to be the formation and propagation of a single crack. Controlled experiments are needed to obtain the post peak part of the response in the t-C-T region. Application of fracture mechanics can be explored to explain the behavior of concrete in the post peak part of the response.

A distinction is generally made between low-cycle, high-amplitude fatigue and high-cycle, low-amplitude fatigue. The former involves a few load cycles of high stress (earthquakes, storms, etc.) while the latter is characterized by a greater number of cycles of low stress (wind and wave loading). A comprehensive review of the fatigue behavior of concrete subjected to uniaxial fatigue is provided in committee reports of RILEM and ACI (7,8,9). Few experimental results on the response of concrete subjected to repeated biaxial loading are available in the literature. All the results in the literature pertain to cyclic biaxial compression (where both principal stresses are compressive) (10,11,12). From their experimental investigation considering four different principal stress ratios, Su and Hsu [1988] established the S-N curve for concrete subjected to biaxial compression. The S-N relationship was found to be non-linear instead of being a straight line. The observed S-N curve can be idealized by two straight lines that have significantly different slopes. The slope for the low-cycle fatigue was seen to be several times higher than that for high cycle fatigue. Su and Hsu also concluded that the fatigue strength of concrete under biaxial compression is greater than that under uniaxial compression for any given number of cycles. Observation of the failure modes indicated that the failure mode under monotonic and fatigue loadings was identical. The S-N curve was also observed to be a curve in the case of high strength concrete (12). No reference concerning fatigue behavior of concrete subjected to fatigue loading in the C-T region was found in the literature.

## OBJECTIVES AND SCOPE

An experimental investigation to characterize the material response in biaxial t-C-T region is conducted in this paper. The main objective is to understand the response of concrete subjected to high-amplitude biaxial fatigue loading in the tensile-compression-tension (t-C-T) stress space. Characterization of the material response under quasi-static loading and establishment of the mechanism of failure under such loading is the essential first step in the process. A predictive model for response of concrete under fatigue loading is also proposed.

## EXPERIMENTAL AND ANALYTICAL PROCEDURES

In the experimental program two different test configurations were used. Concrete beam specimens were tested in a three-point bend configuration to obtain the tensile material properties. The experimental setup for applying biaxial t-C-T stresses consisted of applying combined torsional and axial loading to hollow cylindrical concrete specimens. Torsion introduces a state of pure shear stress in the material which, when resolved in terms of principal stresses corresponds to a biaxial state of stress wherein the two principal stresses are equal in magnitude but of opposite signs. The ratio of the two principal stresses can be changed by superposition of axial load over the torsional load. The entire C-T region of the biaxial stress space can be spanned by varying the magnitude of the applied axial load, the limit corresponding to uniaxial tension and compression. The entire process is illustrated schematically in Figure 1.

Experimental evidence suggested that the decrease in stiffness of a specimen tested in the t-C-T region is associated with the growth of a single crack. Further, the decrease in the stiffness of the specimen in constant amplitude fatigue loading in the t-C-T region was also observed to be due to an increase in the crack length.

To understand the material response to t-C-T loading, it is hence important to obtain information about the crack growth resulting from load application. The problem can then be framed in terms of fracture mechanics, and a fracture-based crack growth criterion can be established for such loading. The evolution of damage under fatigue loading can then be studied and a fracture based fatigue damage law can be developed.

In concrete, the non-linear process zone ahead of crack tip is relatively large and its effects cannot be neglected as in linear elastic fracture mechanics (LEFM). One approach to account for the non-linear fracture process zone ahead of the crack tip is using the concept of effective crack. It is a first order correction to LEFM. In this approximation, the crack in a concrete structure surrounded by a large non-linear zone is replaced with an effective elastic crack that gives the same compliance of the structure as the actual crack. Effective crack is the traction free crack that gives the same compliance as the true crack. Crack growth in the concrete structure is framed in terms of the effective traction free crack and principles of LEFM are applied to study crack propagation (13).

Closed form solutions that relate the crack length with the compliance of a concrete beam are available in the literature (14). For a notched beam specimen tested in three-point bend configuration, the Young's modulus (E) of the material is calculated as

$$E = \frac{6Sa_0 V_1(\alpha)}{C_i bd^2} \quad \text{where } V_1(\alpha) = 0.76 - 2.26\alpha + 3.87\alpha^2 - 2.04\alpha^3 + \frac{0.66}{(1-\alpha)^2} \quad (1)$$

$C_i$ is the compliance (inverse of the stiffness) calculated from the load-crack mouth opening displacement (CMOD) curve, $a_0$ the initial notch length, S the span of the beam, b and d are the thickness and depth of the beam respectively and $\alpha = a_0/d$. The procedure for determining the crack length from the observed change in compliance associated with crack growth using Equation 1 is described by Jenq and Shah (14). The Mode I stress intensity factor ($K_I$), at a given point on the quasi-static envelope and for the given crack length can be computed using equation 2

$$K_I = 3P \frac{S(\pi a_e)^{1/2} F(\alpha)}{2bd^2} \quad \text{where} \quad F(\alpha) = \frac{1.99 - \alpha(1-\alpha)(2.15 - 3.93\alpha + 2.7\alpha^2)}{\pi^{1/2}(1+2\alpha)(1-\alpha)^{3/2}} \quad (2)$$

$\alpha = a_e/d$ and P is the load at the given point in the post-peak response.

The effective crack length in a cylinder subjected to torsion can also be similarly estimated from the change in rotational stiffness (or compliance) of the specimen. Unfortunately, closed-form solutions for prediction of the stiffness as a function of crack length are not available for this particular specimen geometry. Numerical analyses were performed for a cylinder with cracks of different lengths to determine the influence of crack length on the rotational stiffness (inverse of compliance) and the stress intensity factors at the crack tip (15).

From the results of the numerical analysis it was observed that the mode I stress intensity factor ($K_I$) is considerably larger than the stress intensity factors for modes II and III ($K_{II}$ and $K_{III}$) at all non-trivial crack lengths (15). Through regression analysis of the numerical data the expression for computing $K_I$ for a given crack length and applied torque is given as

$$K_I = (2.6296 \times 10^{-6}(a) + 7.2064 \times 10^{-6}) * \text{Torque} \quad (3)$$

where $K_I$ is in N/mm$^{3/2}$, Torque is in N-mm, and a is in mm. Similarly, from the numerical analysis an expression that relates the decrease in rotational stiffness of the cylinder with the increase in crack length was obtained. The expression given in Equation 4 that relates the crack length with the observed decrease in rotational stiffness was obtained.

$$S = -0.0353a^2 - 0.0208a \quad (4)$$

where a is the crack length in mm, and S is the percentage decrease in rotational stiffness with respect to an uncracked cylinder.

## EXPERIMENTAL RESULTS

Under quasi-static loading it was observed that a crack initiated in the pre-peak region of the response under both the flexural and torsional loading (15,16,17). The crack propagation under quasi-static loading was studied by unloading and reloading the specimen at several points on the quasi-static load envelope. Typical cyclic-quasi-static test response obtained by unloading the specimen at different points on the load envelope in flexure is shown in Figure 2(a). The crack length at these points in the quasi-static load response was determined from the decrease in unloading stiffness. Using the results of the previous section, the crack lengths were determined by matching the percentage drop in the stiffness with respect to the pristine specimen with the numerical computed change in stiffness associated with an increase in the effective crack length. The $K_I$ corresponding to the given crack length is then determined using the expressions that relate crack length and applied loading to $K_I$ (such as Equations 2 and 3). The value of $K_I$ computed at different points on the post-peak quasi-static load envelope are shown plotted in Figure 2(b). It has been previously established that the value of $K_I$ corresponding to the peak load of the specimen is a material constant referred to as the critical stress intensity factor, $K_{IC}$ (14). This analysis shows that the stress intensity factor during crack growth in the post peak part of the quasi-static response stays constant and is equal to $K_{IC}$. An average value of 40 N/mm$^{3/2}$ is obtained for $K_{IC}$. Similar response was also obtained from the torsional specimens, wherein a constant value of $K_I$ equal to 40 N/mm$^{3/2}$ was obtained in the post-peak part of the torsional response. The crack growth along the quasi-static post-peak envelope can hence be characterized by a fracture condition

$$K_I = critical \text{ value of } K_I = K_{IC} \tag{5}$$

Constant amplitude, fatigue tests were conducted in both flexure and torsion at three different load ranges (15,16,17). In all the fatigue tests the lower load level in a fatigue cycle was kept constant, equal to 5% of the average quasi-static peak load. Three different upper load levels corresponding to 75%, 85%, and 95% of the average quasi-static peak load were used in the study. The crack length during fatigue loading was determined from the observed change in compliance (or stiffness) of the specimen using the compliance calibration curves.

The plot showing the rate of crack growth versus crack length for all the specimens tested in flexure at maximum load in a fatigue cycle equal to 85% of the average quasi-static peak load is shown in Figure 3. Similar responses were obtained at all the load ranges tested and for both flexure and torsional loading. The rate of crack growth follows a two-stage process; a deceleration stage followed by an acceleration stage up to failure. In the deceleration stage there is a decrease in the rate of fatigue crack growth with an increase in crack length. The deceleration stage typically lasts for the first 30-40% of the fatigue life of the specimens. In the acceleration stage the rate of fatigue crack growth increases continuously. There is a distinctive bend-over point ($a_{bendover}$) that corresponds to the crack length where the rate of crack growth changes from deceleration to acceleration.

**Fatigue failure criterion**

A comparison of $K_I$ obtained from post-peak quasi-static response and constant amplitude fatigue loading for flexural and torsional tests is shown in Figure 4. For fatigue loading, the $K_I$ corresponding to crack lengths at failure have been computed using the maximum load in the fatigue cycle. The values of $K_I$ at fatigue failure are seen to compare closely in flexural and torsional loading. Further, comparable critical values of $K_I$ are obtained from quasi-static post-peak and fatigue loadings for flexure and torsion. It can be concluded that the load-deformation response obtained from quasi-static loading therefore acts as the failure envelope curve for fatigue loading when framed in terms of crack lengths. Further, the condition, $K_I = K_{IC}$, can be used to predict fatigue failure.

## ANALYTICAL MODEL

A schematic representation of the results obtained from the quasi-static and fatigue loading of concrete in the t-C-T region is attempted in this section. The load response of a specimen (Figure 5(a)) can be transformed to the equivalent load – effective crack length space, as shown in Figure 5(b). In Figure 5(b), the origin, $O'$, is located at the initial flaw, $a_o$. As the load is increased there is no crack growth till a certain load is reached (point O). The region below the threshold value O in the equivalent load-effective crack length space corresponds with the linear portion of the load response. The region OB represents stable crack growth in the quasi-static monotonic response up to peak load. After peak load, there is a decrease in the load with an increase in crack length (region BD). Region BD corresponds with the post-peak part of the load response.

The fatigue response, which exhibits crack growth due to repeated loading between two fixed load levels, is also shown schematically by line AC in Figure 5(b). From the previous analysis, it has been demonstrated that the quasi-static response acts as the failure envelope to the fatigue response when framed in terms of crack length. Point A represents the crack length at the beginning of the first cycle of fatigue loading and point C represents the crack length at fatigue failure. The post peak part of the quasi-static response (region BD in Figure 5(b)) can be predicted using the failure condition, $K_I = K_{IC}$, given in Equation (5). Using the failure envelope concept in terms of effective crack length, the same conditions also applies to fatigue failure. It is of interest to predict the crack growth in terms of load cycles.

## Mechanistic Understanding of Crack Growth

To predict the fatigue crack growth it is important to understand the various mechanisms that influence the crack growth in concrete. There are two competing mechanisms involved in crack growth process in concrete and are illustrated schematically as the G and R curves in Figure 6. The R-curve of an infinite-sized specimen is shown by the curve OCF and the R-curve for the finite-sized specimen is given by a constant value after the critical crack length, curve OCE. The R curve represents the energy absorbed due the growth of the inelastic fracture process zone during crack growth and the G curve is the crack driving force resulting from elastic unloading of the body during crack growth. The quasi-static load response of the specimen can be approximately predicted using G and the R-curves. The process is shown schematically in Figure 6. For the given specimen, the intersection of the G curves starting at the same initial notch length $a_o$ (for different stress levels) and the R-curve generates the quasi-static response that is shown on load effective crack plot in Figure 6(b). For a given stress level, the G-curve intersects the R-curve at two points. The first point of intersection is in the pre-peak portion while the second point is in the post-peak of the quasi-static response. By varying the stress, the entire load-deformation response of the specimen can be approximately predicted from the intersection of G and R-curves. On increasing $\sigma$, there comes a point where the G- curve intersects the R-curve tangentially. This corresponds to the peak load of the specimen and the crack length ($a_{crit}$). The constant value of the R-curve after $a_{crit}$ corresponds to the observed constant value of $K_{IC}$ in the post-peak part of the quasi-static response.

A simple phenomenological model for fatigue-crack growth is described in this section. Constant amplitude fatigue is a process of stable crack growth between two fixed load (stress) levels up to failure. The upper load (stress) level in the load cycle governs the energy release rate for unit-crack extension during this process. For a given upper load (stress) level during fatigue loading, (say $\sigma_1$ in Figure 6), the energy release rate with crack growth follows the G-curve from point A to point E in Figure 6(a); Point A represents the crack length during the first cycle of loading and point E represents crack length at failure given by the failure condition, $K_I = K_{IC}$. With an increase in crack length there is an increase in the stress intensity factor, $K_I$, at the crack tip, which provides a larger crack driving force. However, as the crack length increases the resistance to crack propagation (energy required for crack propagation) also increases along the path ACE on the R-curve (Figure 6(a)). The resistance to crack growth increases up to point C and then remains constant along CE. The point C corresponds to the critical crack length ($a_{crit}$) at the peak load of the quasi-static response.

The fatigue crack growth experiences increasing resistance up to $a_{crit}$. It can be therefore be hypothesized that the deceleration stage in the fatigue crack growth ends at $a_{crit}$. A comparison of the critical crack length at the peak load of the quasi-static response ($a_{crit}$) and the crack length at the bend-over point, where the rate of crack growth changes from deceleration to acceleration ($a_{bendover}$) is shown in Figures 7 (a) and (b) for flexure and torsion specimens respectively. Each point on the graph corresponds to one specimen tested in fatigue. The X-axis corresponds to the maximum applied load in a fatigue-load cycle as a percentage of the average maximum quasi-static monotonic peak load. The Y-axis of the graph is the ratio of $a_{bendover}$ to the average $a_{crit}$ obtained by testing six specimens under cyclic quasi-static loading. From the figure it can be seen that $a_{bendover}$ in the fatigue response at load ranges used in this study is the same magnitude as $a_{crit}$ of the quasi-static monotonic response. Hence, $a_{bendover}$ in constant-amplitude fatigue, for the load ranges considered, can be predicted from $a_{crit}$ obtained from the quasi-static response.

## PROPOSED MODEL FOR FATIGUE CRACK GROWTH

A proposed model for fatigue-crack growth rate is based on the following assumptions:
1. The crack growth rate in the deceleration stage is governed only by the increasing resistance, R.
2. The crack growth rate in the acceleration stage is governed only by the stress intensity factor, $K_I$.

In the deceleration stage, increasing $K_I$ with increasing crack has a minor influence. This effect is ignored in this simplified model. In the deceleration stage, the reduction in the crack growth rate per unit crack advance is assumed to be proportional to the increasing resistance (rising R-curve) experienced by the crack. The resistance is proportional to the total increase in crack length from the beginning of loading (Da = a – $a_o$). For unnotched specimens $a_o$ can be taken equal to 2mm (18). It is proposed that the rate of crack growth in this stage can be represented by the expression

$$\frac{\Delta a}{\Delta N} = C_1 (Da)^{n_1} \tag{6}$$

where $C_1$ and $n_1$ are the constant and the exponent respectively, to be determined from the experimental data.
In the acceleration stage, it is assumed that the crack growth rate can be predicted by Paris law, rewritten here as

$$\frac{\Delta a}{\Delta N} = C_2 (\Delta K_1)^{n_2} \tag{7}$$

where $C_2$ and $n_2$ are the constant and exponent of the Paris law respectively, to be determined from the experimental data.

From the fatigue test data of three-point bending configuration, the obtained values of the coefficients, $Log(C_1)$ and $n_1$ of Equation 6, were –2.05 and –1.7 respectively, when the maximum stress in the fatigue cycle is below 95% of the peak quasi-static stress. For the very high maximum fatigue load case the coefficients were found to be –1.0 and –1.5 respectively. This difference in the very high load ranges was attributed to the significant influence of increasing $K_I$ with increasing crack length. Similarly, the Paris law exponents of Equation 7, $Log(C_2)$ and $n_2$, were determined from the flexural tests to be 28.6 and 17.33 respectively.

Equations 6 and 7 therefore provide a convenient approach for modeling fatigue crack growth that can be implemented relatively easily; material constants determined from the uniaxial tests can be used to predict the structural response in the t-C-T region with sufficient accuracy. The fatigue crack growth process simulated using Equations 6 and 7 for flexure and torsion is shown in Figures 8 (a) and (b). The material constants obtained from the flexure tests were used in both simulations.

The analytical curves were obtained by numerically integrating Equations 6 and 7. The first step in the procedure consists of numerically determining $a_{bendover}$. The rate of crack growth predicted by Equations 6 and 7 (shown in Equation 8) are equated and the crack length solved iteratively.

$$Log(C_1) + n_1 Log(Da) = Log(C_2) + n_2 Log(\Delta K_I) \tag{8}$$

The procedure for obtaining the simulated crack fatigue growth now consists of determining computing the number of cycles ($\Delta N_i$) for a chosen crack increment ($\Delta a_i$) and maximum and minimum load levels ($P_{max}$ and $P_{min}$) as shown below

deceleration stage: ($a <= a_{bendover}$)
1.  set $a = a_o$ and $N = 0$
2.  $a = a + \Delta a_i$, $Da = a - a_o$
3.  $\Delta N_i = \dfrac{\Delta a_i}{C_1(\Delta a_i)^{n_1}}$ and $N = N + \Delta N_i$
4.  repeat steps 2-3 till $a = a_{bendover}$

acceleration stage: ($a_{bendover} < a$)
5.  set $a = a_{bendover}$
6.  $a = a + \Delta a_i$, $\Delta K_I = K_I(a, P_{max}) - K_I(a, P_{min})$
7.  $\Delta N_i = \dfrac{\Delta a_i}{C_2(\Delta K_I)^{n_2}}$ and $N = N + \Delta N_I$
8.  repeat steps 6-7 till $\Delta N_i = 0$

Typical experimental results are also plotted in the same figure for comparison. It can be seen that the S-shaped trend observed in the fatigue crack growth is well captured by the simplified model consisting of Equations 6 and 7. Also, it can be seen that the fatigue crack growth in torsion is accurately predicted using coefficients obtained from flexural tests.

## CONCLUSIONS

Based on the results presented in this paper the following conclusions can be drawn.
1.  The load-deformation response obtained from quasi-static loading acts as the failure envelope curve for fatigue loading when framed in terms of crack length. The condition, $K_I = K_{IC}$, can be used to predict failure in high-amplitude fatigue.

2. Fatigue-crack growth is characterized by two distinct stages of crack growth rate: (a) a deceleration stage, where the rate of crack growth decreases as the crack grows; and (b) an acceleration stage, where there is a steady increase in crack growth rate to failure. The acceleration stage follows the deceleration stage.

3. There are two competing mechanisms that influence the crack growth rate during constant amplitude fatigue loading: (a) increasing resistance to crack growth owing to the rising R-curve, and (b) increasing $K_I$ with crack length that accelerates the crack growth. The resistance to crack growth increases up to the crack length at the peak load of the quasi-static response and then assumes a constant value. The crack length where the rate of crack growth changes from deceleration to acceleration can hence be predicted from the crack length at the peak load in quasi-static loading.

## ACKNOWLEDGEMENTS

This paper was prepared from a study conducted in the Center of Excellence for Airport Pavement Research. Funding for the Center of Excellence was provided in part by the Federal Aviation Administration under Research Grant Number 95-C-001. The Center of Excellence is maintained at the University of Illinois at Urbana-Champaign and is in partnership with Northwestern University and the Federal Aviation Administration. The authors would also like to acknowledge support from the NSF Center for Advanced Cement Based Materials, Northwestern University, during the course of this study.

## REFERENCES

1.    Karman, T., "Tests on Materials under Triaxial Compression," Verein Deutscher Ingenieure (Berlin), No. 42. 1911.

2.    Bresler, B., and Pister, K., "Strength of Concrete under Combined Stresses," *ACI Journal Proceedings*, Vol. 55, No. 3, Sep. 1958, pp. 321-345.

3.    Goode, C.D., and Helmy, M.A., " The Strength of Concrete under Combined Shear and Direct Stresses," *Magazine of Concrete Research*, vol. 19, No. 59, June 1967, pp. 105-112.

4.  McHenry D., and Karni J., "Strength of Concrete under Combined Tensile and Compressive Stresses," *ACI Journal Proceedings*, Vol. 54, No. 10, Apr. 1958, pp. 829-840.

5.  Kupfer, H., Hilsdorf, H., and Rusch " Behavior of Concrete Under Biaxial Stresses," *ACI Journal proceedings*, Vol. 66, No. 8, August 1969, pp. 656-666.

6.  Kupfer H. B., and Gerstle K.H., " Behavior of Concrete under Biaxial Stresses", *Journal of Engineering Mechanics*, ASCE, Vol. 99, No. EM4, 1973, pp. 853-866.

7.  Hawkins N.M., and Shah S.P.,: American Concrete Institute Considerations for Fatigue. IABSE Colloqium, "Fatigue of Steel and Concrete Structures," Lousanne, March 1982, *Proceedings IABSE Reports*, Vol. 37, Zurich, 1982, pp. 41-50.

8.  RILEM Committee 36-RDL, "Long Term Random Dynamic Loading of Concrete Structures", *RILEM Materials and Structures*, Vol. 17, No. 97, 1984, pp. 1-28.

9.  ACI Committee 215: Fatigue of Concrete Structures, (Editor Shah, S.P.), *American Concrete Institute*, Publication SP-75, Detroit 1982.

10. Buyukozturk O., and Tseng T., "Concrete in Biaxial Compression", *Journal of Structural Engineering*, ASCE, vol. 110, No. 3, March 1984, pp. 461-476.

11. Su E.C.M., and Hsu T.T.C., "Biaxial Compression Fatigue and Discontinuity of Concrete", *ACI Materials Journal*, May-June 1988, pp. 178-188.

12. Nelson E.L., Carrasquillo R.L., and Fowler D.W., "Behavior and Failure of High-Strength Concrete Subjected to Biaxial-Cyclic Compression Loading", *ACI Material Journal*, July-Aug 1988, pp. 248-253.

13. Shah, S.P., Swartz, S.E., Ouyang, C., "Fracture Mechanics of Concrete", Wiley Intersciense, New York, 1995.

14. Jenq, Y.S., and Shah,S.P., "Two Parameter Fracture Model for Concrete," *Journal of Engineering Mechanics*, ASCE, Vol. 111, No. 10, 1985, pp. 1227-1241.

15. Subramaniam Kolluru V., (1999), "Fatigue of Concrete Subjected to Biaxial Loading in the Tension Region," PhD dissertation, Northwestern University.

16. Subramaniam, K.V., O'Neil, E., Popovics, J.S., and Shah, S.P. (2000), "Flexural Fatigue of Concrete: Experiments and Theoretical Model," *J. Engrg. Mech., ASCE,* vol 126(9), pp.891-898.

17. Subramaniam, K.V., Popovics, J.S., and Shah, S.P., (1999) "Fatigue Behavior of Concrete subjected to Biaxial Stresses in the C-T Region," *Mat. Jrnl., ACI,* Vol 96, No.6, pp. 663-669.

18. Ouyang, C., and Shah, S.P., " Geometry Dependent R-Curve for Quasi-Brittle Materials," *Journal of American Ceramic Society,* Vol 74, No. 11, 1991, pp. 2831-2836.

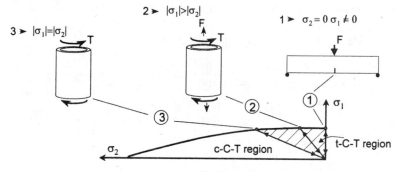

Figure 1: Schematic representation of the test configurations in the biaxial t-C-T stress-state.

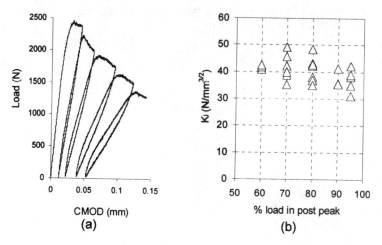

Figure 2: (a) Typical cyclic quasi-static response from a flexure test, (b) The $K_I$ stress intensity factor at different points in the post-peak of quasi-static response.

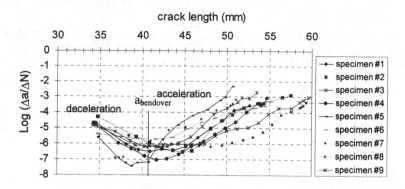

Figure 3: Rate of fatigue crack growth as a function of crack length during constant amplitude flexural fatigue loading.

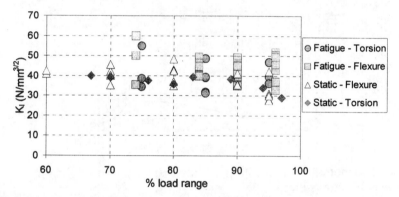

Figure 4: $K_I$ at fatigue failure and in the quasi-static post peak of specimens tested in flexure and torsion.

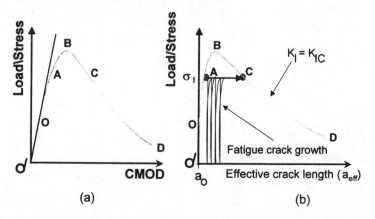

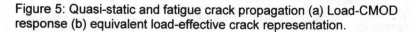

Figure 5: Quasi-static and fatigue crack propagation (a) Load-CMOD response (b) equivalent load-effective crack representation.

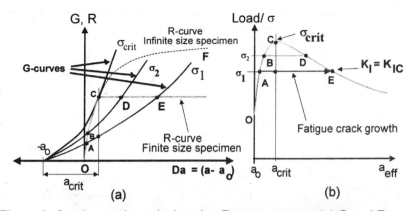

Figure 6: Crack growth analysis using R-curve concept (a) G and R-curves showing the energy release rates at different applied stress levels and the resistance to crack growth with increasing crack length (b) load-effective crack representation of the load-deflection response.

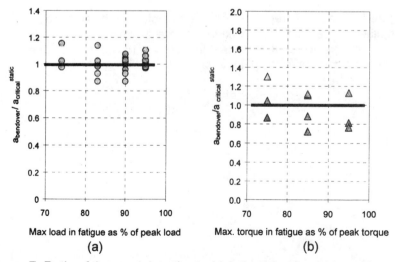

Figure 7: Ratio of the crack length at which the rate of crack growth changes from deceleration to acceleration ($a_{bendover}$) to crack length corresponding to the peak load ($a_{critical}^{static}$) in quasi-static loading (a) flexure (b) torsion.

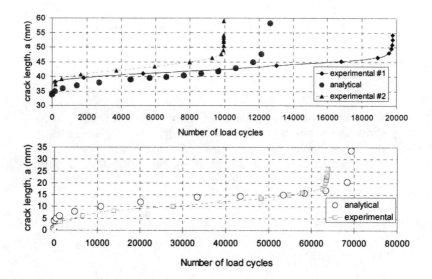

Figure 8: Analytical prediction of fatigue crack growth using the simplified model (a) flexural fatigue with maximum load equal to 95% of quasi-static peak load (b) torsional fatigue with maximum torque equal to 75% of the quasi-static peak torque.

# Analysis of Reinforced Concrete Beams Strengthened with Composites Subjected to Fatigue Loading

by C. G. Papakonstantinou, P. N. Balaguru, and M. F. Petrou

**Synopsis:** Use of high strength composites for repair and rehabilitation of bridges and parking decks is steadily increasing. Since these structures are subjected to fatigue loading, the performance of strengthened beams under this type of loading needs to be evaluated. An analytical procedure that incorporates cyclic creep of concrete and degradation of flexural stiffness is presented. The method is verified by computing cycle dependent deflections and comparing them with experimental results. The results presented in this paper also provide a summary of an experimental investigation, in which reinforced concrete beams were strengthened with glass fabrics (sheets) and subjected to fatigue loading. The comparison shows that the analytical model provides reasonably accurate prediction of deflections, for both reinforced beams and reinforced concrete beams reinforced with composites. Although Glass fiber composites were used for the evaluation, the model is also applicable to other types of fibers.

Keywords: analysis; carbon; deflection; fatigue; fiber reinforced polymer; glass; reinforced concrete; stiffness; strengthening

ACI student member **C.G. Papakonstantinou** is working towards a Ph.D. degree in Structural Engineering at Rutgers, the State University of New Jersey. His research interests include fatigue of reinforced concrete, strengthening of reinforced concrete with high strength composites, use of inorganic polymers in structural elements.

ACI member **P.N. Balaguru** is Professor of the Department of Civil and Environmental Engineering, at Rutgers, the State University of New Jersey. His research interests include fatigue of reinforced concrete, behavior of reinforced concrete and ferrocement, corrosion of reinforced concrete, use of inorganic polymers for composites, development of innovative construction materials.

ACI member **M.F. Petrou** is an Associate Professor and Graduate Director of the Department of Civil and Environmental Engineering, University of South Carolina, Columbia. His research interests include civil engineering materials, behavior of reinforced and prestressed concrete structural elements, structural modeling, laboratory and field-testing of bridges, and corrosion.

## INTRODUCTION

There is an increasing interest in the use of high strength composites for repair and rehabilitation of reinforced concrete elements. Since most of these elements are structural members of bridges or parking garages, there is a need to understand the behavior of strengthened elements under repeated loading. A small number of researchers have investigated the fatigue properties of reinforced concrete beams strengthened with glass and carbon fibers, (1),(2),(3),(4).

Meier (1992), (1), reported two fatigue tests of reinforced concrete beams strengthened with a glass/carbon-hybrid sheet. Barnes and Mays (1999), (2), reported fatigue test results of two reinforced concrete beams and three reinforced concrete beams strengthened with CFRP plates. Papakonstantinou (2000), (4), also reported on fatigue tests of six reinforced concrete beams and eight reinforced concrete beams strengthened with GFRP plates. The results of all these tests indicate that beams failed in a primary flexural mode and initial failure was due to the steel fracture rather than failure of concrete, adhesive failure or fracture of the FRP plates. There were no distinguishable differences in behavior between the strengthened and non-strengthened beams during load cycling. The number of cycles to failure varied according to the stress range of steel reinforcement.

Shahawy and Beitelman (1999), (3), tested six tee-beams under fatigue loading, using a load range of 0-25% of the ultimate load. Two beams were used as control beams, two were fully wrapped with two layers of CFRP and two were wrapped with three layers of CFRP on the tension side. From the results of these tests it was concluded that the failure was based on the fracture of steel, the use of FRP increases the fatigue life of the beams as well as that the increase of the number of layers of CFRP results in a further increase of the fatigue life of the beams.

The aforementioned investigations were conducted mainly to obtain the fatigue strength of the strengthened beams. There is no study so far to address the serviceability aspect of the strengthened beams subjected to fatigue loading. Since concrete beams subjected to a repeated loading will experience an increase in deflections, satisfactory prediction of this time dependent quantity may be important in cases where serviceability criteria govern the design. In addition excessive deflection increase could signal the impending failure. In certain cases there could be change in stress in reinforcement due to creeping of concrete.

In this paper an analytical model is presented which can be used to predict the increase in deflection of reinforced concrete beams strengthened with high strength fiber reinforced composite fabric. The analytical results are compared with available experimental data. The model can also be used to compute any changes in stresses due to cyclic creep of concrete.

## ANALYTICAL PROCEDURE

Balaguru and Shah (1982), (5), reported an analytical model, which can be used to predict the increase of deflections of reinforced concrete beams, subjected to fatigue loading. Based on this model, the two major contributing factors are: (i) cyclic creep of concrete and (ii) degradation of flexural stiffness due to increase in cracking and reduction in modulus of rupture under fatigue loading.

The cyclic creep strain of concrete can be expressed as the sum of two strain components; a mean strain component, based on $\sigma_m = ((\sigma_{max} + \sigma_{min})/2)/fc'$ and a cyclic strain component, based on $\Delta = (\sigma_{max} - \sigma_{min})/fc'$. A regression equation based on experimental results was obtained as:

$$\varepsilon_c = 129 \cdot \sigma_m \cdot t^{1/3} + 17.8 \cdot \sigma_m \cdot \Delta \cdot N^{1/3} \tag{1}$$

where     $\varepsilon_c$ is the cyclic creep strain in micro mm/mm

$\Delta$ is the stress range expressed as a fraction of the compressive strength

$\sigma_m$ is the mean stress expressed as a fraction of the compressive strength

$\sigma_{max}$ is the maximum applied compressive stress in concrete

$\sigma_{min}$ is the minimum applied compressive stress in concrete

N is the number of cycles

t is the time from start of loading in hours.

Once the cyclic creep strain is known, the cycle-dependent secant modulus for concrete in compression, $E_N$ can be expressed using the equation:

$$E_N = \frac{\sigma_{max}}{\dfrac{\sigma_{max}}{E} + \varepsilon_c} \tag{2}$$

where E is the initial secant modulus

$E_N$ is the cyclic modulus after N number of cycles

Use of this cycle dependent modulus for the analysis is presented in the following section.

## Outline of Analytical Procedure for a Rectangular Beam

The analytical procedure is explained using a rectangular section reinforced with steel and strengthened with composites. The details of beam cross section, strain and stress distribution at working loads are shown in Fig. 1.

The depth of the neutral axis, kd can be computed using the force equilibrium equation. The contribution of the composite is considered to be similar to the contribution of steel reinforcement. The area of composite is multiplied by the modular ratio, $n_f$. In addition to the assumptions of classical bending theory, it is assumed that: (i) there is a perfect bond between the composite plate and the beam up to failure and (ii) the behavior of the composite plate is linearly elastic up to failure.

$$b\frac{(kd)^2}{2}+(n-1)A_s{}'(kd-d') = nA_s(d-kd)+n_fA_f(h-kd) \quad (3)$$

where    $A_s$   is the cross sectional area of the reinforcing bars in tension side

$A_f$   is the cross sectional area of the composite plate in tension face

$A_s$'   is the cross sectional area of the reinforcing bars in compression side

b    is the width of the beam

d    is the depth of the reinforcing bars in the tension side

d'    is the depth of the reinforcing bars in the compression side

kd    is the depth of the neutral axis

h    is the thickness of the beam

n    is the modular ratio of steel reinforcement given by the equation: $n = \dfrac{E_s}{E \text{ or } E_N}$

$E_s$    is the modulus of elasticity of the reinforcing bars

$n_f$ is the modular ratio of the composite given by the equation:

$$n_f = \frac{E_f}{E \text{ or } E_N}$$

$E_f$    is the modulus of elasticity of the composite plate

Once the depth of the neutral axis is known, the moment of inertia of cracked section, $I_{cr}$ can be computed using the following equation:

$$I_{cr} = \frac{b(kd)^3}{3}+nA_s(d-kd)^2+n_fA_f(h-kd)^2+(n-1)A_s{}'(kd-d')^2 \quad (4)$$

For the uncracked section, the gross moment of inertia, Ig is given by:

$$I_g = \frac{bh^3}{12} \qquad (5)$$

The stresses at the extreme compression fiber, $\sigma_c$, the composite in the tension fiber, $\sigma_f$, as well as at the reinforcement in the tension, $\sigma_s$, and compression side, $\sigma_s'$, can be calculated using the $I_{cr}$ and the principles of mechanics:

$$\sigma_c = \frac{M}{I_{cr}} kd \qquad (6)$$

$$\sigma_f = \frac{M}{I_{cr}} (h - kd) n_f \qquad (7)$$

$$\sigma_s = \frac{M}{I_{cr}} (d - kd) n \qquad (8)$$

$$\sigma_s' = \frac{M}{I_{cr}} (kd - d') n \qquad (9)$$

where M is the applied moment.

For the first cycle; maximum stress in concrete:

$$\sigma_{max} = \frac{M_{max}}{I_{cr}} kd \qquad (10)$$

while minimum stress in concrete:

$$\sigma_{min} = \frac{M_{min}}{I_{cr}} kd \qquad (11)$$

where $\quad$ $M_{max}$ is the maximum applied moment
$\quad\quad\quad\quad\quad\quad$ $M_{min}$ is the minimum applied moment

Once $\sigma_{max}$ and $\sigma_{min}$ in concrete are known, the cycle dependent modulus $E_N$ for a given number of cycles can be calculated using Eq. 1. and 2.

## Computation of Flexural Stiffness Degradation

In the case of static loading, the contribution of concrete in the tension zone is accounted for by using the effective moment of inertia. The effective moment of inertia can be obtained using the formula from ACI-318 (1999), (6):

$$I_e = I_{cr} + \left(\frac{M_{cr}}{M_a}\right)^3 (I_g - I_{cr}) \qquad (12)$$

where $I_e$, $I_{cr}$, $I_g$ are the effective, cracked and gross moment of inertia, $M_a$ is the maximum value of the applied moment along the beam, and $M_{cr}$ is the cracking moment given by $M_{cr} = I_g f_r / (h - \bar{y})$ where $f_r$ is the modulus of rupture of concrete and $\bar{y}$ is the depth of neutral axis of the uncracked section.

The stiffness of the beam will reduce when subjected to fatigue loading due to the fatigue of concrete in the tension zone. This progressive reduction in the stiffness can be accounted for by using a reduced cycle-dependent modulus of rupture. This relationship can be expressed as:

$$f_{r,N} = f_r \cdot \left( 1 - \frac{\log_{10} N}{10.954} \right) \qquad (13)$$

It should be noted that the increase of the number of cycles will lead to the reduction of the modulus of rupture, cracking moment $M_{cr}$ and consequently the effective moment of inertia $I_e$.

For the static loads, the deflections are calculated using the initial effective moment of inertia ($I_e$) and the first cycle modulus of elasticity of concrete (E), that is:

$$\delta = \frac{f(load, span)}{E \cdot I_e} \qquad (14)$$

where $\delta$ is the instantaneous deflection

f(load, span) is a function for the particular load and span arrangement

and $EI_e$ is the initial effective stiffness of the beam

In the case of fatigue loading equation (14) changes to:

$$\delta = \frac{f(load, span)}{E_N \cdot I_{e,N}} \qquad (15)$$

where $E_N I_{e,N}$ is the reduced stiffness of the beam after N cycles.

$E_N$ is given by equation (2) and $I_{e,N}$ is given by the following equation:

$$I_{e,N} = I_{cr,N} + \left( \frac{M_{cr,N}}{M_a} \right)^3 (I_g - I_{cr,N}) \qquad (16)$$

$$M_{cr,N} = \frac{I_g}{(h - \bar{y})} f_{r,N} \qquad (17)$$

where        $f_{r,N}$ is given by equation (13) and

$I_{cr,N}$ is the moment of inertia calcutated using $E_N$

The tensile stiffeness degradation is accounted by the term $M_{cr,N}$ while the effect of the concrete creep is included in the terms $E_N$ and $I_{cr,N}$.

A step by step procedure and a numerical example are presented in the Appendix.

### DETAILS OF EXPERIMENTAL RESULTS USED FOR EVALUATION OF THE MODEL

The experiments reported by Papakonstantinou (2000), (4), were used for evaluating the model. In this investigation seventeen reinforced concrete beams, nine of them strengthened with a Glass Fiber Reinforced Polymer (GFRP) system were tested in third-point flexure. Four beams were initially tested under static load in order to establish their ultimate load carrying capacity.

The average compressive strength of concrete was 40 MPa whereas the yield strength of the steel was 427 MPa. The beams were 1320 mm long, 152 mm wide and 152 mm thick and were simply supported over a span of 1220 mm. Two 12.7 mm diameter bars were used as flexural tension reinforcement, and two more were placed in the compression region for fabrication ease. For shear reinforcement, 9.5 mm bars were used.    The flexural and shear reinforcement was designed to ensure flexural failure.  Twenty 9.5 mm stirrups were placed in the beam as shear reinforcement at a spacing of 50 mm in the two outer thirds of the beam, and two stirrups were spaced evenly in the center third, zero-shear/constant-moment portion of the beam. There was a 25 mm clear cover of concrete for each side of the stirrups except for the top surface, which had a 13 mm cover.  Specimen details are shown in Fig. 2.

Two designations: N, for non-repaired reinforced concrete beams, and S, for beams strengthened with GFRP composite sheets, were used. Each beam was numbered in a consecutive order.  All the repaired beams were reinforced over their entire tensile face with a GFRP sheet. The GFRP system had a strength of 1,730 MPa and a stiffness of 72.4 GPa. Repaired specimens were allowed to cure at least 10 days after the FRP application.

A summary of the beams and the loading conditions is presented in Table 1. The maximum and minimum loads applied are presented for the fatigue tests. It should be noted that the maximum loads were always determined based on the strain on the steel reinforcement. The percentage of ultimate capacity is determined from the static tests. The minimum applied load was necessary to avoid the instability because the supports provide restraint in only one direction.

The tests were carried out using a 270 kN capacity hydraulic actuator. The actuator was operated under load control for the cycling loading and under stroke displacement control for the static tests. The load was applied at the third points of the beam within the 1220 mm supported span. All contact points along the beam were supported with rollers to ensure that the load was applied evenly during testing (Fig. 2).

Strains on steel, concrete and the GFRP sheet as well as deflections at midspan were recorded during testing in order to study their change under static and fatigue loading conditions. Two electrical resistance strain gages were attached on opposite sides at the mid-span of each longitudinal reinforcing bar to resume the axial strain in the bar. Strain gages were also located on the top concrete surface as well as on the GFRP sheet at selected locations (Fig. 2).

The deflection at mid-span was measured using a "yoke" deflection measuring system. This system, illustrated in Fig. 2, consists of an aluminum frame, bolted on the neutral axis of the beam at the end supports and two LVDT's mounted on the "yoke" at midspan, one on each side of the beam. The LVDT's measure the deflection of the specimen through contact with two aluminum plates, bonded to the beam with epoxy. The average of the two measurements represents the net mid-span deflection

## Experimental Results from Static Tests

Two non-repaired (N-1 and N-2) beams and one repaired (S-4) beam were tested under monotonic loading conditions and were used as control beams for the fatigue testing. The experimental results are shown in Fig. 3. The strengthened beam was stiffer than the control beams before and after reinforcement yielding. FRP increased the ultimate load capacity by about 50% and the reduction in ductility was less than 10%. The failure was initiated by yielding of the reinforcement and subsequent delamination of the GFRP sheets.

## Experimental Results from Fatigue Tests

Fourteen beams were tested under cycling load at different load ranges. Six of them were non-repaired and eight repaired. The maximum load in each case was chosen based on the maximum stress on the steel reinforcement. The minimum load depended on the frequency of the loading and was selected to ensure stability of the test set up. The maximum number of cycles applied was limited to $2 \times 10^6$ cycles. During all fatigue tests, static tests were conducted on a regular basis in order to monitor the damage accumulation and its effect on the deflection and stress values. All tests were conducted at 2 or 3 Hz (Table 1).

The data in the Table 1 include the maximum and minimum applied load and strain on the reinforcing bars and on the concrete (compression), as well as the ratio of applied load to ultimate ($P_{max}/P_{ult}$) and yielding loads ($P_{max}/P_y$), the number of cycles to failure. The ultimate values were determined from the original static control tests.

The change of deflections with the increase of the number of cycles for all beams that failed under cycling loading are presented in Fig. 4. For each beam the deflection was recorded during monotonic loading tests. Note that in all beams there was an initial substantial increase of the deflections, followed by a stable region where the deflection increases were minimal. Just before failure there was another period of substantial deflection increase. It can also be observed that, in all beams that failed after 400,000 cycles, there is a more gradual increase in deflections with increase in number of cycles. This may be caused by a greater number of cracks with increased cycling. Generally, however, the deflection remained the same until approximately 100,000 cycles before failure occurred. This is important, since one can detect an upcoming failure by monitoring the deflections.

## COMPARISON OF ANALYTICAL AND EXPERIMENTAL RESULTS

Comparison of experimental and analytical deflection values are presented in Figs. 5 to 18. These figures show graphs of normalized deflection vs. number of cycles for each beam subjected to fatigue loading. The values of the experimental deflections were normalized using the first value recorded after the beginning of the experiment. The normalized theoretical values were obtained by dividing each of the theoretical values by the value calculated for the corresponding experimental value.

Fig. 5. presents the comparison data for the beam N-3. It is evident that there is a very good correlation between the predicted theoretical and experimental values. The predicted values are also close to the experimental for the beams N-6, N-7 and N-8 (Figs. 8, 9 and 10). There is only some inconsistency between the analytical and the experimental values of beam N-5 (Fig. 7.). In this case the experimental deflections remained constant for almost 1,500,000 cycles and there is a sudden increase at 1,600,000 cycles, whereas the analytical values increase gradually. It should be noted that there is only 5% difference during the last cycles.

Comparison charts for the strengthened beams S-2, S-5, S-6, S-7, S-8, S-9, S-10 and S-11 are shown in Figs. 11, 12, 13, 14, 15, 16, 17 and 18 respectively. These graphs show, that the predicted values are close to the experimental, in almost all of the cases. The difference in all cases is less than 15%. There is only one case where the prediction under-estimates the experimental values. That happened for the beam S-10 (Fig. 17.) and the reason is probably a slippage of the "yoke" on the beam. It should also be noted that the graph for beam S-8 is limited to 20,000 cycles because there was a sudden increase in deflection due to a shear crack. In addition, the maximum applied load was higher than the yielding load. The main reason for using such a high load was to have a big variation of maximum applied loads.

Comparison of strains and stresses at steel and composite is not presented, since the increase in strains was very small in all cases. The analytical model does predict the small increase in strains that happens due to stress redistribution caused by cyclic creep of concrete.

## CONCLUSIONS

Based on the analysis described, the following conclusions can be drawn:

- The proposed analytical model provides reasonable accurate prediction of deflections, for both reinforced concrete beams and reinforced concrete beams strengthened with composites. Although Glass fiber composites were used for the evaluation, the model could also be applicable to other types of fibers.
- Cyclic loading lead to increase in deflections for both control and strengthened beams. The deflection increases are slightly lower for strengthened beams.

## REFERENCES

1. Meier, U., Deuring, M., Meier, H., and Schwegler, G.. "Strengthening of structures with CFRP laminates: Research and applications in Switzerland." *Advanced Composite Materials in Bridges and Structures*, K. W. Neale and P. Labossiere, eds., Canadian Society for Civil Engineers. 1992.

2. Barnes, R.A. and Mays, G.C. "Fatigue Performance of Concrete Beams Strengthened with CFRP Plates." *ASCE Journal of Composites for Construction*, 3(2), 1999, pp. 63-72.

3. Shahawy, M. and Beitelman, T.E. "Static and Fatigue Performance of RC Beams Strengthened with CFRP Laminates." *ASCE Journal of Structural Engineering*, 125(6), 1999, pp. 613-621.

4. Papakonstantinou, C.G. "Fatigue Performance of Reinforced Concrete Beams Strengthened with Glass Fiber Reinforced Polymer Composite Sheets." M.S. Thesis, University of South Carolina, Columbia, SC. 2000.

5. Balaguru, P.N. and Shah, S.P. " A Method of Predicting Crack Widths and Deflections for Fatigue Loading" ACI Special Publication 75-5, Detroit, 1982, pp. 153-175.

6. ACI Committee 318. "Building Code Requirements for Structural Concrete." *American Concrete Institute*, 1999, pp 96-101.

Table 1 Summary of monotonic and fatigue test results.

| Beam | Testing Frequency | Ultimate Applied Load $P_{ult}$ (kN) | Applied Load at Yield $P_y$ (kN) | Applied Load (kN) max | Applied Load (kN) min | $P_{max}/P_{ult}$ | $P_{max}/P_y$ | Strain on Steel (microstrain) max | Strain on Steel (microstrain) min | Strain on Concrete in compression (microstrain) max | Strain on Concrete in compression (microstrain) min | Number of Cycles to Failure N |
|---|---|---|---|---|---|---|---|---|---|---|---|---|
| non-strengthened | | | | | | | | | | | | |

**Non-strengthened reinforced concrete beams**

| Beam | Testing Frequency | $P_{ult}$ (kN) | $P_y$ (kN) | max | min | $P_{max}/P_{ult}$ | $P_{max}/P_y$ | Steel max | Steel min | Conc. max | Conc. min | N |
|---|---|---|---|---|---|---|---|---|---|---|---|---|
| N-1 | monotonic | 73.8 | 58.2 | na | na | na | na | 2400[1] | na | 3000 | na | na |
| N-2 | monotonic | 73.3 | 58.2 | na | na | na | na | 2460[1] | na | 3100 | na | na |
| N-4 | 3 Hz | assumed to be the same as N-1 and N-2 | | 31.2 | 3.3 | 0.42 | 0.53 | 950 | 60 | 742 | 40 | 2,000,000 |
| N-5 | 3 Hz | | | 35.6 | 3.3 | 0.48 | 0.61 | 1100 | 60 | 1500 | 63 | 2,000,000 |
| N-8 | 3 Hz | | | 40.0 | 3.3 | 0.52 | 0.68 | 1493 | 124 | 852 | 71 | 650,000 |
| N-3 | 2 Hz | | | 43.6 | 3.3 | 0.59 | 0.74 | 1701 | 85 | 878 | 44 | 275,000 |
| N-6 | 2 Hz | | | 53.4 | 4.4 | 0.73 | 0.91 | 2010 | 143 | 1090 | 77 | 155,000 |
| N-7 | 2 Hz | | | 62.3 | 3.3 | 0.85 | 1.05 | 2350 | 99 | 1500 | 64 | 80,000 |

**Reinforced concrete beams strengthened with one ply of GFRP**

| Beam | Testing Frequency | $P_{ult}$ (kN) | $P_y$ (kN) | max | min | $P_{max}/P_{ult}$ | $P_{max}/P_y$ | Steel max | Steel min | Conc. max | Conc. min | N |
|---|---|---|---|---|---|---|---|---|---|---|---|---|
| S-4 | monotonic | 109.7 | 80.1 | na | na | na | na | 2500[1] | na | 3500 | na | na |
| S-11 | 3 Hz | assumed to be the same as S-4 | | 40.0 | 3.3 | 0.36 | 0.50 | 1274 | 105 | 950 | 23 | 6,000,000 |
| S-10 | 3 Hz | | | 44.5 | 3.3 | 0.41 | 0.55 | 1419 | 106 | 979 | 73 | 685,000 |
| S-2 | 3 Hz | | | 46.7 | 2.2 | 0.43 | 0.58 | 1410 | 66 | 700 | 33 | 880,020 |
| S-5 | 3 Hz | | | 48.9 | 4.0 | 0.45 | 0.61 | 1477 | 101 | 890 | 61 | 800,000 |
| S-7 | 3 Hz | | | 53.4 | 3.3 | 0.49 | 0.67 | 1580 | 99 | 808 | 51 | 570,000 |
| S-9 | 2 Hz | | | 57.8 | 3.3 | 0.53 | 0.72 | 1825 | 110 | 1090 | 66 | 235,000 |
| S-6 | 2 Hz | | | 64.5 | 4.4 | 0.59 | 0.81 | 2080 | 160 | 1284 | 99 | 126,000 |
| S-8 | 2 Hz | | | 80.1 | 4.0 | 0.73 | 1.0 | 2350 | 135 | 1563 | 90 | 30,500 |

[1] strain in reinforcing steel at yield, $P_y$.

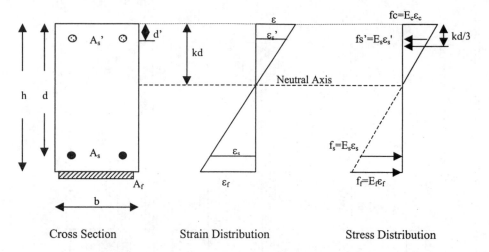

Cross Section          Strain Distribution          Stress Distribution

Fig. 1. Strain and stress distributions of strengthened reinforced concrete section at working load

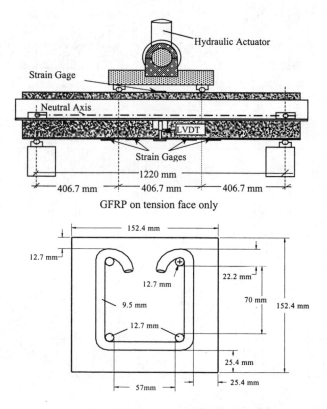

Fig. 2 Test set-up and specimen details.

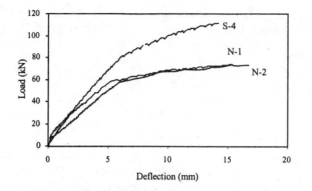

Fig. 3 Load vs Deflection diagram for static tests

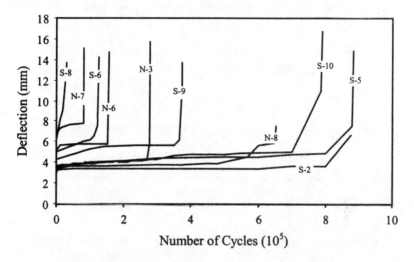

Fig. 4 Deflection vs. number of cycles due to fatigue (5).

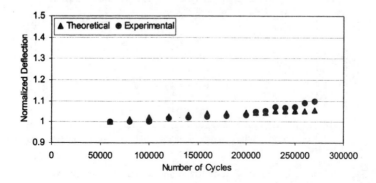

Fig. 5 Experimental and analytical normalized deflection vs. number of cycles (beam N-3)

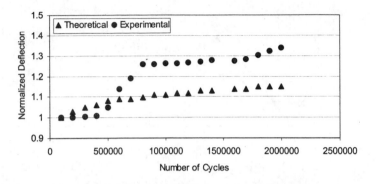

Fig. 6 Experimental and analytical normalized deflection vs. number of cycles (beam N-4)

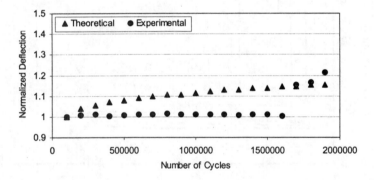

Fig. 7 Experimental and analytical normalized deflection vs. number of cycles (beam N-5)

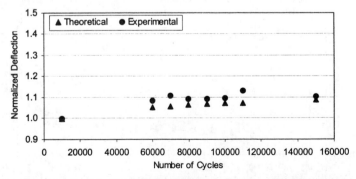

Fig. 8 Experimental and analytical normalized deflection vs. number of cycles (beam N-6)

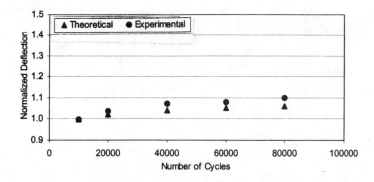

Fig. 9 Experimental and analytical normalized deflection vs. number of cycles (beam N-7)

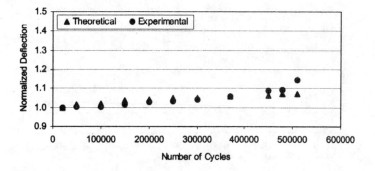

Fig. 10 Experimental and analytical normalized deflection vs. number of cycles (beam N-8)

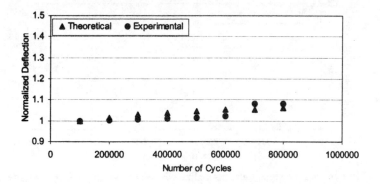

Fig. 11 Experimental and analytical normalized deflection vs. number of cycles (beam S-2)

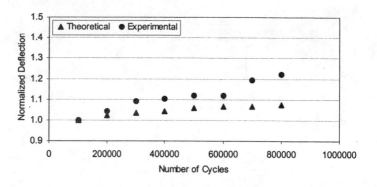

Fig. 12 Experimental and analytical normalized deflection vs. number of cycles (beam S-5)

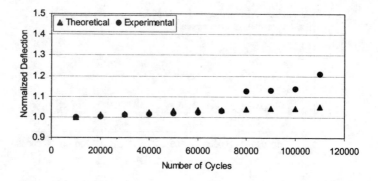

Fig. 13 Experimental and analytical normalized deflection vs. number of cycles (beam S-6)

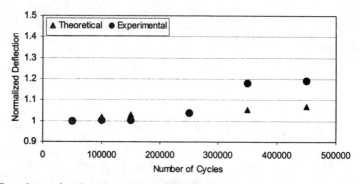

Fig. 14 Experimental and analytical normalized deflection vs. number of cycles (beam S-7)

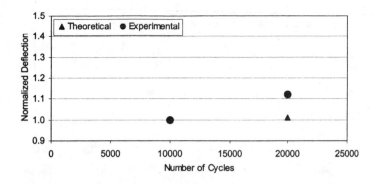

Fig. 15 Experimental and analytical normalized deflection vs. number of cycles (beam S-8)

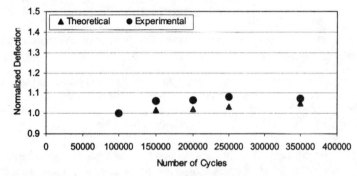

Fig. 16 Experimental and analytical normalized deflection vs. number of cycles (beam S-9)

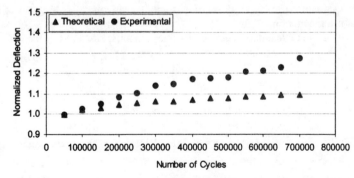

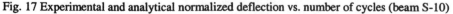

Fig. 17 Experimental and analytical normalized deflection vs. number of cycles (beam S-10)

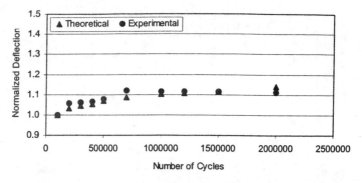

Fig. 18 Experimental and analytical normalized deflection vs. number of cycles (beam S-11)

## APPENDIX

### I. Step-by-Step Procedure for Computation of cycle dependent stresses and deflections.

The procedure described below is for given cross-sectional properties of a reinforced concrete beam strengthened with composite plates, maximum moment, $M_a$, number of cycles, N, and known properties of the materials.

1. Calculate $I_g$ and $\bar{y}$ from the cross sectional dimensions and properties of the material.
2. Calculate the depth of neutral axis kd, and the maximum compressive stress, $\sigma_{max}$, under the static bending moment $M_a$ from the cross sectional properties and the properties of the materials using cracked section analysis
3. Calculate the modulus of rupture of concrete $f_{r,N}$ using equation (13)
4. Calculate $M_{cr,N}$ from equation (17)
5. Calculate the apparent modulus of elasticity of concrete after N cycles of loading, $E_N$ using equation (1) and (2)
6. Calculate the cycle dependent depth of neutral axis, $kd_N$ and $I_{cr,N}$ using $E_N$ and $E_s$
7. Calculate $I_{e,N}$ using equation (16)
8. Calculate the deflection using equation (15)

### II. Numerical Example  (Beam S-2 at 800,000 cycles)

Compressive strength of concrete, fc'= 40 MPa, Yield strength of reinforcement, fy = 427 MPa, Modulus of elasticity of concrete, $E_c$ = 30 GPa, Modulus of elasticity of steel reinforcement, $E_s$ = 200 GPa, Modulus of elasticity of composite, $E_f$= 72.4 MPa, Width of beam, b = 152 mm, Height of beam, h = 152 mm, Span length, l = 1220 mm, depth of the reinforcing bars in the tension side, d = 109 mm, depth of the reinforcing bars in the compression side, d' = 29 mm, $A_s$'= $A_s$ = 258 mm$^2$, $A_f$= 129 mm$^2$, Maximum applied load, P = 46.7 kN

Maximum moment, Ma = 9.45 kNm, Maximum stress on concrete, $\sigma_{max}$ = 21 MPa, Minimum stress on concrete, $\sigma_{min}$ = 1 MPa, Number of cycles, N=800,000

Gross moment of inertia, Ig = bh$^3$/12=152*152$^3$/12=44.5x10$^6$ mm$^4$

$\Delta = (\sigma_{max}-\sigma_{min})/f_c' = (21-1)/40 = 0.5$

$\sigma_{max}= (\sigma_{max}+\sigma_{min})/2f_c' = (21+1)/2*40 = 0.275$

$\varepsilon_c =129 \cdot \sigma_m \cdot t^{1/3}+17.8 \cdot \sigma_m \cdot \Delta \cdot N^{1/3} = 372.3$

$$E_N = \frac{\sigma_{max}}{\frac{\sigma_{max}}{E} + \varepsilon_c} = \frac{21}{\frac{21}{30000}+372.3\times10^{-6}} = 19584 \text{ MPa}$$

$f_r$= 4 MPa, $n = \dfrac{E_s}{E_N} = \dfrac{200}{19.6} = 10.2$, $n_f = \dfrac{E_f}{E_N} = \dfrac{72.4}{19.6} = 3.69$

The depth of the neutral axis at 800,000 cycles, $k_d$ can be computed using the force equilibrium equation:

$$b\frac{k_d^2}{2}+(n-1)A_s'(k_d - d') = nA_s(d - k_d)+n_f A_f(h - k_d) \Rightarrow k_d = 47.3mm$$

Using the $k_d$ calculated above we can calculate the $I_{cr,N}$:

$$I_{cr,N} = \frac{b\,k_d^3}{3}+nA_s(d - k_d)^2 +n_c A_c(h - k_d)^2 +(n-1)A_s(k_d - d')^2 = 21.62\times10^6 mm^4$$

$$f_{r,N} = f_r \cdot \left(1-\frac{\log_{10} N}{10.954}\right)=4 \cdot \left(1-\frac{\log_{10} 800000}{10.954}\right)=1.8 \ MPa$$

$$M_{cr,N} = \frac{I_g}{k_d} f_{r,N} = 1.69kNm$$

$$I_{e,N} = I_{cr,N} +\left(\frac{M_{cr,N}}{M_a}\right)^3 (I_g - I_{cr,N}) = 21.75\times10^6 mm^4$$

$$\delta = \frac{23 \cdot M_a \cdot l^2}{216 \cdot E_N \cdot I_{e,N}} = 3.52mm$$

The recorded value for beam S-2 at 800,000cycles was 3.62 mm

# Fatigue Investigation of the Steel-Free Bridge Deck Slabs

## by A. A. Mufti, A. H. Memon, B. Bakht, and N. Banthia

Synopsis: During its lifetime, a bridge deck slab is subjected to a very large number of wheels of different magnitudes. By contrast, the laboratory investigation of the fatigue resistance of a bridge deck slab is usually conducted under wheel loads of constant magnitude. No current method seems to be available for establishing the equivalence between the actual wheel population on a bridge and fatigue test loads. The design codes are also not explicit in this respect.

Taking the statistics of wheel loads from a Japanese survey, an upper-bound data-base is prepared for expected wheel population in Canada. There is close correspondence between the maximum wheel loads observed in Japan and Canada. The data-base is likely to be applicable to the USA as well.

A fairly simple mathematical model has been developed to determine the number of passes of one wheel load so that the damage induced by it is equivalent to the damage caused by a given number of passes of another load of known magnitude. The model, which does not necessarily assume a linear relationship between accumulated damage and the number of passes, can be used to determine the number of passes of any test wheel with respect to any predetermined wheel load statistics. The method can also be used to formulate specific fatigue load requirements for bridge deck slabs.

Two sets of tests on full-scale models of both cast-in-place and precast deck steel-free slabs have been described briefly. It is concluded that both these slabs have more fatigue strength than required to sustain the projected population of wheel loads.

Keywords: bridge deck slab; deck slab; fatigue resistance; steel-free deck slab; wheel loads

ACI member Aftab A. Mufti is Professor of Civil Engineering at the University of Manitoba, Canada, and President of ISIS Canada. He has been involved in structural engineering research for more than 35 years, and is one of the inventors of the steel-free deck slab. He has received many awards for his research work.

Amjad H. Memon is a Ph.D. student in the Department of Civil Engineering at the University of Manitoba, Canada; he is also a member of the ISIS Canada research team.

ACI member Baidar Bakht is President of JMBT Structures Research Inc., Toronto, Canada. He has been involved in bridge engineering research for about 35 years, and is one of the inventors of the steel-free deck slab. He has received many awards for his research work.

ACI member Nemkumar Banthia is Professor of civil engineering at the University of British Columbia, Canada. He serves on ACI, RILEM, CSA, and CSCE committees. He has authored or coauthored more than 60 papers on concrete and fiber reinforced concrete. He has received many awards for his research work.

## INTRODUCTION

During its lifetime, a bridge deck slab is subjected to several hundred million truck wheels, the loads on which range from the lightest to the heaviest. Passes of lighter wheels are very large in number, whereas those of the heavier wheels are fewer. By contrast, the laboratory investigation of the fatigue resistance of a deck slab is usually conducted under a test load of constant magnitude. The time available for such investigations is necessarily much smaller than the lifetime of a bridge. Consequently, the test loads are kept large so that the number of passes required to fail the slab in fatigue are manageably small. To authors' knowledge, no method is currently available to correlate the actual wheel loads with the fatigue test loads. The design codes (e.g. AASHTO (1) and CHBDC (2)) are also not explicit with respect to the design fatigue loads on deck slabs. An analytical method is presented in this paper for establishing the equivalence between fatigue test loads and a given population of wheel loads. While the method is general enough to be applicable to all deck slabs of concrete construction, it is developed especially for steel-free deck slabs (e.g., (3), (4), (5) and (6)), which are relatively new and do not have a long track-record of field performance.

## WHEEL LOADS DATA

Commentary Clause C3.6.1.4.2 of the AASHTO Specifications (1998) notes that the Average Daily Traffic (ADT) in a lane is physically limited to 20,000 vehicles, and the maximum fraction of trucks in traffic is 0.20. Thus the maximum number of trucks per day in one direction ($ADTT$) is 4,000. When two lanes are available to trucks, the number of trucks per day in a single lane, averaged over the design life, ($ADTT_{SL}$) is found by multiplying $ADTT$ with 0.85, giving $ADTT_{SL} = 3,400$. It is assumed that the average number of axles per truck is four (a conservative assumption), and that the life of a bridge is 75 years. The maximum number of axles that a bridge deck would experience in one lane during its lifetime = $3400 \times 4 \times 365 \times 75 = 372$ million. A well-confined deck slab under a wheel load fails in the highly-localized punching shear mode. Accordingly, the consideration of wheel loads in more than one lane is not necessary.

The Calibration Report in the Commentary to the Canadian Highway Bridge Design Code (CHBDC, 2000) is based on vehicle weight surveys in four Canadian provinces; from this report, it can be calculated that the expected annual maximum axle loads in the Canadian provinces of Ontario, Alberta, Saskatchewan and Quebec are 314, 150, 134 and 159 kN, respectively. The expected maximum lifetime axle loads are about 10 % larger than the annual maximum loads (7), thus leading to the maximum lifetime axle load anywhere in Canada being 345 kN. As noted by Matsui et al. (8), the maximum wheel load observed in Japan is 32 t, or 313 kN. The close correspondence between the expected annual maximum axle weight in Canada and the maximum observed axle load in Japan indicates similarity between the axle loads in the two countries. Matsui (9) have also provided a histogram of axle weights observed on 12 bridges in Japan. In the absence of data on Canadian trucks, this histogram was used to construct the wheel load statistics, which are shown in Table 1. This table also includes the numbers of wheels of various magnitudes, corresponding to a total of 372 million wheel passes. Any fatigue test load on a bridge deck slab should induce the same damage in the slab as the damage induced by all the wheel loads included in this or similar table.

## NUMBER OF CYCLES VERSUS FAILURE LOAD

A given number of cycles $N$ of a load $P$ can be equated to $N_e$ cycles of an experimental load $P_e$ only on the basis of an established relationship between $P$ and $N$. Matsui and his colleagues in Japan are the only researchers (8) who have provided a $P$-$N$ relationship based on rolling wheel tests on full-scale models of both reinforced concrete and reinforcement-free deck slabs; their conclusions are quantified by the following equation, which is applicable to both reinforced and unreinforced slabs.

$$\log (P/P_s) = -0.07835 \times \log (N) + \log (1.52) \qquad (1)$$

where $P_s$ is the static failure load. Equation (1) gives $P/P_s$ greater than 1.0 for $N$ smaller than about 500. Matsui contends that this equation is valid only for $N$ greater than 10,000.

North American Researchers (10) and (11)) and Korean Researchers (12) have also presented similar relationships. However, their results are based on tests on small-scale models (1-6.6 and 1-3.3, respectively). The slight differences between the Matsui et al. (8) relationship and those by the above researchers could be attributed to the effect of scale in the models.

Several standard cylindrical specimens of 35 MPa concrete have recently been tested in the University of Manitoba under compressive fatigue loads. The cylinders can be seen in Fig. 1, and the test set up in Fig. 2. Notwithstanding the inconclusive nature of results of tests on some specimens, it was observed that the concrete cylinders do fail in fatigue under compressive loads; the fatigue failure loads are smaller than the 'static' failure loads. Specimens under higher loads fail under smaller number of cycles. Similar observations have also been made by others (13). An intuitive interpretation of the results of tests at the University of Manitoba led to the following variation of the Matsui et al. equation, i.e. Equation (1).

$$P/P_s = 1.0 - \ln (N) / 30 \qquad (2)$$

As can be seen in Fig. 3, for $N$ greater than 10,000, this simple equation gives nearly the same results as Equation (1). It also gives the correct result for $N = 1$. Current studies on full-scale models of steel-free deck slabs at Dalhousie University and the University of Manitoba are expected to confirm the validity of this equation. In the absence of a more-reliable relationship, the above equation is used to determine the equivalent number of cycles of various loads. The following notation is introduced.

$P_1$ and $P_2$ are two different wheel loads; $n_1$ and $n_2$ are the corresponding number of passes of $P_1$ and $P_2$, respectively, so that the two loads have the same damaging effect; $N_1$ and $N_2$ are the limiting number of passes corresponding to $P_1$ and $P_2$, respectively; $R_1 = P_1/ P_s$; and $R_2 = P_2/ P_s$.

It is assumed that for ratio $R_i$, the cumulative damage to the deck slab is proportional to $(n_i / N_i)^m$, where $n_i$ is the number of passes of the load, $N_i$ is the limiting number of passes, and $m$ is any value larger than or equal to 1.

It can be shown that the following relationship holds true for any value of $m$.

$$N_2/N_1 = n_1/n_2 \quad (3)$$

The significance of the above relationship is that it is equally applicable to linear and non-linear models of damage. Equations (2) and (3) lead to the following equation for $n_2$

$$n_2 = n_1 \times e^S \quad (4)$$

where

$$S = (R_1 - R_2) \times 30 \quad (5)$$

Consider a steel-free deck slab that has a static failure load ($P_s$) of 100 t (979 kN). By using Equations (4) and (5), it can be shown that for this deck slab, the wheel loads of Table 1 are equivalent to 105 million passes of a 7.5 t (73 kN) wheel, or 49 million passes of a 10 t (98 kN) wheel, or 6115 passes of a 40 t (391 kN) wheel, or 510 passes of a 60 t (587 kN) wheel, and so on. By using this procedure and a representative deck slab, it is a straightforward matter to formulate design requirements for fatigue loads of deck slabs.

## FATIGUE TESTS ON STEEL-FREE DECK SLABS

A full-scale ArchDeck slab is currently being tested in Dalhousie University under simulated rolling wheel loads. The static failure load ($P_s$) of a similar slab was found to be 986 kN. This slab has already withstood more than 50,000 passes of a 394 kN load, which is 40% of $P_s$. About 6000 passes of this test load are equivalent to all the wheel loads that the deck slab is likely to sustain during its lifetime. It can be concluded that the slab under consideration has considerably more fatigue resistance than required.

The full-scale model of a non-composite precast steel-free deck slab is currently being tested for fatigue at the University of Manitoba. Being tested at 33% of the static failure load, it has already sustained without failure more number of passes than those, which correspond to the wheel loads of Table 1. Test setup is shown in Fig. 4.

## CONCLUSIONS

During its lifetime, a bridge deck slab is subjected to different numbers of wheels of different magnitudes. Statistics for a sample population of wheels are developed. A method has been presented to obtain the number of passes of a test load of fixed magnitude, so that the fatigue damage induced by it in a deck slab is equivalent to the cumulative damage induced by a given population of wheels. It has been shown that both cast-in-place and precast steel-free deck slab can sustain more damage than induced by the projected population of wheels that a bridge deck slab is expected sustain during its lifetime.

## ACKNOWLEDGEMENTS

Research funding provided by ISIS Canada, A Network of Centers of Excellence, is gratefully acknowledged. Special thanks to Moray Mcvey and Grant Whiteside for setting up of the experiment, Liting Han for setting up the Data Acquisition System and to ISIS Canada administrative staff for their support.

## REFERENCES

1.    AASHTO, 1998, *LRFD Bridge Design Specifications*, American Association of State Highway and Transportation Officials, Washington,  D.C.

2.    CHBDC, 2000, "Canadian Highway Bridge Design Code, "Canadian Standards Association International, Toronto

3.    Mufti, A.A.; Jaeger, L.G.; Bakht, B.; and Wegner, L.D., 1993, "Experimental Investigation of FRC Slabs Without Internal Steel Reinforcement," *Canadian Journal of Civil Engineering*, V. 20 No.3, pp. 398-406.

4.    Bakht, B., and Mufti, A.A., 1998, "Five Steel-Free Bridge Deck Slabs in Canada," *Journal of the International Association for Bridge and Structural Engineering (IABSE)*, V. 8, No. 3, pp. 196-200.

5.    Newhook, J.P.; Bakht, B.; Mufti, A.A.; and Tadros, G., 2001, "Monitoring of Hall's Harbour Wharf," *Proceedings, 6th Annual Conference of the International Society for Optical Engineering*, Newport, California.

6.    Mufti, A.A.; Banthia, N.; and Bakht, B., 2001, "Fatigue Testing of Precast ArchPanels," *Proceedings, Third International Conference on Concrete under Severe Conditions*, Vancouver, Vol. 1, pp. 1033-1041.

7.    Agarwal, A.C., 2002, Private communication.

8.    Matsui, S.; Tokai, D.; Higashiyama, H.; and Mizukoshi, M., 2001, "Fatigue Durability of Fiber   Reinforced Concrete Decks Under Running Wheel Load," *Proceedings, Third International Conference on Concrete under Severe Conditions*, Vancouver, V. 1, pp. 982-991.

9.    Matsui, S., 2001. Private communication

10.   Petrou, M.F., Perdikaris, C.P, and Wang, A., 1993, "Fatigue Behavior on Noncomposite Reinforced Concrete Bridge Deck Models," Transportation Research Record 1460, pp. 73-80, Transportation Research Board, Washington, D.C.

11.   Perdikaris, C.P., and Beim, S., 1988, "RC Bridge Decks under Pulsating and Moving Loads," ASCE Journal of Structural Engineering, V. 114, No. 3, pp. 591-607.

12.   Youn, S.-G., Chang, S.-P., "Behavior of Composite Bridge Decks Subjected to Static and Fatigue Loading," ACI Structural Journal, V. 95, No. 3, pp. 249-258.

13.   Dyduch, K., Szerszen, M., 1994, "Experimental Investigation of the Fatigue Strength of Plain Concrete Under High Compressive Loading," Materials and Structures, 27, pp.505–509.

TABLES

Table 1. Statistics of wheels loads for a total of 372
million wheels, adopted from Matsui (9)

| Wheel weight, tonnes | Percentage of total | No. of wheels, in millions |
|:---:|:---:|:---:|
| 1 | 21.25 | 79.05 |
| 2 | 32.06 | 119.26 |
| 3 | 21.61 | 80.39 |
| 4 | 12.60 | 46.87 |
| 5 | 6.48 | 24.11 |
| 6 | 3.24 | 12.05 |
| 7 | 1.44 | 5.37 |
| 8 | 0.54 | 2.01 |
| 9 | 0.32 | 1.19 |
| 10 | 0.18 | 0.67 |
| 11 | 0.11 | 0.41 |
| 12 | 0.07 | 0.26 |
| 13 | 0.04 | 0.15 |
| 14 | 0.02 | 0.07 |
| 15 | 0.01 | 0.04 |
| 16 | 0.004 | 0.01 |

Fig. 1: Casting of the cylinders

Fig. 2: Test Setup and Cyclic Loading

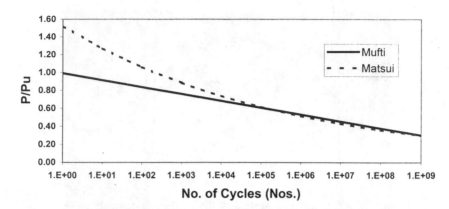

Fig. 3: Comparison of S-N Curves

Fig. 4: View of Test Setup of Non-composite precast
Steel-free Deck Slab

SP 206—5

# Mechanical Behavior of the Interface Between Substrate and Repair Material

by P. Paramasivam, K. C. G. Ong, and W. Xu

**Synopsis:** The service performance of repaired structures depends mainly on the mechanical properties of the substrate and repair materials and the mechanical behavior of the interface between them. However, in most studies in the literature, only bond strength is used to evaluate the repair and the deformation behavior of the interface is usually neglected. In this study, three types of tests were performed to evaluate the bond strength and mechanical behavior of the interface between substrate and repair materials. A total of 4 slabs (800×850 mm) were cast using conventional concrete as a substrate. The substrate surface was roughened with jackhammers. A layer of 50- to 60-mm repair materials were cast by shotcreting or by hand. The test specimens were cored or cut to the required size. Direct tensile and compression tests were performed to evaluate the bond strength and mechanical behavior of the interface between substrate and repair materials. The test results indicate that the bond strength was affected by the mix proportions and independent on the casting method and the inclusion of steel fibers. However the casting methods had a strong influence on the mechanical behavior of the interface between substrate and repair material.

**Keywords:** bond strength; interface; mechanical behavior; repair material; repair method; steel fiber

P. Paramasivam, FACI, is professor of civil engineering at the National University of Singapore. He has been active in R&D applications of ferrocement and fibre reinforced cement composites. He has received Maurice P. Van Buren Structural Engineering Award in 1989 for innovative use of ferrocement sunscreens in high-rise flats and ACI Award in 1997 for his outstanding and sustained contributions in concrete technology in Singapore.

K. C. G. Ong, MACI, is an associate professor of civil engineering and Director, Centre for Construction Materials and Technology at the National University of Singapore. His research interests include composite materials, durability and assessment of concrete structures, and retrofitting. He is currently President of the Singapore Concrete Institute and a Past President of the American Concrete Institute - Singapore Chapter.

W. Xu is a former Ph.D student at the National University of Singapore. He obtained his MEng from China Institute of Water Resources and Hydropower Research in 1988. His fields of interest are durability of concrete, repair materials and technologies.

## INTRODUCTION

A large number of concrete structures constructed several decades ago have exhibited severe degradation in the form of spalling of concrete cover which is caused by corrosion of embedded reinforcing steel due to ingress of chloride ions and/or carbonation of concrete. This has been attributed to the inadequate thickness of the cover, cover concrete of high porosity and poor quality, and the presence of cracks caused by shrinkage, temperature effects and accidental overload. Therefore, repairs are needed to improve aesthetic appearance, reduce permeability, protect reinforcement and if necessary to restore the design load-bearing capacity of a damaged member or to increase the load-carrying capacity of an under-designed member.

One of the major advantages of shotcrete is the usually excellent bond it has with the substrate materials produced by shotcreting process (1-4). This has made the shotcrete process particularly well-suited for the repair of vertical and overhead surfaces. Shotcrete has thus been the selected system for repair of a wide variety of concrete structures (5-8).

This study presents the results of bond tests performed to evaluate the bond strength and mechanical behavior of the interface between substrate and repair materials. Slabs (800×850 mm) made of conventional concrete were cast and cured in ambient laboratory environment. The substrate surface was roughened with jackhammers. The repair materials were cast by shotcreting or cast by hand. Direct tensile and compression tests, performed after 2 and 6 months, to evaluate the bond strength and mechanical behavior of the interface between substrate and repair materials.

## RESEARCH SIGNIFICANCE

Limited information is available concerning the mechanical behavior of the interface between the repair material and substrate under different stress states, particularly the stress-strain relationship at the interface of composite specimens under tension. The investigation presented in this paper indicated that the presence of the interface might greatly affect the mechanical behavior and durability of the repaired structures.

## LITERATURE REVIEW

When a repair material is cast against a surface of old substrate concrete, the contact of fresh cement paste with a hardened concrete surface (aggregate or old cement paste) results in the formulation of an interfacial transition zone. In this zone, the absence of aggregate interlocking action between the old substrate concrete and the "new" applied repair material simply results in a mechanical plane of weakness although this could to a certain extent be mitigated by roughening the surface of the old concrete substrate. In fact, the interface is not only the preferential zone of rupture, but it may be the most highly stressed zone of the composite structure as it comprises two different materials that are linked together. Although the transition zone between the "new" cement paste and aggregate is well documented (9-11), the transition zone between "new" applied repair material and old concrete substrate has been less often investigated.

An important property of all repair materials is the bond strength between the repair material and its substrate. Without an adequate bond, a repair layer is ineffective no matter what the quality of the repair material itself. Schrader and Kaden (4) stated that the bond between shotcrete and an old sound concrete is generally very good, probably due to the shotcrete compaction effect and the normally low water/cement ratio of the repair material, particularly in dry-mix shotcretes. Parker (1) mentioned that when shotcrete first impacts on the receiving surface only the cement paste bonds to the surface, and the other components rebound until a sufficient thickness of paste is obtained. A well-compacted layer of cement paste with a low water/cement ratio is thus formed at the bond interface and acts rather like a bonding agent.

Opsahl (2) reported bond strength for fiber reinforced wet-mix shotcrete of 0.8 to 2.5 MPa, measured from 60-mm diameter core pull-off tests. Opsahl (3) also reported bond strengths of a wet-mix shotcrete to a concrete substrate in the range of 0.4 to 2.2 MPa, the lowest values being obtained when shotcreting on deteriorated concrete or when using high dosages of accelerator in the wet-mix shotcrete.

Talbot (12) evaluated the influence of surface preparation on long-term bonding of shotcrete by bonding test. The test results indicate that the type of surface preparation has a strong influence on the strength and durability of the bonding, and that hydrodemolition is probably the best type of surface preparation. The shotcrete mix composition, however, was found to have relatively little influence on bonding durability.

However, very few investigators have analyzed the durability of the bond between a new concrete layer and an old concrete surface. Saucier (13) studied the mechanisms by which this bond can be slowly damaged. He found that the microstructure of the interface was highly influenced by the characteristics of the cement in the new concrete. He also observed that microcracking of the cover layer due to shrinkage was generally the main cause of the reduction of the bond strength, and that the influence of other effects such as freezing and thawing cycles was less pronounced.

Direct tensile tests have been widely used to evaluate the properties of shotcrete and other repair materials (12, 14-17). However, reported information from the direct tensile test has concentrated on the bond strength between repair materials and substrate. The information on the deformation behavior of the interface between repair materials and substrate is relatively limited.

In order to understand the mechanical properties of the interface between repaired materials and substrate, three innovative tests were carried out to investigate the deformation behavior of the interface under different stress states.

## ANALYTICAL CONSIDERATION

### Load Direction Perpendicular to the Interface

Assuming a uniform uniaxial stress at the interface, the deflection (elongation if in a state of tension) of the composite specimen will depend on the deflection of substrate, deflection of repair material and the deflection of the interfacial transition zone between the substrate and repair material. If the bond between the substrate and repair material is perfect and the thickness of the interfacial transition zone assumed to be very small, the deflection of the composite specimen under a uniform state of uniaxial stress can be calculated according to the two-phase material model shown in Fig. 1.

For a given composite specimen, its total length can be expressed as:
$$L = L_r + L_c \tag{1}$$
where: $L$, $L_r$ and $L_c$ are the lengths of the composite specimen, the repair material and the substrate respectively.

Under a uniform uniaxial stress, $\sigma$, the total deformation, $\Delta L$, can be expressed as:
$$\Delta L = L \bullet \varepsilon = L_r \bullet \varepsilon_r + L_c \bullet \varepsilon_c \tag{2}$$

where: $\varepsilon$, $\varepsilon_r$ and $\varepsilon_c$ are the strains in the composite specimen, the repair material and the substrate respectively.

The modulus of elasticity of the composite specimen, $E$, can be written as

$$\frac{1}{E} = \frac{1}{L}\left(\frac{L_r}{E_r} + \frac{L_c}{E_c}\right) \qquad (3)$$

Where: $E_r$ and $E_c$ are the modulus of elasticity of the repair material and substrate respectively.

According to this model, the modulus of elasticity of the composite specimen $E$ should be fall in between $E_r$ and $E_c$. If $L_r = L_c$, then

$$\frac{1}{E} = \frac{1}{2}\left(\frac{1}{E_r} + \frac{1}{E_c}\right) \qquad (3a)$$

If the values of $E_r$, $E_c$, $L_r$ and $L_c$ are known, the value of elastic modulus ($E$) of the composite specimen can be calculated using Equation (3a).

## Load applied parallel to the interface

When the load applied on to a composite prism specimen is as shown in Fig. 2(a) and assuming a uniform strain, $\varepsilon$, sustained across a cross section of the specimen, the stresses in the substrate ($\sigma_c$) and the repair material ($\sigma_r$) can be expressed as follows:

$$\sigma_c = \varepsilon \cdot E_c; \quad \text{and} \quad \sigma_r = \varepsilon \cdot E_r \qquad (4)$$

where: $E_r$ is the modulus of elasticity of the repair material;

$E_c$ is the modulus of elasticity of the substrate.

The transverse strain and Poisson's ratio of the composite specimen can be expresses as:

$$\varepsilon_t = \frac{\varepsilon_{tr}b_r + \varepsilon_{tc}b_c}{b} \qquad (5)$$

$$\mu = \frac{\mu_r b_r + \mu_c b_c}{b} \qquad (6)$$

where: $b$, $b_r$ and $b_c$ are the widths of the composite specimen, the repair material and the substrate respectively.

For the special case with $b_r = b_c$, Equations (5) and (6) can be written as:

$$\varepsilon_t = \frac{\varepsilon_{tr} + \varepsilon_{tc}}{2} \qquad (5a)$$

$$\mu = \frac{\mu_r + \mu_c}{2} \qquad (6a)$$

where: $\varepsilon_t$, $\varepsilon_{tr}$, $\varepsilon_{tc}$ are the transverse strains of the composite specimen, the repair material and the substrate, respectively; $\mu$, $\mu_r$, $\mu_c$ are the Poisson's ratio of the composite specimen, the repair material and the substrate, respectively.

## EXPERIMENTAL PROGRAM

### Material

Ordinary Portland cement, natural sand, 10-mm maximum size crushed granite and steel fibers were used in the preparation of substrate and repair materials. The gradations of the aggregates have important influence on the properties of fresh and hardened concrete, especially for wet-mix shotcrete. The aggregates were tested in accordance to ASTM standards and compared with the requirements of ASTM C33-01 and BS 882. In addition, the gradation of the aggregate of pre-packed shotcrete material was also tested by washing out the cement particles with water. The test results indicated that the gradation of the fine aggregates not only conforms mostly to the requirements of ASTM and BS standards, but also to the specifications of ACI 506R-90 (18).

The fiber used in the present test program is straight steel fiber of 0.16 mm diameter and 6-mm long with an aspect ratio 37.5.

### Mix Design of Repair Material

In order to investigate the effects of constituents of shotcrete, steel fibers and casting method on the bond strength and the mechanical behavior of the interface between the repair material and substrate, the mix proportions shown in Table 1 were employed for the investigation reported herein.

### Preparation of Specimens

The substrates were cast in wooden molds. The mix proportions by weight of the concrete used as substrates were as follows: Cement : Sand : Coarse Aggregate : Water = 1 : 2.11 : 2.68 : 0.55. Nine 100 mm cubes and nine dia.100 mm by 200 mm high cylinders were cast accompanying each batch of substrate specimens to determine the compressive strength and compressive elastic modulus of the concrete at the desired age. All the test substrates were air-cured at ambient indoor laboratory conditions (27 ± 2 °C and 80 ± 10 % relative humidity). After 28 days of air-curing, the bonding surfaces of the concrete substrates were prepared by roughening using mechanical chipping. Normally a 3-mm surface roughness can be obtained using this method. The thickness of the substrate was controlled to 92 ± 2 mm. Dust and loose materials were cleaned using a steel-wire brush and flushed with clean water prior to shotcreting or casting.

The substrate panels were soaked so as to obtain a "saturated surface dry" condition prior to shotcreting or casting to form the composite panels. All the composite panels were demolded 3 days later and again air-cured at ambient indoor at least for a further 28 days before they were cut or cored to obtain the various types of composite specimens for testing.

The designation used to describe the specimens is as follows: Letters to classify specimen type (TS - composite cylindrical specimens for direct tensile test; CS - composite cylindrical specimens for compressive test; PS - composite prism specimens for compressive test) followed by a number to indicate the Mix No. of repair material.

## Test Procedure

The cylindrical specimens under direct tension were designed to evaluate the bond strength and the tensile behavior of the interface between repair material and substrate. The specimens were cored from the composite panels and then trimmed to level the top and bottom surfaces. Two steel discs were attached to the top and the bottom surfaces of the specimens with epoxy resin. Six strain gages (30 mm) were used to measure the strain in the repair material, the strain across the interface and the strain in the substrate. Tests were conducted using a servo-controlled universal testing machine as shown in Fig. 3. Load was applied at a constant displacement rate of 0.1 mm/minute (0.004 in./minute). During each test, the ultimate strength and the final failure mode were monitored.

The cylindrical specimens under compression herein were designed to assess the mechanical behavior of the interface between repair material and substrate under compression. The preparation method used to obtain compression specimens was similar to the direct tensile specimens. Strain gages were also used to monitor the changes in strain in the repair material and substrate portion and across the interface.

The prism specimens under compression herein were designed to evaluate the mechanical behavior of the composite prism under compression and the influence of the existence of the interface. The specimens were cut from the composite panels. The dimensions of the tested prism are 100×100×200 mm with the thickness of repair material 28 mm. Strain gages were used to monitor the changes in longitudinal and transverse strain in the repair material and substrate portion, and transverse strain across the interface as shown in Fig. 4. The failure strength and failure mode were also monitored during each test.

## TEST RESULTS AND DISCUSSIONS

### Cylindrical Composite Specimen Under Tension

The test results of the four kinds of composite specimens under tension are shown in Table 2 and the typical stress-strain curves for TS1 and TS3 are shown in Figs. 5 and 6 respectively.

It could be observed from the test results that shotcreted mortar had higher bond strength than shotcreted pre-packed material (on average by about 45%). This showed that the mix composition of the repair material had significant influence on the bond of shotcreted repair material to its substrate. A higher bond strength could be achieved by optimizing the mix composition through proper selection of repair material. However, the casting method used herein seemed to have little effect on the bond between the repair material and substrate. In other words, TS1 specimens could achieve the same direct tensile strength as TS4 specimens. Furthermore, the present test results also revealed that the incorporation of steel fibers had no direct influence on the bonding strength.

The strain monitored across the interface was always much higher than either the strain monitored in the repair material or in the substrate and behaved in a non-linear manner, especially at high tensile stress levels (i.e. near to the ultimate failure stress). This revealed that of the three zones, the interface between the repair material and substrate is the weakest, probably due to the presence of defects, such as pores and microcracks. These defects lead to higher stress concentration, especially at high stress levels.

Unlike the strain measured by the strain gauge that straddled the interface, the strain monitored in the repair material or substrate may not be the maximum within the two latter materials and is in any case not expected to be. This is probably due to the relative lower length/diameter ratio of the repair material portion (about 0.6) and the substrate portion (about 0.9) of the composite specimens tested, and the epoxy resin adhesive system used at the steel disc. Therefore, the strain monitored within the repair material portion or the substrate portion of the composite specimen during the test was lower than that monitored when the modulus of elasticity tests were carried out on each individual material by itself. The comparison of the modulus of elasticity monitored from different tests and theoretical analysis is presented in Table 3 and the changes of modulus of elasticity across the interface at different tensile stress levels are shown in Fig. 7.

Figure 8 shows the effect of repair methods and material compositions on the mechanical behavior of the interface. The results indicated that the ultimate bond strength depended mainly on the material compositions. The strain at failure depended mainly on the casting methods. Compared to that monitored across the interface of TS1 specimens, the strain at failure monitored across the interface of TS3 and TS4 specimens increased by about 50%. This perhaps showed that the interfacial transition zone formed when shotcreting was used had lower deformation capacity than that formed when casting was used. This lower deformation capacity is probably due to the likely scenario that shotcreting produces an interfacial transition zone of lower thickness and higher density. Furthermore, the test results also demonstrated that the incorporation of steel fibers in the repair material seemed to have no significant effect on deformation behavior and the ultimate failure strain achieved across the interfacial transition zone.

Compared to that at 60 days, the bond strength of TS1 and TS2 specimens decreased by about 8 % and 21 % respectively after 180 days. The strain at failure monitored across the interface decreased by 14 % for TS1 specimens and remained the same for TS2 specimens. It may therefore be concluded that ageing from 60 to 180 days under ambient outdoor conditions had a significant influence on the bonding of the repair material to substrate. This phenomenon was also observed by other researchers (12), and is most often linked to shrinkage and its associated damage.

The failure modes of TS1, TS2, TS3 and TS4 specimens were all very similar. All specimens failed mainly at the interface and partly at the substrate. The failure modes also revealed that the weakest zone seemed to be the interface. Thus it is possible to improve bond by selecting an appropriate method for preparing the receiving substrate surface and proper choice of repair materials or mix proportions of repair material used.

## Cylindrical Composite Specimen under Compression

The test results indicated that the strains monitored by the strain gauge straddling the interfacial transition zone were higher than that in the repair material or substrate portions for all the specimens tested and, especially at low stress levels as shown in Fig. 9. This should not be the case if the two materials were perfectly bonded. This also confirmed the observation that there must be the presence of some defects, such as air voids or microcracks, present at the interface transition zone, which could have caused the relatively higher deformation monitored.

The modulus of elasticity measured across the interface was lower than that of the repair material and substrate portions and that calculated using Equation 3 (a) as shown in Table 4. If the two materials were perfectly bonded together, the modulus of elasticity measured across the interface should assume a value that fell between that of the two materials alone, thus providing evidence for considering IZT material as being viewed as a $3^{rd}$ phase. The failure mode observed was similar to that of normal cementitious materials.

## Prism Composite Specimens under Compression

The ultimate strength, the slope of the stress-strain curves of the repair material and substrate at a load applied corresponding to one-third of the maximum strength (modulus of elasticity), the Poisson's ratio of the repair material and substrate and the failure mode of the specimens tested are summarized in Table 5. Typical stress-strain curves of composite specimens PS1, and PS3 in compression are shown in Figs. 10 and 11 respectively.

From the test results obtained, it may be concluded as follows:

- The transverse strain monitored across the interfacial transition zone was higher than that in the repair material or substrate portions for all the specimens tested, especially at high stress levels. This revealed that the existence of a clearly defined interfacial transition zone which had a significant effect on the mechanical behavior of such composite specimens. High deformation at the interface may lead to premature failure of the bond.
- The transverse strain monitored across the interface was affected by the mix proportions of repair material used. Compared to that of PS1 specimens, the transverse strain at the interface of PS2 specimens decreased by about 30%. This was perhaps caused by the better bond between the substrate and repair materials and the higher modulus of elasticity across the interface for PS2

specimens. However, the cast method used herein seemed to have little effect on the transverse strain across the interface.

- If the Poisson's ratio was used to assess the bond, the values were found to be less than 0.2 for the substrate and repair material. This value was consistent with most concrete materials (19). The Poisson's ratio across the interface, however, was higher than the values calculating using Equation 6(a) for all the specimens tested. Furthermore, the Poisson's ratios were higher than 0.2 for most of specimens tested except PS2, which registered a value of 0.18. These values were unexpected since Poisson's ratio should be lower than 0.2.

- The test results obtained showed that Poisson's ratio was lower for specimens with relatively high-bond strength at the interface and higher for specimens with a lower bond strength at the interface.

- Bond failure was observed for some of the specimens tested as shown in Fig. 11.

## CONCLUSION AND RECOMMENDATIONS

Three innovative tests were performed to evaluate the bond strength and mechanical behavior of the interface between substrate and repair materials. The test results indicated that the bonding strength between repair material and substrate was affected mainly by the composition of the repair material. No significant difference was observed between the bonding strength of concrete cast by hand or by using shotcreting. However, casting methods seemed to have a significant effect on the mechanical behavior of the interface under tension. Further research is needed to find the reasons.

All the test results seemed to suggest that the interface was a weak zone that had significant effects on the mechanical behavior of composite specimens. The higher strain was perhaps caused by the presence of microcracks and high porosity at the interfacial zone. These microcracks perhaps existed partly in the substrate, which was caused by the mechanical roughening of the surface when prepared the specimens and partly at the interface, which was caused by the differential shrinkage between the already shrunk old concrete and newly cast repair material. The high porosity and presence of microcracks at the interfacial zone was also of great importance in the durability of composite structures. Further research is recommended to analyze the influence of the interface on the durability of the composite structures.

## ACKNOWLEDGMENT

The authors would like to acknowledge the financial support provided by the National University of Singapore. Thanks are also due to Master Builders Technologies (Singapore) Pte Ltd for supplying the pre-packed material and Harvest Resources Contracts Pte Ltd for supplying the spraying equipment.

## REFERENCES

1. Parker, H. W. A practical new approach to shotcrete for rebound losses. *ASCE and ACI SP 54*, Shotcrete for Ground Support, Detroit, **1977**, 149-187.

2. Opahl, O. A. Steel fibre reinforced shotcrete for rock support. Report for Royal Norwegian Council for Scientific and Industrial Research, *NTNF Project* 1053.09511, 33. **1982**.

3. Opahl, O. A. Study of a wet-process shotcreting method - *Volume 1. Division of Building Materials*, University of Trondheim, Report No. BML 85.101, November, **1985**.

4. Schrader, E. K. and Kaden, R.A. Durability of Shotcrete. *Concrete Durability*. (SP-100), American Concrete Institute, Detroit, **1987**, 1071-110.

5. Heneghan, J.I. Shotcrete Repair of Concrete Structures in Marine Environment, *Performance of Concrete in Marine Environment,* Edited by V. M.Malhotra, ACI SP-65, **1980**, 509-526.

6. Gilbride, P., Morgan, D.R., and Bremner, T.W. Deterioration and Rehabilitation of Berth Faces in Tidal Zone at the Port of Saint John. American Concrete Institute, Detroit, Concrete in Marine Environment, SP-109, **1988**, 199-226.

7. Morgan, D.R. and Neill, J.N. Durability of Shotcrete Rehabilitation Treatments of Bridges, *Transportation Association of Canada, Annual Conference*, Winnipeg, Manitoba, September, **1991**, 36.

8. Morgan, D.R. New Developments in Shotcrete for Repairs and Rehabilitation, *CANMET International Symposium on Advances in Concrete Technology*, Athens, Greece, May, **1992**, 675-720.

9. Chatterji, S. and Jensen, A. D. Formation and Development of Interfacial Zones between Aggregates and Portland Cement Paste in Cement-based Materials. Edited by J. C. Maso. Interfaces in Cementitious Composites, RILEM, Proceedings 18, Toulouse, France, **1992**, 3 - 12.

10. Goldman, A. and Bentur, A. Effects of Pozzolanic and Non-reactive Microfillers on the Transition Zone in High Strength Concretes. Edited by J. C. Maso. Interfaces in Cementitious Composites, RILEM, Proceedings 18, Toulouse, France, **1992**, 53 - 61.

11. Nilsen, U., Sandberg, P. and Folliard, K. Influence of Mineral Admixtures on the Transition Zone in Concrete. Edited by J. C. Maso. Interfaces in

Cementitious Composites, *RILEM, Proceedings 18*, Toulouse, France, **1992**, 65 - 70.

12. Talbot, C., Pigeon, M., Beaupre, D., and Morgan, D.R. Influence of Surface Preparation on Long-Term Bonding of Shotcrete. *ACI Materials Journal*, **1994**, V. 91, No. 6, 560 - 566.

13. Saucier, F., and Pigeon, M. Durability of New-to-Old Concrete Bondings. *Proceeding of the ACI International Conference on Evaluation and Rehabilitition of Concrete Structures and Innovations in Design*, SP-128, American Concrete Institute, Detroit, 1991, pp.689-707.

14. Knab, L. I. and Spring, C. B. Evaluation of Testing Methods for Measuring the Bond Strength of Portland Cement Based Repair Materials to Concrete. *Cement, Concrete and Aggregate*, Vol. 11, No.1, Summer, **1989**, 3-14.

15. Silfwerbrand, J., Improving Concrete Bond in Repaired Bridge Decks. *Concrete International*. Vol.12, No.9, Sept, **1990**, 61-66.

16. Kuhlmann, L. A., Styrene-Butadiene Latex-Modified Concrete: The Ideal Concrete Repair Material. *Concrete International: Design & Construction*, Vol. 12, No.10, Oct, **1990**, 59-65.

17. Robin, P. J., and Austin, S. A., A Unified Failure Envelope from the Evaluation of Concrete Repair Bond Tests. *Magazine of Concrete Research*, Vol. 47, No. 170, Mar., **1995**, 57-68.

18. ACI Committee 506 Guide to Shotcrete. *ACI Manual of Concrete Practice, Part 5*, American Concrete Institute, Detroit, **1990**, 41.

19. Mehta, P. K. and Monteiro, P. *Concrete: Structure, Properties, and Materials*. Prentice-Hall Inc, 1993.

Table 1 --- Mix proportions of repair materials

| Mix No. | Description | Repair method |
|---|---|---|
| 1 | Pre-packed shotcrete material(shotpatch)<br>Shotpatch : Water = 25 : 3.75 | spraying |
| 2 | Ordinary mortar.<br>Cement : Sand : Water = 3.0 : 1 : 0.5 | spraying |
| 3 | Pre-packed shotcrete material (shotpatch) and steel fibre.<br>Shotpatch : Fibre : Water = 25 : 0.75 : 3.75. | casting |
| 4 | Pre-packed shotcrete material (shotpatch).<br>Shotpatch : Water = 25 : 0.75 : 3.75. | casting |

Table 2 --- Direct tensile test results of composite specimens

| Specimen No. | | Tensile strength (MPa) | Ultimate strain of repair material $(\times 10^{-6})$ | Ultimate strain across interface $(\times 10^{-6})$ | Ultimate strain of substrate $(\times 10^{-6})$ |
|---|---|---|---|---|---|
| TS1 (60 days) | Mean | 1.37 | 15.0 | 133 | 23.0 |
| | St. Dev. | 0.11 | 2.3 | 5.0 | 1.00 |
| TS1 (180 days) | Mean | 1.26 | - | 114 | - |
| | St. Dev. | 0.05 | - | 25.4 | - |
| TS2 (60 days) | Mean | 1.98 | 29.5 | 113 | 44.0 |
| | St. Dev. | 0.10 | 1.8 | 13.7 | 4.00 |
| TS2 (180 days) | Mean | 1.57 | - | 113 | - |
| | St. Dev. | 0.13 | - | 8.8 | - |
| TS3 (60 days) | Mean | 1.38 | 15.0 | 210 | 25.5 |
| | St. Dev. | 0.14 | 0 | 34.0 | 5.44 |
| TS4 (60 days) | Mean | 1.39 | 15.3 | 203 | 31.2 |
| | St. Dev. | 0.11 | 2.3 | 13.6 | 0.82 |

Note: - Not available.

Table 3 --- Comparison of Modulus of Elasticity from Tensile Test

| No. of specimen | $E_R$ (from modulus test) (GPa) | $E_S$ (from modulus test) (GPa) | $E_R^*$ (from tensile test) (GPa) | $E_S^*$ (from tensile test) (GPa) | $E^*$ (from tensile test) (GPa) | E (from analysis) (GPa) |
|---|---|---|---|---|---|---|
| TS1 | 22.7 | 28.6 | 79.4 | 48.4 | 14.1 | 25.7 |
| TS2 | 22.6 | 30.7 | 61.0 | 43.2 | 20.3 | 26.7 |
| TS3 | 25.8 | 29.1 | 66.1 | 48.0 | 12.3 | 27.5 |
| TS4 | 24.2 | 29.7 | 70.7 | 35.6 | 11.4 | 26.9 |

*: based on the values at 2/3 tensile strength.

Table 4 --- Comparison of Modulus of Elasticity from Compressive Test

| No. of specimen | $E_R$ (from modulus test) (GPa) | $E_S$ (from modulus test) (GPa) | $E_R^*$ (from Compressive test) (GPa) | $E_S^*$ (from compressive test) (GPa) | $E^*$ (from compressive test) (GPa) | E (from analysis) (GPa) |
|---|---|---|---|---|---|---|
| CS1 | 22.74 | 28.6 | 28.4 | 32.1 | 14.6 | 25.7 |
| CS2 | 22.6 | 30.7 | 26.0 | 37.6 | 19.1 | 26.7 |
| CS3 | 25.8 | 29.09 | 26.0 | 27.7 | 17.8 | 27.5 |
| CS4 | 24.21 | 29.66 | 25.0 | 24.1 | 19.6 | 26.9 |

*: based on the values at 1/3 compressive strength.

Table 5 --- Compressive Test Results of Composite Prism Specimens

| Specimen No. | PS1 | PS2 | PS3 | PS4 |
|---|---|---|---|---|
| Compressive stress (MPa) | 15.07 | 15.09 | 15.06 | 15.07 |
| Transverse strain in Repair material/Substrate ( T – R/S ) ($\times 10^{-6}$) | 86/76 | 71/70.5 | 84/91 | 92/77 |
| Transverse strain across Interface ( T – I ) ($\times 10^{-6}$) | 168 | 90.5 | 120.5 | 129 |
| Longitudinal strain in Repair material/Substrate ( L – R/S ) ($\times 10^{-6}$) | 688/422 | 509/510 | 456/538 | 509/557 |
| Failure strength (MPa) | 41.2 | 45.8 | 44.1 | 41.9 |
| Poisson's ratio of repair material | 0.13 | 0.14 | 0.18 | 0.18 |
| Poisson's ratio of substrate | 0.17 | 0.14 | 0.17 | 0.14 |
| Poisson's ratio across the interface | 0.23 | 0.18 | 0.24 | 0.24 |
| Poisson's ratio across the interface (Using equation 6(a)) | 0.15 | 0.14 | 0.18 | 0.16 |

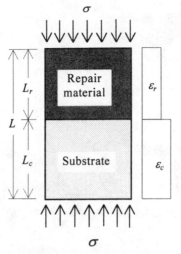

Fig. 1 --- Composite specimen under stress perpendicular to interface

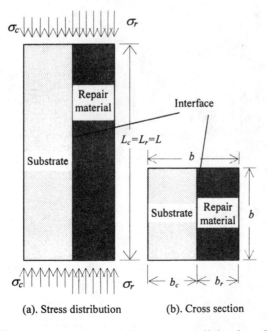

(a). Stress distribution          (b). Cross section

Fig. 2 --- Composite specimen under stress parallel to interface

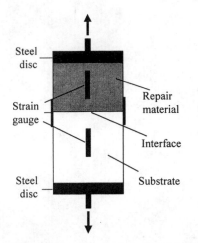

Fig. 3 --- Direct tensile test set-up for cylindrical specimens

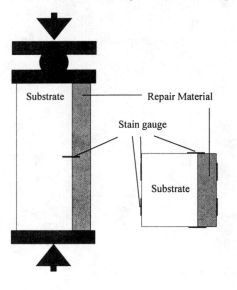

Fig. 4 --- Compressive test set-up for prism specimens

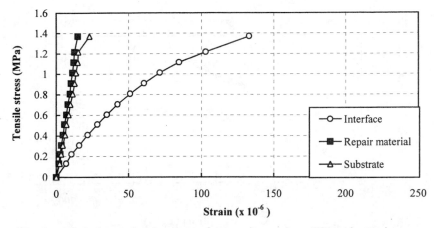

Fig. 5 --- Typical stress-strain curves of composite specimen TS1 under tension

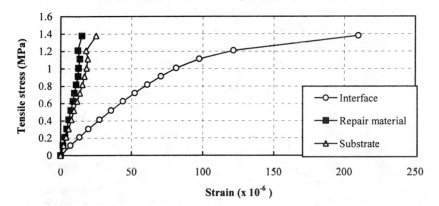

Fig. 6 --- Typical stress-strain curves of composite specimen TS3 under tension

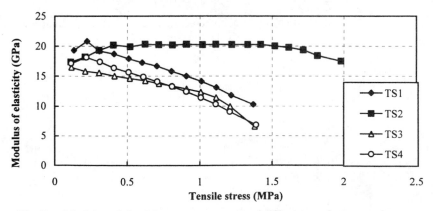

Fig. 7 --- Modulus of elasticity vs. stress curves of different specimens

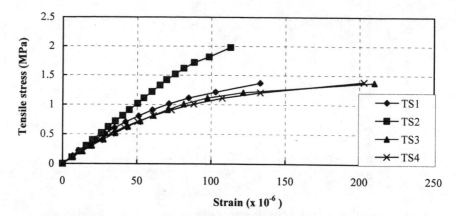

Fig. 8 --- Influence of casting method and repair materials on the mechanical properties
of interface

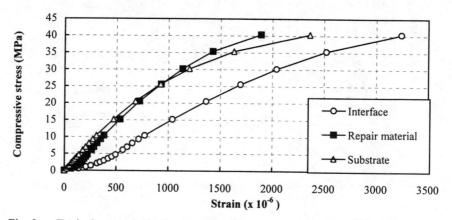

Fig. 9 --- Typical stress-strain curves of composite specimen CS1 under compression

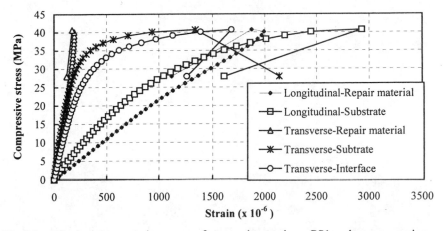

Fig. 10 --- Typical stress-strain curves of composite specimen PS1 under compression

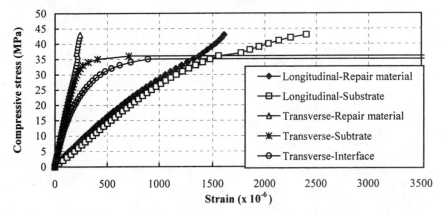

Fig. 11 --- Typical stress-strain curves of composite specimen PS3 under compression

# Fatigue Characteristics of Fiber Reinforced Concrete for Pavement Applications

## by V. S. Gopalaratnam and T. Cherian

Synopsis: Results from flexural fatigue tests on plain concrete and two fiber reinforced concrete (FRC) mixes (hooked-end steel fiber and polyolefin fiber) are presented and discussed. The specimens were made using the same concrete materials used for the MoDOT field test program. MoDOT's RDT Division was responsible for field implementation, which included design and construction of unbonded fiber-concrete overlays in the southbound lanes of Interstate 29 in Atchison County, Missouri, between Route A and US 136. The fatigue performance of both the FRC mixes investigated in this study were superior to that of the Control mix. Crack widths in the Steel Fiber Reinforced Concrete (SFRC) specimens were typically smaller than those in the Polyolefin Fiber Reinforced Concrete (PFRC) specimens under comparable levels of fatigue loading (stress level as well as number of fatigue cycles). This property influences the long-term durability of the material for pavement applications. The difference between the FRC mixes and the Control mix becomes readily apparent at the higher levels of upper limit of fatigue stress. Fatigue failure in FRC can be characterized by a three-stage process. In the first stage fatigue damage is accumulated in the concrete matrix. Rapid growth in net-deflection occurs with increasing fatigue cycles. The second stage is characterized by little or no growth in net-deflections, attributable to stable and steady growth of damage along fiber-matrix interfaces. Only when this damage reaches a threshold level does the third stage begin. The third stage is characterized by a rapid growth in net-deflections resulting in fiber-pull out and/or fractures at the critical cross-section and associated catastrophic growth of the main matrix crack.

**Keywords:** fatigue loading, fiber reinforced concrete; pavements; polymeric fibers; steel fibers; testing

**Vellore S. Gopalaratnam** is Professor of Civil Engineering at the University of Missouri-Columbia. A Fellow of the American Concrete Institute, he has served as the Chairman of ACI Committee 446 Fracture Mechanics and Secretary of ACI 544 Fiber Reinforced Concrete. His research interests include interface mechanics, rate effects, fracture, and failure of concrete materials.

**Thomas Cherian** is Project Engineer with Black and Veatch, Kansas City. The work reported here was completed when he was a Master of Science student at the Department of Civil and Environmental Engineering at the University of Missouri-Columbia.

# INTRODUCTION

### Background Information

The incorporation of fibers in concrete has been known to enhance the flexural toughness, crack propagation resistance, fatigue performance and freeze-thaw resistance of the resulting composite (ACI Committee 544, 1988). These attributes make it ideally suited for pavement applications. It is possible to extend the service life as a result of incorporation of fibers in concrete pavements. Joints can be spaced further apart thus requiring less maintenance than conventional unreinforced pavements. Pavement thickness can also be reduced compared to unreinforced pavements. It is with this motivation that Missouri Department of Transportation (MoDOT) initiated an experimental investigation to study the performance of FRC for unbonded overlays for concrete pavements. The study was a collaborative effort between MoDOT's Research, Development and Technology (RDT) Division and the Department of Civil and Environmental Engineering, University of Missouri-Columbia.

The larger investigation (Gopalaratnam and Cherian, 2001) comprised two components. Phase I was a laboratory study to characterize the flexural toughness response of fourteen fiber reinforced concrete mixes and one unreinforced concrete mix. One steel and one polymeric fiber reinforced mix were chosen based on their flexural toughness characteristics for test sections to be placed as a part of the MoDOT field test program. MoDOT's RDT Division was responsible for field implementation, which included design and construction of unbonded fiber-concrete overlays in the southbound lanes of Interstate 29 in Atchison County, Missouri, between Route A and US 136. MoDOT has also conducted regular pavement surveys since the pavement sections were put down during Summer 98 (Chojnacki 2000). Flexural specimens for Phase II of the study involving more systematic laboratory fatigue study were made by MoDOT under field conditions using the same concrete materials used for the construction of the test sections of the unbonded overlay. Some of the significant fatigue test results from Phase II are described here.

**Previous Studies on FRC Fatigue**

There are no test standards specifically intended for establishing the flexural fatigue response of FRC materials (Ramakrishnan et al. 1996). However, procedures used for fatigue properties of plain concrete are usually adapted for tests on FRC. Additionally earlier investigators like Batson (1972) have also successfully used test procedures, loading configurations and specimen sizes employed for static flexural tests of FRC. Generally, flexural fatigue tests reported for FRC involve constant amplitude non-reversed sinusoidal loading (Ramakrishnan et al. 1987, 1996). The loading frequency used varies from a low value of a few Hertz to the more commonly used value of 20 Hertz. (Ramakrishna et al., 1996, Paskova and Meyer, 1997, Naaman and Hammoud, 1998). The works by Ramakrishnan et al (1987, 1996) provide some of the more exhaustive parametric studies of fatigue performance of FRC (varying fiber type, volume fraction and stress range). Typically fatigue tests are conducted in a load-controlled mode with specified lower and upper limit fatigue stress levels. The parameters selected for the current fatigue-testing program are based in part on some of these earlier studies. Unlike many previous fatigue test programs, one of the goals of this study was to investigate the progressive degradation of specimens stiffness and resultant changes in hysteresis loops and associated energy absorption.

**Experimental Program**

Three types of materials were tested in flexural fatigue during Phase II of this investigation, namely Control (unreinforced concrete), PFRC (polyolefin reinforced concrete – 50 mm monofilament 3M polyolefin fibers at 25 lb/yd$^3$), and SFRC (steel fiber reinforced concrete – 60 mm hooked-end Bekaert steel fibers at 75 lb/yd$^3$). Smaller specimen size for the Control mix (4.5" x 3.5" x 16") and larger specimen sizes (6" x 6" x 20") for the FRC specimens (per ASTM C-1018 to account dimensional constraints for fiber lengths in excess of 30 mm) were used. Tests were conducted with the lower limit flexural fatigue stress fixed at 10% of the static modulus of rupture for each of the materials studied. The upper limit stress level was varied in the four series of tests to 60%, 70%, 80% and 90% of the static modulus of rupture for the materials studied. Up to six replicate specimens were tested for each data point (a total of 72 fatigue specimens tested during Phase II). A 20-Hz sinusoidal wave was used as the command signal. Tests were conducted under load-controlled conditions using a closed-loop electro-hydraulic testing machine. A custom developed LabView program, described in the next section, accomplished the task of test control and data acquisition. Fatigue cycling at 20Hz was interrupted at regular intervals, to incorporate three cycles of slow loading /unloading at 1 cycle per minute (0.0167 Hz.) to allow for a study of progressive degradation of the material. Tests were started with three slow cycles (0.0167-Hz) to establish the virgin load-deflection response of each fatigue specimen. Load from the load cell (see Figure 1), net-deflection, ram displacement and number of cycles of fatigue loading were recorded regularly during the fatigue tests.

**LabView Program Details**

LabView software and associated National Instruments data acquisition board (12-bit analog to digital converter at a 40 KHz throughput) were used for the investigation. Several subprograms (also called as VIs or Virtual Instruments) were written to handle the test control and data acquisition features of the fatigue test. When the program is executed, the user is first asked to enter test information on three user-friendly screens. Then the user is asked to select filenames for the two output files. Thirdly the main testing screen reappears and waits for the start button to be activated. Once the test has begun the program switches back and forth between two separate parts of the software. The first part of the test consists of three slow cycles at a frequency of 0.0167-Hz. This part of the program allows the load-deflection response, and more importantly the stiffness, to be monitored as the test progresses. The second part of the test is the fast cyclic loading conducted at 20-Hz. An outer loop controls how many fast cycles the test completes between the slow cycle subprograms. The test stops when either the stop button is activated or a prescribed limiting ram displacement is achieved.

The main testing screens consists of three graphs, readouts of the current upper limits of all channels, a listing of what is being graphed, the number of cycles that has been completed, and three buttons that allow the user to start/restart, pause and stop the test. For all three graphs the x-axis is set to auto scale. The graphing limits on the y-axis whether it is set to graph ram displacement, net deflection, or load can be adjusted during the test so that the relevant data can always be displayed on the screen. During the fast cycles part of the test the program reads in a group of data points and then computes the mean and amplitude of the sample set. This mean and amplitude are added to obtain the upper limit, which is displayed on the screen for each channel. The number of fatigue cycles recorded is cumulative and includes both the slow and fast cycles.

The outer loop that causes the fatigue test sequence structure to repeat can be terminated in three ways. If the ram displacement exceeds the prescribed limiting deflection (0.125" in this investigation) the program terminates the main do-loop. Inside the slow cycles subprogram a similar check is made and if that check fails then again the main do-loop is terminated. The third option to stop the main do-loop is by activating the Stop button on the testing screen.

The *slow cycles* subprogram facilitates control during the three cycles of loading/unloading at 0.0167-Hz. Data is recorded every second to a hysteresis loop data file. This subprogram brings up a testing screen similar to the main one.

The subprogram *test control* sends properly formatted commands through the serial port to the MTS Microprofiler (function generator for the command signal to the testing machine). If the Microprofiler does not receive, or returns an error signal, another subprogram is called which displays a summary of the problem on the screen.

As the specimen suffers stiffness degradation due to accumulated fatigue damage, the computer-controlled servo-hydraulic testing machine adjusts ram displacements appropriately so as to maintain prescribed levels of lower and upper limit load. This feature allows one to characterize the nature of progressive fatigue failure in FRC materials during the various stages of the test.

**Results and Discussions from the Fatigue Tests**

Figure 2 shows plots of maximum flexural load (upper limit load) versus number of cycles to failure for the three mixes tested with a view to highlight the extent of scatter typical in fatigue tests. While it is not common to plot flexural load on the y-axis, particularly when different specimen sizes were used, as in this investigation, this graph offers an excellent overview of the complete set of fatigue test results. Arrows at the far right indicate that the tests were stopped after 2 million cycles of loading at which time failure had not occurred (failure is defined in this investigation as a prescribed ram displacement of 0.125 in., given the ductility of the fiber mixes). Control specimens generally fracture into two halves well before this limit. All fatigue tests are stopped at 2 million cycles if failure as defined above does not precede this limit of fatigue life.

Figure 3 shows a plot where the maximum fatigue stress (upper fatigue limit) is normalized with respect to the static flexural strength of the material. This plot removes the influence of specimen size inherent in Figure 2. It can be clearly observed from the plot that both the fiber concrete mixes perform much better under flexural fatigue loading than the control mix.

The improved performance of the fiber mixes is more pronounced at the higher stress levels (difference between Control and FRC mixes is larger at 90% stress range (or also termed 0.9 stress range) than at the 60% stress range. Although tests were stopped at the 2 million-cycles, it appears that there is no well-defined endurance limit (independent of number of cycles to failure, like that observed in some ferrous alloys). For FRC however several researchers (Ramakrishnan et al., 1996) have chosen to define "endurance limit" as the maximum stress level (expressed as a fraction of the static flexural strength) that a material can sustain without failure at 2 million cycles. If this definition is used, then the FRC mixes exhibit an endurance limit of between 55-70% of the static flexural strength while the Control mix does not seem to exhibit an endurance limit even while using this restricted definition of "endurance limit".

Results from all the fatigue tests conducted during Phase II of the study are summarized in Table 1. It is relevant to note here that it is possible to impose 2 million cycles on a flexural specimen in less than 2-days when fatigue tests are conducted at a frequency of 20-Hz. From Table 1 it is clear that many of the Control specimens failed during the first set of slow cycles at the 80% and 90% stress range. All of the SFRC specimens tested at the 60% stress range survived the 2 million cycles without failure. Some of these SFRC specimens had no visible cracks on the surface. When statically tested to failure after surviving 2 million fatigue cycles a few exhibited strengths in excess of the virgin flexural strength. Such strengthening has been reported earlier as well (Ramakrishnan, et al., 1996). Of the two FRC mixes, clearly the SFRC mix performed significantly better than the PFRC mix (it should be noted that fiber lengths and volume fractions are different for these two mixes and are based on what the manufacturers recommended to MoDOT as ideal for pavement applications). The steel fiber mix had the inherent advantage of larger fiber length, hooked-end fibers, and a higher fiber modulus. The polyolefin fiber mix had an advantage with regard to larger fiber volume fraction (even though the fiber weight fraction was smaller than the SFRC mix).

Figures 4-6 are plots showing the hysteresis loops during selected slow-cycles for the Control, PFRC and SFRC specimens, respectively at an upper limit stress level of 60% of the corresponding static flexural strengths. In general, the following can be observed at the low level of upper limit stress (60% of static flexural strength): (1) there appears to be only a small amount of degradation of stiffness (as determined from the average slopes of the load-deflection response) in any of the three mixes tested (Table 2), (2) the shapes of the hysteresis loops themselves do not change as a function of the number of fatigue cycles while there is some increase in the energy absorbed within each hysteretic loop with fatigue cycling (Table 2), and (3) there is no significant difference in the shapes of the loops between Control and the two fiber mixes.

Figures 7-9 are plots showing the hysteresis loops during selected slow-cycles for the Control, PFRC and SFRC specimens, respectively at an upper limit stress level of 80% of the corresponding static flexural strengths.

In general, the following can be observed at the high level of upper limit stress (80% of static flexural strength): (1) the control specimen (Figure 7) fails during the first slow cycle of hysteresis (within 3 cycles of loading/unloading) with accumulation of deformation (shift in the hysteresis loops to the right), (2) the PFRC specimen exhibits both a degradation in stiffness with accumulated fatigue cycles (slope of load-deflection plot becomes shallower), as well as increased energy absorption in the hysteretic loops with fatigue cycling (later loops are more rounded and larger in area than the earlier loops, Figure 8), (3) the shape of the hysteresis loops of the PFRC specimen do not show significant change with fatigue cycling, (4) the SFRC specimen, like the PFRC specimen exhibits both a stiffness degradation with accumulated fatigue cycles and a significant increase in the energy absorbed in the hysteresis loops with fatigue cycling, and (5) unlike the PFRC specimen, the SFRC specimen exhibits a noticeable change in the shape of the hysteresis loops with fatigue cycling, Figure 9. The earlier loops have a rounded shape similar to fatigue cycling at a lower level of upper fatigue stress level (eg. 60% of static flexural strength). The later loops have a shape that appear pinched at the top and bottom ends of the loop. Table 3 includes numerical details with regard to stiffness degradation and increased hysteretic energy absorption with fatigue cycling at the higher level of upper fatigue stress (80% of static flexural strength) for the specimen responses shown in Figures 7-9. Clearly at the higher level of upper fatigue stress, the fiber mixes behave differently from the control mix. This can be attributed to the fact that at these stress levels, the fiber matrix interface is subjected to progressive damage resulting in the greater levels of stiffness degradation and hysteretic energy absorption.

It is clear comparing Figures 4-9 and Tables 2-3, that the SFRC mix is superior to the PFRC mix as far as overall fatigue performance is concerned. Among the parameters used in this investigation to judge superior fatigue performance include: the number of fatigue cycles to failure at comparably prescribed stress level (preferably as a fraction of the respective static flexural strength), ability to accumulate damage without failure (hysteretic energy absorption capacity), ability to sustain significant degradation in stiffness without failure (stiffness degradation). In addition, the ability to resist crack opening is also inherently superior for the SFRC mix as compared to the PFRC mix. Higher modulus fibers have an advantage in this department. Although no systematic crack-width measurements were made at the various stages of fatigue performance, typically, cracks widths in SFRC specimens were smaller than those in the PFRC specimens. The PFRC and SFRC specimens in most of the tests did not fracture into two halves (specimen integrity was maintained even if failure was due to limiting prescribed ram displacement of 0.125") unlike the Control specimens, Figure 10. In several cases where the test was stopped at 2 million fatigue cycles without failure, the SFRC specimens exhibited little or no signs of surface cracking or distress.

Figure 10 and 11 show plots of growth of net-deflection versus number of fatigue cycles for the three types of specimens tested at two different levels of upper fatigue stress (80% and 90% of static flexural strength, respectively). The inset in Figure 10 shows the response of control and PFRC specimen using an enlarged scale to highlight the responses. This type of three-stage response was fairly typical, particularly at the higher levels of upper fatigue stress, and provides some interesting information with regard to mechanisms of accumulation of fatigue damage in FRC materials.

The first stage of the response is similar for the control and FRC mixes where the net deflection grows rapidly with the number of fatigue cycles. This phenomenon may be associated with accumulation of fatigue damage in the concrete matrix. In the control mix, when the accumulated damage reaches some threshold amount, there is catastrophic crack growth resulting in fracture of the specimen in two halves. In the fiber mixes, this threshold damage initiates a role for the fibers in load transfer and energy dissipation. The second and third stages of response appear to be unique to the fiber mixes. The second stage involves accumulation of fatigue cycles with little additional growth in net-deflection. It is possible that during this stage, the fiber-matrix interface is responsible for much of the energy dissipation, with progressive damage to the interfaces of the fibers bridging cracks. The third and final stage again sees a rapid growth in net-deflection with additional fatigue cycles. This is most likely due to failure of fiber-matrix interface bonds and resultant rapid growth in the main matrix crack. What can also be observed is the fact that typically the SFRC mix can sustain significantly larger deflections prior to failure then the PFRC or control mixes.    The PFRC mix however sustains significantly larger deflections prior to failure compared to the Control mix.

**Conclusions**
1.  The custom-developed LabView program greatly facilitated test control and data acquisition of a fairly complicated flexural fatigue testing protocol which combined fast-cycle fatigue testing (at 20 Hz) where limited data was stored (amplitude and mean values of all parameters) with regular slow-cycle fatigue loading (at 0.0167 Hz) where more detailed data on the complete load-deflection response was stored.
2.  The LabView programs developed allowed monitoring of progressive stiffness degradation and increase in hysteretic energy absorption for the three mixes including unreinforced concrete Control, PFRC and SFRC. This approach to fatigue testing has provided unique insight to mechanisms of fatigue damage accumulation in FRC materials discussed below.
3.  The fatigue performances of both the FRC mixes investigated in this study were superior to that of the Control mix. The fatigue performance of the SFRC mix investigated was significantly better than the PFRC mix. It would not however not be appropriate to conclude that in general steel fiber mixes will perform better than polyolefin fiber mixes, as fatigue performance is depended also on other important reinforcing parameters (fiber type, fiber aspect ratio, fiber content and fiber-matrix interface bond).
4.  Crack widths in the SFRC specimens were typically smaller than those in the PFRC specimens under comparable levels of fatigue loading (stress level as well as number of fatigue cycles). This property will influence the long-term durability of the material for pavement applications.
5.  At a limiting number of cycles of 2 million, the Control mix did not appear to have an endurance limit. All of the SFRC specimens and a couple of the PFRC specimens tested at an upper fatigue stress level of 60% survived without failure at 2 million cycles (failure defined here as complete fracture or a limiting ram displacement of 0.125").
6.  The difference between the FRC mixes and the Control mix becomes readily apparent at the higher levels of upper fatigue stress. In addition to improved fatigue life, the FRC specimens are also able to sustain larger deflections and absorb more energy in each fatigue cycle compared to the Control mix.
7.  Fatigue failure in FRC can be characterized by a three-stage process. In the first stage fatigue damage is accumulated in the concrete matrix. Rapid growth in net-deflection occurs with increasing fatigue cycles. The second stage is characterized by little or no growth in net-deflections, attributable to steady growth of damage along fiber-matrix interfaces. Only when this damage reaches a threshold level does the third stage begin. The third stage is characterized by a rapid growth in net-deflections resulting in fiber-pull out and/or fractures at the critical cross-section and associated catastrophic growth of the main matrix crack.

**Acknowledgments**
        Support from Missouri Department of Transportation (MoDOT) under research contracts RI97-006/SPR 1997-59 and RI97-015/SPR 1998-85 are gratefully acknowledged.    Discussions on the projects with Ms. Patty Lemongelli and Mr. Tim Chojnacki, both of the Research, Development and Technology (RDT) of MoDOT were invaluable.    Preliminary testing during Phase I of the study and development of the Lab View programs used in this investigation was undertaken by Mr. Matthew Eatherton (former graduate student).    Instrumentation support from Mr. Rich Oberto and machining support from Mr. Rick Wells are greatfully acknowledged.    The authors would also like to thank Mr. Dave Barrett (undergraduate honors project) for his assistance with some of the data analysis.

**References**
1. ACI Committee 544, "Measurements of Properties of Fiber Reinforced Concrete," Report 544 2R-88, ACI Materials Journal, Nov.-Dec. 1988.
2. Batson, G.B., "Flexural Fatigue Strength of Steel Fiber Reinforced Concrete," *ACI Journal*, Vol. 69, November 1972.
3. Chojnacki, T., "Evaluation of Fiber-Reinforced Unbonded Overlay," Missouri Department of Transportation, Report RDT 00-015, Dec. 2000.
4. Gopalaratnam, V.S. and Cherian, T., "Flexural Toughness and Flexural Fatigue Characterization of Fiber Reinforced Concrete Pavement Mixes," Report to Missouri Department of Transportation, April 2001, 44 pp.
5. Naaman, A E. Hammoud, H., "Fatigue Characteristics of High Performance Fiber-Reinforced Concrete", *Cement and Concrete Composites*, Vol. 20, No. 5, Oct 1998, pp 353-363.
6. Paskova, T., Meyer, C., "Low-cycle Fatigue of Plain and Fiber-Reinforced Concrete", *ACI Materials Journal*, Vol.94, Jul-Aug 1997, pp. 273 - 285.
7. Ramakrishnan, V., Gollopudi, S., and Zellers, R., "Performance Characteristics and Fatigue Strength of Polypropylene Fiber Reinforced Concrete", SP-105: ACI, Detroit, MI, 1987, pp. 159-178.
8. Ramakrishnan, V., Meyer, C., Naaman, A. E., Zhao, G., Fang, L., "Cyclic Behavior, Fatigue Strength, Endurance Limit and Models for Fatigue Behavior of FRC", HPFRCC 2, RILEM Proc. 31, Ed.: Naaman, A. E., and Reinhardt, H. W., 1996, Chapman and Hall, London.

**Table 1   Summary of fatigue life at the various maximum fatigue stress levels for the three mixes tested**

| Specimen | Maximum Fatigue Stress Level | | | | Remarks on Failure Mode |
|---|---|---|---|---|---|
| | 60% | 70% | 80% | 90% | |
| Control | 4,274 | 27,880 | 1 | 1 | Single large |
| | 338,420 | 42,181 | 2 | 1 | crack, Beam |
| | 1,079,019 | 62,333 | 2 | 2 | fractures |
| | 1,813,616 | 69,167 | 2 | 2 | into two |
| | 2,000,020 | - | - | 2 | halves |
| PFRC | 1,100,042 | 141,451 | 1,701 | 200 | Single large |
| | 1,530,037 | 177.942 | 2,600 | 440 | crack with |
| | 1,70499 | 243,717 | 8,271 | 640 | finer crack |
| | 1,960,039 | 640,757 | 17,654 | - | branches |
| | 2,000,000 | - | 19,720 | - | |
| SFRC | 2,000,000 | 288,699 | 4,321 | 400 | Single |
| | 2,000,020 | 545,824 | 7,096 | 441 | narrow |
| | 2,000,020 | 691,239 | 9,753 | 722 | crack with |
| | 2,000,034 | 1329858 | 65,500 | 1,787 | finer |
| | 2,032,598 | 2,000,020 | 2,000,020 | 2,000 | hairline branches |

**Table 2**   **Progressive degradation in stiffness and increased hysteretic energy absorption during each loading/unloading loop of fatigue (upper fatigue stress level of 60% of static flexural strength)**

| Specimen | Fatigue life | Cycle number | Stiffness ratio | Energy absorbed ratio |
|---|---|---|---|---|
| Control | 1,079,019 | 3 | 1.00 | 1.00 |
| | | 3,000 | 0.99 | 1.63 |
| | | 33,000 | 0.99 | 1.54 |
| | | 110,000 | 0.97 | 1.25 |
| | | 235,000 | 0.94 | 1.36 |
| | | 500,000 | 0.90 | 1.50 |
| PFRC | 1,704,991 | 3 | 1.00 | 1.00 |
| | | 3,000 | 0.88 | 1.28 |
| | | 33,000 | 0.87 | 1.39 |
| | | 110,000 | 0.84 | 1.43 |
| | | 235,000 | 0.83 | 1.50 |
| | | 750,000 | 0.82 | 1.48 |
| | | 1,500,000 | 0.80 | 2.24 |
| SFRC | 2,000,020 | 3 | 1.00 | 1.00 |
| | | 3,000 | 0.99 | 1.20 |
| | | 33,000 | 0.96 | 1.26 |
| | | 235,000 | 0.91 | 1.31 |
| | | 750,000 | 0.89 | 1.72 |
| | | 1,500,000 | 0.85 | 1.94 |
| | | 2,00,020 | 0.82 | 2.26 |

**Table 3**   **Progressive degradation in stiffness and increased hysteretic energy absorption during each loading/unloading loop of fatigue (upper fatigue stress level of 80% of static flexural strength)**

| Specimen | Fatigue life | Cycle number | Stiffness ratio | Energy absorbed ratio |
|---|---|---|---|---|
| Control | 1 | 1 | 1.00 | 1.00 |
| PFRC | 2,600 | 3 | 1.00 | 1.00 |
| | | 500 | 0.88 | 1.02 |
| | | 1,000 | 0.83 | 1.11 |
| | | 2,000 | 0.76 | 1.42 |
| SFRC | 2,000,020 | 3 | 1.00 | 1.00 |
| | | 500 | 0.99 | 1.23 |
| | | 1,000 | 0.97 | 1.34 |
| | | 21,000 | 0.93 | 1.45 |
| | | 1,100,000 | 0.26 | 5.07 |
| | | 1,500,000 | 0.19 | 8.48 |
| | | 2,000,020 | 0.16 | 10.88 |

**Fig. 1      Overall view of the flexural fatigue test set-up**

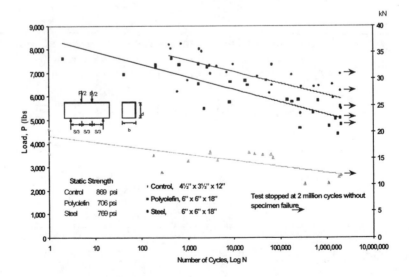

**Fig. 2      Upper limit fatigue versus number of fatigue cycles**

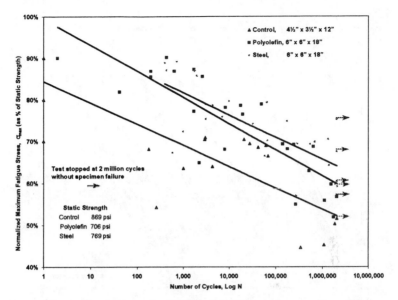

**Fig. 3** Normalized upper limit fatigue stress versus number of fatigue cycles

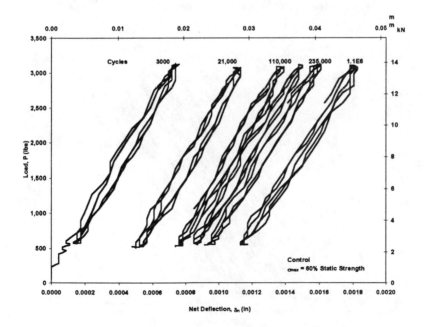

**Fig. 4** Hysterisis loops during slow cycle fatigue at an upper limit fatigue stress of 60% for control specimens

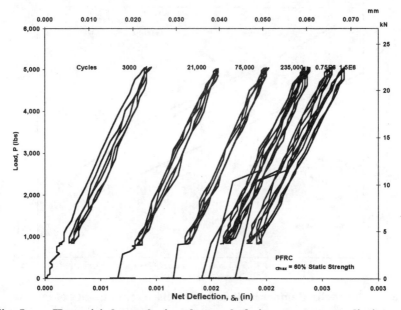

**Fig. 5    Hysterisis loops during slow cycle fatigue at an upper limit fatigue stress of 60% for PFRC specimens**

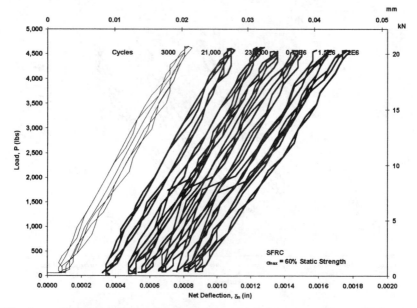

**Fig. 6    Hysterisis loops during slow cycle fatigue at an upper limit fatigue stress of 60% for SFRC specimens**

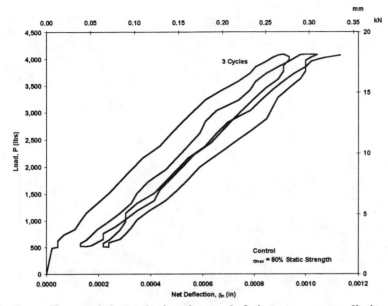

**Fig. 7**    **Hysteresis loops during slow cycle fatigue at an upper limit fatigue stress of 80% for control specimens**

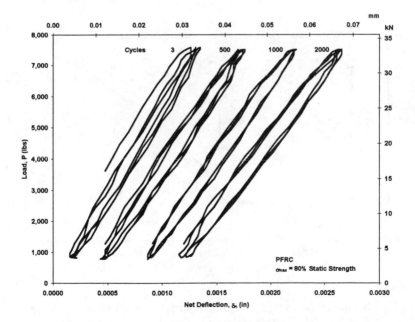

**Fig. 8**    **Hysteresis loops during slow cycle fatigue at an upper limit fatigue stress of 80% for PFRC specimens**

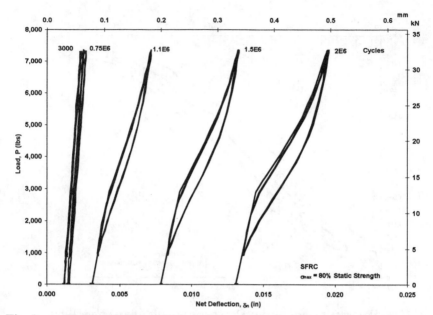

**Fig. 9    Hysteresis loops during slow cycle fatigue at an upper limit fatigue stress of 80% for SFRC specimens**

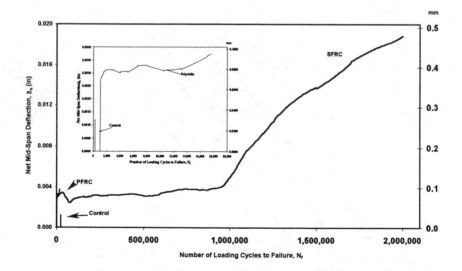

**Fig. 10    Net-deflection versus number of fatigue cycles showing the three stages in fatigue damage growth (80 % upper limit fatigue stress). Inset shows enlarged scale plots of control and PFRC response**

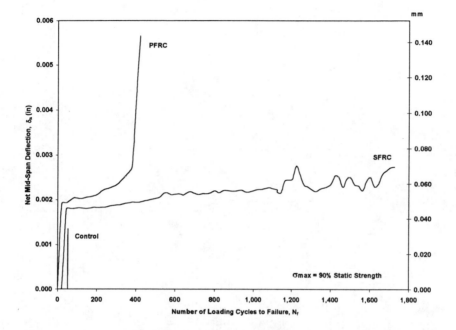

**Fig. 11**   **Net-deflection versus number of fatigue cycles showing the three stages in fatigue damage growth (90 % upper limit fatigue stress)**

# Assessment of Stresses in Reinforcement of Kishwaukee River Bridge

by J. Halvonik and M. W. Wang

**Synopsis**: Southbound bridge, as a first part of two Kishwaukee River Bridges, suffers from extensive cracking in the webs since its construction. Eight segments on either side of each pier are heavily cracked. The first part of the article deals with reasons and conditions of this damage.

For further investigation it was necessary to determine the most damaged webs of the bridge. The magnitude of steel stresses in shear reinforcement was assumed to be a major criterion for the severity of damage.

Varying shear resistance of the joints during construction and after retrofit caused a complex flow of internal forces. Ordinary models for steel stress calculations have become unworkable. Therefore, a model for assessment of stresses in shear reinforcement based on measured crack width has been developed. The model is based on CEB-FIP model for crack width prediction, that was modified for inclined cracks.

**Keywords:** bond; bridge; crack width; shear cracks; shear reinforcement; shear stresses

Jaroslav Halvonik is Associated Professor at Department of Concrete Structures and Bridges, Faculty of Civil Engineering at  Slovak University of Technology in Bratislava, Slovak Republic,  E-mail: halvonik@svf.stuba.sk, He received his ME (1988), and Ph.D (1993) from the Slovak University of Technology in Bratislava. His current work deals with theory of design of structural elements made from structural concrete.

Ming  L.  Wang   is Professor and Director of Bridge Research Center, at Department of Civil and Material Engineering, University of Illinois at Chicago, E-mail: mlwang@uic.edu

## INTRODUCTION

In 1980 and then in 1982, the southbound bridge and the northbound bridge respectively, opened to traffic, as two separate structures of the Kishwaukee River Bridge [1]  in Rockford, Illinois (as shown in Fig.1). Bridges have post-tensioned precast segmental box–girder decks. Single-cell segments have one shear key in each web. In contrast to the northbound bridge the deck of the southbound bridge suffers from extensive cracking in the webs since its construction. The reason is more or less known. After completion of the bridge, it was found out that epoxy glue did not harden properly in most of the joints. The epoxy was not able to carry any shear stresses and instead was acting as a lubricant that caused reduction of friction coefficient in the joints. A substantial part of the shear forces was concentrated at the shear keys. The rest was transferred across the joints by friction mostly in lower part of segments where high compressive axial stresses were present.

Severe cracking was not only a consequence of defective epoxy. Thirteen days after the bridge's completion the male shear key of segment SB1-N1 crushed and the inner surface spalled [1,2]. It resulted in slip of the web at the joint. Relative vertical displacement between the pier segment and segment #1 was 16 mm ($^5/_8$ in).

After  the SB1-N1 shear key failed, all defective  joints were  repaired by  steel pins [2]. The smooth contact surfaces became indented (toothed) and have substantially improved transfer of shear stresses across the joints, especially for loads imposed after retrofit (barriers, wearing surface and vehicular load). The steel pins have also enhanced shear resistance of the joints. The retrofit have not been completed due to the fact that the structure had not been activated before. Redundant steel stresses in shear reinforcement that had developed due to the failed epoxy, have not been removed and cracks have not been sealed.

The question that arises then, is how high are the stresses in the shear reinforcement now and is there any threat of failure? Furthermore, several in-situ load tests were proposed and health monitoring system was designed for the bridge. For this purpose an identification of the most damaged webs, was required. The magnitude of steel stresses in shear reinforcement was assumed to be a major criterion for the severity of damages.

The assessment of steel stresses is a complex task in the southbound bridge, because many conditions have influenced a flow of internal forces in the webs during and after construction, e.g. the shear keys were exposed to high shear forces with magnitude depending on the shear resistance of the joints. A different inclination of the shear cracks, even within one segment, shows different ability of the joints to transfer shear stresses during construction. The flow has significantly changed for loads imposed after insertion of the steel pins. Furthermore, stresses induced by self-weight before retrofit, have been changing due to creep of concrete. All these conditions mentioned did not allow stresses in reinforcement to be determined through using ordinary models. Finally, the idea of utilizing measured cracks width for the assessment of steel stresses led to the development of engineering model, designed to predict inclined crack widths in RC elements.

**Description of the Bridge**

The deck of the bridge has five spans with lengths of 51.8 m + 3 x 76.2 m + 51.8 m (170 ft + 3 x 250 ft +170 ft), see Fig.2. The overall length of the decks is 334 m (1096 ft) .
Precast segmental decks were built with the balanced cantilever method. Each cantilever consisted of seventeen single-cell segments 2150 mm (7'-$^3/_5$") long and one pier segment 1067 mm (3'-6") long. Cast-in-place closures have a length of 984 mm (3'-2$^3/_4$").

Each of the two bridges support two 3.66 m (12 ft) wide traffic lanes, two shoulders, curbs and parapets, see Fig.3. The total width of the structure is 12.8 m (42 ft).
The cross-section of segments is constant except for the first five segments from the pier, where thickness of the bottom slab changes from 203 mm to 457 mm (8 in–18 in). The web width 356 mm (14 in) is constant all over the structure. The maximum depth of the girder is 3550 mm (11'-7$^3/_4$"). Segments have match-cast epoxy joints with one shear key in each web. The design compressive strength of concrete was 37.9 MPa (5500 psi).

The decks are entirely prestressed by Dywidag high-strength threaded bars with diameter 32 mm (1$^1/_4$ in) located in the top and bottom slabs (no draped tendons are present). In longitudinal direction, the number of bars in the top slab (cantilever bars) ranges from 100 in pier segments to 8 bars in segments #17. The number of bars in bottom slab (continuity bars) ranges from 36 bars in segments #17 to six bars in segments #7. Top slab of each segment is also prestressed by 3 bars in transverse direction.

Segments are reinforced by mild steel reinforcement grade 60, see Fig.4. Each web contains eight stirrups at 10 in (254 mm) spacing. Stirrups were made from two #7 bars ($2\phi22.2$ mm) in the first three segments next to the piers, two #6 bars ($2\phi19.1$ mm) in further three segments and in the others from two #5 bars ($2\phi15.9$ mm). Longitudinal reinforcement consists of #4 bars ($\phi12.7$ mm) spaced at 254 mm (10 in) at both surfaces.

**Damages of the Bridge**

As mentioned above the webs of southbound bridge suffer from extensive cracking. Eight segments on either side of each pier are heavily cracked. The crack pattern is very different even within one segment (east and west web). The angle of the cracks varies from 10° to 42°. The widest cracks are sloping at 15°, usually propagating from the bottom part of the female key (crack "A" in Fig.5) towards the next segment. In many segments, it can be observed that crack propagates from bottom part of the male key (crack "B" in Fig.5). These cracks are shorter and less wide than the former ones. In most of the segments, cracks end at the opposite end without continuing to the adjoining segment. The widest cracks are located next to the female key, having an average width of 0.75 mm (0.03 in). In the middle of some segments, it was found to be 0.65 mm (0.026 in). The most frequently observed crack width is 0.40 mm (0.016 in).

**ASSESSMENT OF STEEL STRESSES IN SHEAR REINFORCEMENT**

It is known that the crack width is influenced by many parameters e.g. stresses in reinforcement, bond between concrete and steel, the reinforcement ratio, the loading configuration and the way of loading, long-term properties of concrete and etc. The idea is to determine steel stresses by comparing measured "$w_{meas}$" and predicted "$w_{cal}$" crack widths. The problem is to find proper model for crack width calculation, because most of them only predict crack widths that cross reinforcement perpendicularly or those under angles less 75°, e.g. ACI. CEB-FIP Model Code 90 [3] introduces one of the engineering models that can predict the width of inclined (shear) cracks.

## Model for Crack Width Calculation

The MC90's model is based on calculation of average steel $\varepsilon_{sm}$ and concrete $\varepsilon_{cm}$ strains along so called transmission length $l_t$ which is a length over which slip between steel and concrete occurs, see Fig.6. A difference between the average strains of steel and concrete multiplied by transmission length is a slip $s_{cr}$ of reinforcement at the crack. The maximum crack width is twice the slip at the crack. In RC-elements with quasi-constant stress field over cracked area the probability that parallel cracks will have a constant spacing of $2 \times l_t$, is very small. Therefore, the term average crack width have been introduced.

The maximum crack width can be determined by eq. (1) and average crack width by eq. (2). An influence of shrinkage on crack width is expressed by strain $\varepsilon_{cs}$.

$$w_{max} = s_{r,max}\left[\left(\varepsilon_{sm} - \varepsilon_{cm}\right) - \varepsilon_{cs}\right] \tag{1}$$

$$w_{avr} = s_r\left[\left(\varepsilon_{sm} - \varepsilon_{cm}\right)_t - \varepsilon_{cs}\right] \tag{2}$$

The maximum spacing between the cracks is as a twice of the transmission length.
$$s_{r,max} = 2 l_t \tag{3}$$

While average spacing can be taken: $s_r = \dfrac{2}{3} s_{r,max} = \dfrac{4}{3} l_t$ (4)

If stabilized cracking is assumed, the transmission length can be determined by:

$$l_t = \frac{\sigma_{sr} - \sigma_{sE}}{4\tau_{bm}}\phi_s = \frac{F_{cr}/A_s - \alpha_e F_{cr}/\left(A_{c,eff} + \alpha_e A_s\right)}{4 \times 2.25 f_{ctm}} = \frac{\phi_s}{9\rho_{s,eff}} \tag{5}$$

According to [3], mean value of the average bond stress in eq. (5) can be expressed by: $\tau_{bm} = 2.25 f_{ctm}$ (6)

Cracking force by: $F_{cr} = f_{ctm}\cdot\left(A_{c,eff} + \alpha_e A_s\right)$ (7)

Modular ratio by: $\alpha_e = E_s / E_{ci}$ (8)

Effective reinforcement ratio: $\rho_{s,eff} = A_s / A_{c,eff} = A_s /\left(d_{c,eff} l_{c,eff}\right)$, see Fig.7 (9)

The difference of average strains for maximum crack width calculation can be determined by:

$$\left(\varepsilon_{sm} - \varepsilon_{cm}\right) = \left[\varepsilon_s - \beta\left(\varepsilon_{sr} - \varepsilon_{sE}\right)\right] - \beta\,\varepsilon_{sE} = \varepsilon_s - \beta\,\varepsilon_{sr} \tag{10}$$

MC90's model for tension stiffening with empirical integration factor $\beta$ for the steel strain along the transmission length is introduced in eq. (10). The factor $\beta$ depends on distribution of tensile stresses in concrete under tension (bending, pure tension), on bond properties of the steel bars (plane, deformed) and also on type of load (short term, long term, cyclic).

In our case, the webs are reinforced by deformed bars, the distribution of tensile stresses before cracking was similar with pure tension (principal stresses) and permanent actions are loading the structure for more than 22 years. Therefore, factor $\beta = 0.38$ for long term and cyclic load was used. The factor $\beta$ for long term load includes also an influence of creep. For short term load $\beta = 0.6$ can be taken. The calculation of strain difference for average crack width is analogous only integration factor of $\beta_t = \frac{2}{3}\beta$ has to be applied in eq. (10).

## Model for Inclined Crack Width Calculation

Since inclined cracks cross reinforcement with an angle other than 90° and reinforcement is orientated in two directions, the MC90's model has to be modified.

Steel strains $\varepsilon_{srL}$ and $\varepsilon_{srt}$ under cracking force can be determined by:

$$F_{crL} = f_{ctm}\left(A_{cL,eff} + \alpha_e A_{sL}\right)\sin\theta \Rightarrow \sigma_{srL} = F_{crL}/A_{sL} \Rightarrow \varepsilon_{srL} = \sigma_{srL}/E_s \qquad (11)$$

$$F_{crt} = f_{ctm}\left(A_{ct,eff} + \alpha_e A_{st}\right)\cos\theta \Rightarrow \sigma_{srt} = F_{crt}/A_{st} \Rightarrow \varepsilon_{srt} = \sigma_{srt}/E_s \qquad (12)$$

The effective concrete area in eq. (11) and (12) is defined by:

$$A_{cL,eff} = 2.5 d_{c,eff}\, l_{cL,eff} \qquad (13)$$

$$A_{ct,eff} = 2.5 d_{c,eff}\, l_{ct,eff} \qquad (14)$$

The length of effective concrete area for longitudinal reinforcement $l_{cL,eff}$ is a lesser value of $15\phi_{sL}$ and $s_L/\sin\theta$, for transverse reinforcement $l_{ct,eff}$ lesser value of $15\phi_{st}$ and $s_t/\cos\theta$. Where $\phi_{sL}$ is diameter of reinforcement in horizontal direction and $\phi_{st}$ in vertical direction, respectively.

The depth of the effective concrete area was assumed as an average of depths for vertical and horizontal direction: $d_{c,eff} = 2.5\left[c + 0.5\left(\phi_{st} + \phi_{sL}\right)\right]$ \qquad (15)

The widths of inclined cracks can be determined by:

$$w_\theta = \varepsilon_{smt}\, s_{rt}\cos\theta + \varepsilon_{smL}\, s_{rL}\sin\theta - s_{r\theta}\left(\varepsilon_{crm} + \varepsilon_{cs}\right) \qquad (16)$$

In order to determine notional spacing between cracks $s_{rL}$ and $s_{rt}$ in horizontal and vertical directions, transmission lengths are needed. As mentioned above the transmission length is a length along which slip between concrete and steel occurs. From definition following formulae were derived:

$$l_{tt} = \frac{\left(\sigma_{srt} - \sigma_{sEt}\right)}{4\tau_{bm}} \phi_{st} \quad \Rightarrow \quad s_{rt} = 2l_{tt} = \left(\sigma_{srt} - \alpha_e \sigma_{ct}\right)\frac{\phi_{st}}{4.5 f_{cm}} \qquad (17)$$

$$s_{rL} = \left(\sigma_{srL} - \alpha_e \sigma_{cL}\right)\frac{\phi_{sL}}{4.5 f_{ctm}} \qquad (18)$$

Stresses $\sigma_{cL(t)}$, in eq. (17) and (18) are axial stresses in the concrete element prior to cracking. The stress state in concrete, when the first crack occur, can be determined by equations for stress transformation (19) - (20) [7] with $\sigma_{cr} = f_{ctm}$.

$$\sigma_L = \sigma_{cd} \cos^2\theta + \sigma_{cr} \sin^2\theta + \rho_{sL}\sigma_{sL} \qquad (19)$$

$$\sigma_t = \sigma_{cd} \sin^2\theta + \sigma_{cr} \cos^2\theta + \rho_{st}\sigma_{st} \qquad (20)$$

$$\tau_{Lt} = \left(-\sigma_{cd} + \sigma_{cr}\right)\sin\theta \cos\theta \qquad (21)$$

where:

$\rho_{sL(t)}$ = reinforcement ratios in horizontal (vertical) direction.

If stresses $\sigma_L$ due to imposed load and inclination of the crack $\theta$ are known following procedure can be used for description of stress field in the RC element prior to cracking.

1. Axial stresses in concrete in longitudinal direction are:

$$\sigma_{cL} = \sigma_L / \left(1 + \alpha_e \rho_{sL}\right) \qquad (22)$$

2. Principal compressive stresses prior to cracking:

$$\sigma_{cL} = \sigma_{cd} \cos^2\theta + f_{ctm} \sin^2\theta \quad \Rightarrow \quad \sigma_{cd} \qquad (23)$$

3. Axial stresses in transverse direction: $\sigma_{ct} = \sigma_{cd} \sin^2\theta + f_{ctm} \cos^2\theta \qquad (24)$

Average concrete strains $\varepsilon_{crm}$ in principal direction "r" (see Fig.8), can be determined using the assumption that principal tensile stresses $\sigma_{cr}$ in the center of compressive diagonal (concrete between two parallel inclined cracks) equal to $f_{ctm}$ and principal tensile stresses $\sigma_{cr}$ in the crack to zero (see Fig.9).

$$\varepsilon_{crm} = \beta \frac{f_{ctm}}{E_{ci}} \qquad (25)$$

The maximum spacing between two parallel inclined cracks $s_{r\theta}$ measured in principal direction "r" is the greater value between $s_{rt} \cos\theta$ and $s_{rL} \sin\theta$.

Average strains in reinforcement along the transmission length can be determined by:

$$\varepsilon_{smL} = \left[\sigma_{sL} - \beta\left(\sigma_{srL} - \alpha_e\sigma_{cL}\right)\right] / E_s \qquad (26)$$

$$\varepsilon_{smt} = \left[\sigma_{st} - \beta\left(\sigma_{srt} - \alpha_e \sigma_{ct}\right)\right] / E_s \tag{27}$$

Steel stresses $\sigma_{st}$ and $\sigma_{sL}$ in eq. (26) and (27) ensure stress equilibrium in the crack for acting load. The stresses can be computed by eq. (19) – (21) with a principal tensile stress $\sigma_{cr} = 0$.

Cracking is always accompanied with a redistribution of internal forces (stresses) in the real structures and therefore equilibrium conditions for principal stresses that are identical with orientation of newly created cracks do not valid more. Furthermore, concrete has an ability to transfer shear stresses across the cracks, concrete has some residual shear strength in the crack, which can be explained by aggregate interlocking.

According to [8], concrete with nominal strength 37.9 MPa (5500 psi) and maximum used aggregate size 16 mm (0.63 in) is able to transfer shear stresses $\tau_{ci} = 1.46$ MPa (232 psi) across the crack 0.60 mm (0.024 in) wide. It means that further growth of stresses in reinforcement as well as in concrete is very difficult to predict at least at the level of service loads.

## Material Properties of Kishwaukee River Bridge

The concrete's strength properties play an important role in the assessment. Unfortunately, actual compressive and tensile strength of concrete are unavailable. The bridge is still in service and destructive tests (core samples) are not permitted. Required material properties have been derived from 28-days uniaxial design compressive strength of concrete $f_c' = 37.9$ MPa (5.50 ksi). To avoid an underestimation the calculation was carried out by **mean values** of all material properties including bond strength. The mean value of concrete compressive strength was assessed based on coefficient of variation $V_c = 0.15$.

- Assumed mean value of compressive strength of concrete (50% fractile)

$$f_{cm} = f_c' \left(1 - 1.34 x V_c\right)^{-1} = 47.43 \text{ MPa (6.88 ksi)}$$

- Characteristic compressive strength of concrete (5% fractile):

$$f_{ck} = f_{cm} \left(1 - 1.64 x V_c\right) = 35.76 \text{ MPa (5.19 ksi)}$$

- Mean value of axial tensile strength of concrete (50% fractile):

$$f_{ctm} = 1.40 \left(f_{ck}/10 MPa\right)^{2/3} = 3.274 \text{ MPa (475 psi)}$$

- Modulus of elasticity (mean value): $E_{ci} = 21500 \left[f_{cm}/10 MPa\right]^{1/3} = 36124$ MPa (5232 ksi)

Steel stresses $\sigma_{st}$ in shear reinforcement were acquired by comparison of predicted and measured crack widths $w_{cal} = w_{meas}$, where $w_{cal} = w_{max}(\sigma_{st})$ for single crack pattern and $w_{cal} = w_{avr}(\sigma_{st})$ for multiple crack pattern. Single crack pattern means a system of parallel wide cracks with constant spacing. Multiple crack pattern is a system where space between major cracks is filled by secondary cracks.

Besides $\sigma_{st}$, steel stresses in longitudinal reinforcement $\sigma_{sL}$ are also unknown. Because only one equation is available, an assumption of constant stress ratio $k_\sigma = \sigma_{srL}/\sigma_{srt}$ was applied and stresses $\sigma_{sL}$ were determined by:

$$\sigma_{sL} = k_\sigma \sigma_{st} \tag{28}$$

Long-term properties of concrete were figured out based on MC90's model for shrinkage prediction. Following parameters were assumed in calculation: relative humidity of ambient atmosphere 75%, notational size of member 356 mm, normal hardening cement, 28-days mean compressive strength of concrete 47.4 MPa. Computed average strain due to shrinkage was $\varepsilon_{cs} = -2.5 \times 10^{-4}$.

Axial stresses $\sigma_L$ in longitudinal direction were restored for construction stage (after completion of a cantilever unit) at the centroid of assumed concrete cross-section. Axial stresses due to prestressing were calculated assuming an average effective stress in prestressing bars $\sigma_p = 600$ MPa (87 ksi), see Tab.2.

Assessed steel stresses for the most damaged webs are introduced in Tab.3. The single crack pattern is present in all assumed webs. Reinforcement of assumed webs is in Tab.1.

## CONCLUSIONS

The MC90's model for crack width calculations was created base on large number of laboratory tests, e.g. integration factors $\beta$ are fully empirical values. However, the test results–measured crack widths are statistical data that show quite large scatter for the similar or the same conditions during test's execution. Though the MC90's model is deterministic, the computed values have statistical dimension, so there are always deviation from measured values. Therefore, we can expect some differences between actual and computed stresses in reinforcement.

On the other hand, introduced engineering model made it possible to determine the most damaged webs regarding the magnitude of stresses in shear reinforcement and focus attention to them.

Larger crack widths do not always imply higher stresses in shear reinforcement, e.g. in our case, segments are reinforced by bars with different bond properties (different bar diameters $\phi_s$), different reinforcement ratios and inclination of the cracks. Furthermore axial stresses due to prestressing as well as shear forces are varying from segment to segment. All these parameters finally influenced actual values of the crack widths. The model gives a tool to take into account all those parameters. The model can also be used for predicting inclining crack width in the webs of deep beams or girders. In many cases, the model allows us to check for a minimum shear and longitudinal reinforcement in the webs.

## REFERENCES

1. Nair, S.R. - Iverson, J.K.: "Design and Construction of the Kishwaukee River Bridge." Special Report, *PCI Journal*, Vol.27, No.6, pp.22-47, 1982

2. Shiu, K.N. - Daniel, J.L. - Russel, H.G.: "Time-Dependent Behavior of Segmental Cantilever Concrete Bridges" 1983

3. CEB-FIP Model Code 1990, Design Code, Thomas Telford Services, 1993

4. Serviceability Models: Progress report, *CEB in Bulletin d'Information*, 1997

5. prEN 1992-1 (2nd draft): Design of Concrete Structures - Part 1: General Rules and Rules for Buildings, 2001

6. ENV 1992-2: Design of Concrete Structures - Part 2: Concrete Bridges, 1995

7. Hsu, T.C.: "Unified Theory of Reinforced Concrete." CRC Press, 1993

8. Collins, M. P. – Mitchel, D.: "Prestressed Concrete Structures." Prentice Hall, Englewood, New Jersey 1991.

Table 1: Reinforcement in the webs of segments

| Segment | $\phi_{st}$ | | $s_t$ | | $\phi_{sL}$ | | $s_L$ | |
|---|---|---|---|---|---|---|---|---|
| # | # | [mm] | [in] | [mm] | # | [mm] | [in] | [mm] |
| S1 -S3 | #7 | 22.22 | | | | | | |
| S4 -S6 | #6 | 19.05 | 10 | 254 | #4 | 12.70 | 10 | 254 |
| S7- S17 | #5 | 15.88 | | | | | | |

Table 2: Axial stresses at centriod of the segment's cross-section

| Web | $A_c$ | $A_{p1}$ | $n_{top}$ | $n_{bott}$ | $A_p$ | $N_p$ | $\sigma_L$ | $\sigma_L$ |
|-----|-------|----------|-----------|------------|-------|-------|------------|------------|
|     | [m²]  | [cm²]    | [mm]      | [MPa]      | [cm²] | [MN]  | [MPa]      | [psi]      |
| S1-E | 8.688 |       | 96 | - | 772.0 | -46.3 | -5.33 | -733 |
| N2-E | 8.400 |       | 86 | - | 691.6 | -41.5 | -4.94 | -717 |
| N4-E | 7.859 | 8.04  | 70 | - | 562.9 | -33.8 | -4.29 | -624 |
| S8-E | 7.480 |       | 40 | (10) | 402.1 | -24.1 | -3.22 | -468 |

Table 3: Assessed steel stresses in reinforcement

| Web | Reinf. | $w_\theta$ | $w_\theta$ | $\theta$ | $\sigma_L$ | $\sigma_{srL}$ | $\sigma_{srt}$ | $\sigma_{sL}$ | $\sigma_{st}$ |
|-----|--------|-----------|-----------|----------|-----------|----------------|----------------|---------------|---------------|
|     |        | [in]      | [mm]      |          | [MPa]     | [MPa]          | [MPa]          | [MPa]         | [MPa]         |
| S1-E | 2ϕ22.2 | 0.024 | 0.60 | 39° | -5.22 | 344 | 244 | 408 | **290** |
| N2-E | 2ϕ22.2 | 0.023 | 0.58 | 19° | -4.94 | 178 | 247 | 267 | **371** |
| N4-E | 2ϕ19.1 | 0.024 | 0.60 | 15° | -4.29 | 136 | 319 | 167 | **391** |
| S8-E | 2ϕ15.9 | 0.023 | 0.58 | 17° | -3.22 | 234 | 391 | 255 | **396** |

Figure 1 – Outlook of the Kishwaukee Bridge

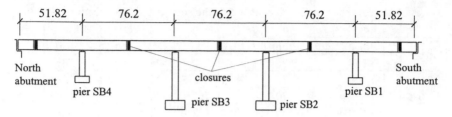

Figure 2 – Longitudinal layout of the Kishwaukee Bridge

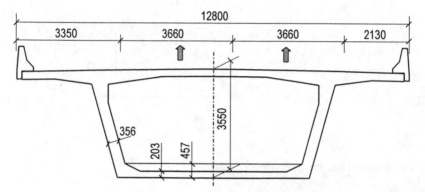

Figure 3 – Cross-section of the bridge

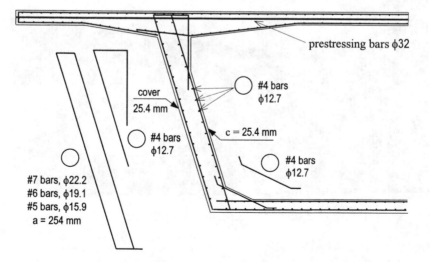

Figure 4 – Ordinary reinforcement in typical segment

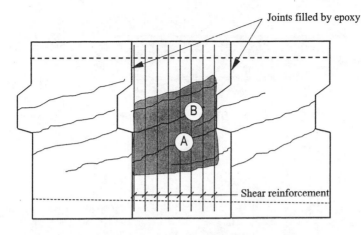

Figure 5 –Typical crack pattern

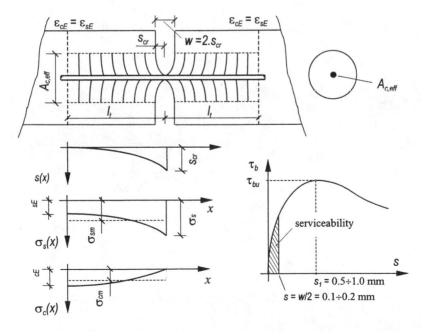

Figure 6 –Model for crack width calculations and bond-slip diagram

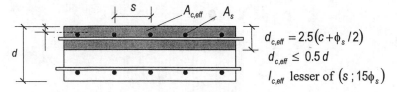

Figure 7–Definition of effective concrete area $A_{c,eff}$ (pure tension)

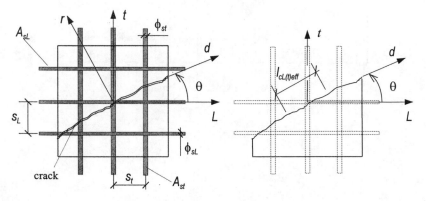

Figure 8–Cracked RC plate element

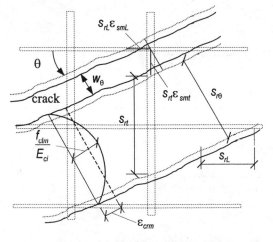

Figure 9 – Modeled inclined crack

# Some Australian Code Developments in the Design of Concrete Structures

## by B. V. Rangan

**Synopsis:** The paper presents a review of some of the Australian code developments in the design of concrete structures. Topics such as shear design of beams, deflection control, punching shear strength of slabs and shear strength of walls are covered.

Keywords: beams; deflection; design; punching; shear; slabs; walls

ACI Fellow B. Vijaya Rangan is Professor of Civil Engineering and Head of School of Engineering and Dean of Engineering, Curtin University of Technology, Perth, Australia. He is a co-author of a text book on concrete structures and has received many awards for his research. Dr Rangan is a member of several ACI committees and a member of Standards Australia Committee on Concrete Structures.

## INTRODUCTION

In the past thirty years, there have been a number of significant developments in the utilisation and application of concrete in civil and structural engineering works. Concrete technology has reached a high degree of sophistication and the design methods have become more precise and rational than before. The benefits of research and how the outcomes have influenced the design codes are substantial. It is an onerous task to fully review all developments. The paper therefore presents a review of some of the developments in the Australian code in the design of concrete structures. Wherever possible other codes are also compared.

## DESIGN OF BEAMS

### Flexural Strength

The flexural strength of a beam is customarily calculated by assuming a linear strain distribution over the depth of the section and considering the equilibrium of forces and moment (1). Although the ultimate compressive strain varies with concrete strengths, a value equal to 0.003 represents the test results satisfactorily. The scatter of test data does not justify a variation of the strain with the compressive strength (2). This value is specified in several codes (3,4,5). In the Canadian Standard (6), $\varepsilon_{cu}$ is taken as 0.0035.

In the codes and standards, the actual distribution of compressive stresses in the concrete is replaced by an equivalent rectangular stress block. The stress block has a uniform stress of $0.85 f_c'$ and a depth less than the neutral axis depth. It is generally accepted that the uniform stress should be smaller than $0.85 f_c'$ for $f_c' > 50$ MPa. In the New Zealand Standard (3), the depth of equivalent rectangular stress block is taken $\gamma$ times the depth of neutral axis and the uniform stress is taken as $\alpha f_c'$

where:

$$\gamma = 0.85 - 0.008\left(f_c' - 30\right) \tag{1}$$

within the limits of $0.65 \leq \gamma \leq 0.85$   and

$$\alpha = 0.85 - 0.004\left(f_c' - 55\right) \tag{2}$$

within the limits of $0.75 \leq \alpha \leq 0.85$.

Note that $\alpha = 0.85$ when $f_c' \leq 55$ MPa and $\alpha = 0.75$ when $f_c' \geq 80$ MPa.

In the Canadian Standard (6), $\alpha$ and $\gamma$ are given by the following:

$$\alpha = \left(0.85 - 0.0015 f_c'\right) \geq 0.67 \tag{3}$$
$$\gamma = \left(0.97 - 0.0025 f_c'\right) \geq 0.67 \tag{4}$$

The minimum values of 0.67 do not apply until $f_c' > 125$ MPa. In Eqs. 1 to 4, $f_c'$ must be substituted in MPa.

Beams designed in practice are under-reinforced and their flexural strength is controlled by the yield force in the tensile steel. The values of the rectangular stress block parameters will therefore have insignificant effect on the design calculation. From this viewpoint, any of the above proposals are suitable.

To prevent a brittle failure at first cracking, the tensile steel ratio should not be less than a minimum. In the Canadian Standard (6), the minimum area of tensile steel $A_{st\,min}$ is given by:

$$A_{st\,min} = 0.2\sqrt{f_c'}\, b_t D / f_y \tag{5}$$

where, $f_c'$ is the cylinder compressive strength in MPa, $b_t$ is the width of the tension zone of the section under consideration and $D$ is the overall depth of beam.

In the New Zealand Standard (3), $A_{st\,min}$ is given by:

$$A_{st\,min} / b_w d = p_{min} = \left(0.25\sqrt{f_c'}\right) / f_y \tag{6}$$

where $b_w$ is the width of the web. Both Eq. 7 and 8 are similar and acceptable.

## Shear Strength

Extensive research has been carried out on the behaviour and shear strength of reinforced concrete beams. Numerous tests on beams have been conducted and reported in the literature. Many theories for predicting the shear strength have been advanced (7). Of these, the strut-and-tie approach and the plasticity have received wider acceptance by the profession. The shear design provisions given by various codes and standards (3-6) are largely based on these approaches, but empirically modified to fit test trends.

As an example, the approach taken in the Australian Standard (4) is influenced by a re-evaluation of the conventional $45^0$ truss model in the light of test results. It was found that the truss angle was not always $45^0$, but smaller especially for beams that contain light shear reinforcement. Because most practical beams contain only light shear reinforcement, it is appropriate to adopt a limited variable angle truss model which takes advantage of the situation. Accordingly, in the Australian Standard, the shear strength $V_u$ of a beam containing vertical stirrups is therefore given by:

$$V_u = V_{uc} + A_{sv} f_y \frac{d_0}{s} \cot \theta \le V_{u\,max} \tag{7}$$

where the angle $\theta$ may be taken to vary linearly from $30^0$ when $V^* = \phi V_{u\,min}$ to $45^0$ when $V^* = \phi V_{u\,max}$. Here $V^*$ is the (factored) design shear force, $\phi$ is the strength reduction factor taken as 0.7, $V_{u\,min}$ is the shear strength of a beam provided with minimum shear reinforcement and $V_{u\,max}$ is the shear strength limited by web crushing and is given by

$$V_{u\,max} = 0.2 f_c' b_v d_0 \tag{8}$$

In Eq. 7, the shear strength contributed by the concrete $V_{uc}$ in the case of a reinforced concrete beam is given by

$$V_{uc} = \beta_1 \beta_2 \beta_3 b_v d_0 \left[ \frac{A_{st}}{b_v d_0} f_c' \right]^{1/3} \tag{9}$$

The factors for determining $V_{uc}$ in Eq. 9 are given by

$$\beta_1 = 1.1 \left( 1.6 - \frac{d_0}{1000} \right) \ge 1.1 \, (d_0 \text{ in mm}) \tag{10}$$

$$\beta_2 = 1.0 \text{ (where no axial load exists)} \tag{11}$$

$$\beta_3 = \frac{2 d_0}{a_v} \, (1.0 \le \beta_3 \le 2.0) \tag{12}$$

where $a_v$ is the distance of the concentrated load from the nearest support.

In the Canadian Standard (6), the minimum amount of shear reinforcement is given by

$$A_{sv\,min} = \frac{0.06\sqrt{f_c'}\,b_v s}{f_y} \tag{13}$$

so that the shear design method may be applied to concretes with compressive strengths up to 100 MPa.

If we take $A_{sv\,min}$ as given by Eq 13, from Eq 7, $V_{u\,min}$ becomes

$$V_{u\,min} = V_{uc} + \left[\frac{0.06\sqrt{f_c'}\,b_v s}{f_y}\right] f_y \frac{d_0}{s} \cot 30^0 \tag{14}$$

or

$$V_{u\,min} = V_{uc} + 0.10\sqrt{f_c'}\,b_v d_0 \tag{15}$$

Bending Stiffness

In codes of practice, the bending stiffness is usually calculated using simple expressions. Both the ACI Code (5) and the Australian Standard (4) use the expression proposed by Branson (8). Accordingly, the effective moment of inertia $I_e$ is given by

$$I_e = \left(\frac{M_{cr}}{M}\right)^3 \left(I_g - I_{cr}\right) + I_{cr} \tag{16}$$

but not greater than $I_g$ , where $M_{cr}$ is the flexural cracking moment and $M$ is the maximum (unfactored) bending moment at the load stage and the section for which the bending stiffness is being calculated.

To account rationally for the breakdown of tension stiffening due to shrinkage induced tension, Gilbert (4) proposed a simple method to modify Eq. 16. Based on this modification, Eq. 16 in the Australian Standard (4) has been amended as follows:

$$I_e = \left(\frac{M_{cr}}{M}\right)^3 \left(I_g - I_{cr}\right) + I_{cr} \qquad \leq I_{e,max} \tag{17}$$

where

$$I_{e,max} = I_g \text{ when p} \geq 0.005 \tag{18}$$
$$= 0.6 I_g \text{ when p} < 0.005 \tag{19}$$

$$M_{cr} = \frac{\left(f_{cf}' - f_{cs}\right) I_g}{y_t} \geq 0 \tag{20}$$

$$f_{cf}' = 0.6\sqrt{f_c'} \text{ when } f_c' \text{ is in MPa} \tag{21}$$

$$f_{cs} = \frac{1.5p}{1+50p} E_s \varepsilon_{cs} \tag{22}$$

$p$ is the tensile steel ratio, $\varepsilon_{cs}$ is the design shrinkage strain and $E_s$ is the modulus of elasticity of reinforcing steel taken as equal to $200 \times 10^3$ MPa. The Australian Standard (4) gives data to estimate $\varepsilon_{cs}$, with typical values in the range of 500 to 1000 microstrains.

It is prudent that the calculation of $I_e$ is based on full service load conditions so as to make some allowance for the loss of bending stiffness due to cracking produced by the construction loads. Information available in the literature indicates that the construction loads can be as large as full service loads. Therefore, in Eq. 17 the value of $M$ is equal to the bending moment at the section when full service loads are acting.

## DESIGN OF SLABS

The behaviour of a structural concrete slab is both nonlinear and inelastic. As loading on the slab increases, the main bending moments migrate from the location dictated by an ealstic analysis towards the location where the designer has provided most reinforcement. Provided that failure due to some cause other than flexure does not supervene, *the distribution of moments at advanced load is governed by the distribution of the reinforcement.*

Slabs and floor systems are very susceptible to cracking caused by shrinkage of concrete and temperature movements due to their large surface area when compared to the volume. As slabs usually contain only light reinforcement, their cracked bending stiffness is significantly smaller than the uncracked value. The cracked bending stiffness of a typical slab may only be in the order of one-fifth of the uncracked stiffness. In addition the creep and shrinkage of concrete cause degradation of the bending stiffness of a slab over a period of time. All these factors make deflection control the primary criterion in the design of slabs and floor systems. It is prudent to select an adequate thickness for the slab in the preliminary design in order to avoid excessive deflection problems later.

### Punching Shear Strength

The region of a slab in the vicinity of a column may fail in shear by developing a failure surface in the form of a truncated cone or pyramid. This type of failure, called a punching failure, is usually the source of collapse of flat plate and flat slab buildings. Adequate design of this region of slab is therefore of paramount importance.

Several methods are available in the literature to calculate the punching shear strength of slabs. Simmonds and Alexander (10) used three-dimensional space truss model to explain the flow of forces in the slab-column connection subjected to combined shear and moment. It has been demonstrated that the yield strength and the position of flexural reinforcement in the column vicinity play an important role in determining shear strength of the slab-column connection. Lim and Rangan (11) extended this truss model for slabs containing stud shear reinforcement. The truss model shows considerable promise for development of code design methods in the future.

The punching shear strength $V_{u0}$, where moment transfer is zero may be expressed as:

$$V_{u0} = u d f_{cv} \tag{23}$$

In Eq. 23, $u$ is the length of the critical shear parameter defined by a line geometrically similar to the boundary of the column or support and located at a distance of $d/2$ therefrom, $d$ is the effective depth of the slab averaged around $u$, and $f_{cv}$ is given by

$$f_{cv} = 0.17\left[1 + \frac{2}{\beta_c}\right]\sqrt{f_c'} \le 0.34\sqrt{f_c'} \tag{24}$$

where $\beta_c$ is the ratio of the longest overall dimension of the column or support Y to the overall dimension X at right angles to Y, and $f_c'$ is the cylinder strength of concrete expressed in terms of MPa.

Based on large-scale tests, the author (12) developed a method for design of slabs to guard against punching shear failure. The method covers flat slab floors with or without closed ties. The punching shear provisions contained in the Australian Standard (4) are based on this method.

Accordingly, the punching shear strength of a slab $V_u$ is given by a set of expressions for previous cases. When there are no shear reinforcements in the slab, $V_u$ is given by:

$$V_u = \frac{V_{u0}}{1.0 + \left(u M_v^* / 8 V^* a d\right)} \tag{25}$$

where a is the width of the torsion strip (4).

## SHEAR STRENGTH OF WALLS

In building codes (4,5), the shear strength of a wall is calculated by summing the shear resisted by the concrete and the shear carried by the shear reinforcement. In the ACI Code (5), the horizontal reinforcement is considered as the shear reinforcement and the effect of vertical reinforcement is ignored. In the Australian Standard (4), the shear reinforcement is either the horizontal reinforcement or the vertical reinforcement depending on the height-to-length ratio of walls. In both ACI318 and AS3600, the shear resisted by the concrete is a very significant component of the shear strength.

The above code methods are empirical extrapolations of similar methods used in the calculation of shear strength of slender beams and has very little relevance to the actual behaviour of a structural wall. The effect of reinforcements and the concrete strength are therefore not realistically represented by the code expressions.

Therefore, the shear strengths calculated by the code expressions usually show a large scatter with respect to the test data.

The shear strength of a wall subjected to in-plane vertical and horizontal forces may be calculated using a strut-and-tie approach. In practice, the central panel of a wall is usually provided with uniform reinforcement, i.e., bars of the same diameter at equal spacing, in both vertical and horizontal directions. A typical element in the cracked central panel comprises a concrete strut tied together by reinforcing bars in the vertical and horizontal directions.

For design purposes, the author (13) simplified the stress analysis of the element and obtained the following expression for shear strength $V_u$ of a structural wall subjected to in-plane vertical and horizontal forces:

$$V_u = t_w d_w \left( p_l f_y + \frac{N^*}{A_g} \right) \tan \theta \tag{26}$$

where $p_l = A_l / t_w L_w$, $A_l$ is the area of vertical steel in the wall on both faces in length $L_w$ and $d_w$ is the effective horizontal length of the wall between centres of end elements. In the absence of end elements, $d_w$ is assumed to be equal to 0.8 times the wall length $L_w$.

In Eq. 26, $N^*$ is design axial compressive force on the wall and $A_g$ is the gross concrete wall cross-section. The strut angle $\theta$ is given by:

$$\tan\theta = \frac{d_w}{H_w} \tag{27}$$

where $H_w$ is the height of the wall. For design purposes, the value of $\theta$ is limited such that:

$$30^0 \le \theta \le 60^0 \tag{28}$$

To ensure that the vertical steel yields prior to failure, $V_u$ is limited to a maximum value given by:

$$V_{u\,max} = \frac{k_3 f_c' t_w d_w \sin\theta \cos\theta}{1.14 + 0.68\cot^2\theta} \tag{29}$$

In Eq. 29, the factor $k_3$ accounts for the difference in the compressive strengths of in-situ concrete in the wall and the concrete test cylinder and, is given by:

$$k_3 = \left(0.6 + \frac{10}{f_c'}\right) \le 0.85 \tag{30}$$

Note that Eq. 26 is in terms of vertical steel in the wall. Both analytical and experimental studies have shown that vertical steel is an important parameter in the calculation of shear strength provided that the wall contained a minimum amount of horizontal steel. For adequate control of cracking due to shrinkage and temperature effects, the minimum value of horizontal steel ratio $p_t$ may be taken as $1.4/f_{sy}$ (4).

## CONCLUSIONS

Some of the recommendations by the author in the design of concrete structural members may be summarised as follows:

1. The flexural strength design of beams may be refined by using modified rectangular stress block parameters given by Eqs. 1 to 4. The minimum value of tensile steel ratio ise given by Eq 5 or 6.

2. The shear strength of a reinforced concrete beam may be calculated by Eq. 7. The strut angle is not always equal to 45 degrees as originally assumed and may be varied from 30 to 45 degrees as proposed in the Australian Standard.

3. The bending stiffness of beams and slabs may be calculated by Eqs. 17 to 22. It is important to include the effect of shrinkage in these calculations especially for members with tensile steel ratio less than or equal to 0.008. When the tensile steel ratio exceeds 0.008, the bending stiffness may be taken as equal to the cracked moment of inertia.

4. Slabs are capable of accommodating considerable redistribution of internal forces and the distribution of bending moments at advanced loads is governed by the distribution of the reinforcement. Therefore, the flexural strength of slabs can be achieved by using simple analysis methods rather than embarking on elaborate calculation procedures.

5. The punching shear strength of slabs is dependent on the flexural reinforcement details in the vicinity of columns and the space truss model available in the literature shows considerable promise for future developments. The code provisions are semi-empirical and those given in the Australian Standard are comprehensive.

6. Walls are extremely useful to carry lateral forces especially due to seismic effects. In order to ensure ductile behaviour of walls, a shear failure must be avoided. The code provisions for the calculation of shear strength of walls are empirical and rational methods must be developed. The strut-and-tie model may be a good tool to achieve this goal.

## REFERENCES

1. Warner, R. F., Rangan, B. V., Hall, A. S. and Faulkes, K. A., *Concrete structures*, Addison Wesley Longman, Melbourne, 1998, 975 pp.

2. ACI Committee 363, "State-of-the-Art Report on High Strength Concrete," *ACI Journal*, V. 81, No. 4, July-August 1984, pp. 364-411.

3. Standards Association of New Zealand, "Concrete structures NZS 3101 - Part 1: Design," 1995.

4. "Australian Standard for Concrete Structures, AS 3600," Standards Australia, Sydney, 2001, 176 pp.

5. ACI Committee 318, "Building Code Requirements for Structural Concrete", American Concrete Institute, Farmington Hills, 2000, 392 pp.

6. Standards Council for Canada, "Design of Concrete structures for Buildings," CAN3-A23.3-M94, Canadian Standards Association, Rexdale (Toronto), Canada, December 1994, 199 pp.

7. ACI-ASCE Committee 445, "Recent Advances to Shear Design of Structural Concrete," *Report ACI 445R-99*, American Concrete Institute, Farmington Hills, March 2000, 55pp.

8.  Branson, D. E., "Instantaneous and Time-Dependent Deflections of Simple and Continuos Reinforced Concrete Beams," *HPR Report* No. 7, Part 1, Albama Highway Department, Bureau of Public Roads, August 1963, pp. 1-78.

9.  Gilbert, R. I., "Serviceability Considerations and Requirements for High Performance Reinforced Concrete Slabs," *Proceedings of the International Conference on High Performance High Strength Concrete*, Perth, Australia, August 1998, pp. 425-439.

10. Simmonds, S. H. and Alexander, S. D. B., "Truss Models for Edge Columns - Slab Connection," *ACI Structural Journal*, V. 84, No. 4, July-August 1987, pp 296-303.

11. Lim, F. K. and Rangan, B. V., "Studies on Concrete Slabs with Stud Reinforcement in the Vicinity of Edge and Corner Columns," *ACI Structural Journal*, V. 92, No. 5, Sept-Oct 1993, pp. 515-525.

12. Rangan, B. V., "Punching Shear Strengths of Reinforced Concrete Slabs," *Civil Engineering Transactions, The Institution of Engineers Australia*, V. CE29, No. 2, 1987, pp. 71-78.

13. Rangan, B. V., "Rational Design of Structural Walls," *ACI Concrete International*, V. 19, 1997, pp. 29-33.

# CREEP, SHRINKAGE, AND AND EARLY-AGE CRACKING

# Holistic Approach to Corrosion of Steel in Concrete

## by N. S. Berke, M. C. Hicks, J. J. Malone, and K.-A. Rieder

Synopsis:     Concrete is the most widely used construction material in the world with over 8 billion tons of it being produced yearly.  Much of this concrete is steel reinforced since the concrete/steel composite has improved ductility over concrete alone, and the concrete provides a protective environment for the steel.  However, reinforced concrete must be used in severe corrosive environments such as found in marine and deicing salt applications. The ingress of chloride leads to corrosion of the steel resulting in early repairs of the structure.  The subsequent costs are over $50 billion/year in the United States, and represent a major drain on infrastructure resources throughout the world.  In this paper the use of improved concrete designs to control corrosion of steel in concrete are addressed.  These designs incorporate the use of low permeability concrete, corrosion inhibiting admixtures, reduced shrinkage and increased toughness with fiber reinforcement.  It is demonstrated that this holistic approach to the concrete design provides a lower life-cycle cost.

Keywords:  chloride ingress; concrete; corrosion; durability; fibers; life-cycle cost; shrinkage reducing admixture

ACI member, Neal S. Berke is a Principal Scientist, Grace Construction Products, Cambridge, MA. He has written and presented more than 50 papers on his research activities, and has several U.S. patents. He has a BA in Physics from the University of Chicago and a PhD in Metallurgical Engineering from the University of Illinois.

Maria C. Hicks is a Senior Research Engineer, Grace Construction Products, Cambridge, MA, where her work encompasses corrosion of steel reinforcing in concrete, studies of chloride diffusion in mortar and concrete and rehabilitation of concrete reinforcement in an advanced state of corrosion. She has a BS in Chemical Engineering from the University of Lisbon, Portugal.

James J. Malone is a Research Technologist, Grace Construction Products, Cambridge, MA, working on shrinkage reducing admixtures. He has a BSBA from Eastern Nazarene College.

ACI member Klaus-A. Rieder is a Senior Research Physicist at Grace Construction products, Cambridge, MA, U.S.A. He is a member of ACI Committees 209, Creep and Shrinkage; 215, Fatigue of Concrete; 446, Fracture Mechanics; and 544, Fiber Reinforcement. His research interests include all aspects of durability and toughening mechanism of concrete.

## INTRODUCTION

Steel reinforced concrete is extensively used due to its durability and cost advantages over other structural materials. However, due to the aggressive use of deicing salts, or marine salt exposures, often coupled with high ambient temperatures, premature corrosion of embedded reinforcement occurs. These adverse conditions are well documented (1-4). Carbonation of the concrete can lead to corrosion of embedded steel, also. It is not specifically addressed in this paper since the methods to reduce chloride induced attack will prevent carbonation corrosion.

The economic costs of corrosion of steel in concrete are substantial. For example, over half of the Federal bridges in the United States are in need of major repairs, often due to corrosion (5). This amounts to over $50 billion and represents only a small portion of the deteriorating infrastructure. Repair costs for reinforced concrete structures run at values in excess of $200/m² of exposed surface area, not accounting for loss of use or traffic control. Thus, for the billions of square meters of reinforced concrete, the costs of major repairs before the desired end of use is a large potential financial burden to all societies.

Concrete designs used for corrosion protection act in one or more of the following ways:

- reduce the ingress of chloride into the concrete (low permeability concrete)
- reduce corrosion in the presence of chloride next to the steel (corrosion inhibitors)

These systems will essentially have one or more of three effects. These are to:
1. reduce the ingress of chloride
2. increase the chloride level at which corrosion initiates
3. reduce the corrosion rate of the active corrosion.

Accelerated and long-term field and laboratory tests are needed to estimate the performance of a protection system as related to the three effects. The evaluation of the performance of these systems is not trivial and requires an understanding of corrosion mechanisms and of the protection mechanisms. In addition, the same accelerated testing techniques can not be used for all methods.

Once the performance of a corrosion protection system can be quantified, it is possible to perform a life-cycle cost analysis to select the most cost effective system. Since there is usually an up-front cost associated with corrosion protection, a net present value analysis looking at repair costs and loss of use need to be considered. An example of this analysis is included.

## MEANS TO IMPROVE THE ABILITY OF CONCRETE TO PROTECT STEEL REINFORCEMENT BY MODIFYING THE CONCRETE DESIGN

### Reducing the Ingress of Chloride

Low Permeability Concrete -- Traditionally the first steps in improving the durability of reinforced steel involved improving concrete quality. For example, ACI 318 and ACI 357 recommend water-to-cementitious [w/(c+p)] ratios of 0.4 or under and minimum covers of 50 mm for non-marine and 68 mm in marine exposures (6,7). Other codes such as Eurocode 2 and Norwegian Code N5 3474 are less stringent (8,9).

Additional means of reducing the ingress of chloride into concrete involve the addition of pozzolans or ground blast furnace slag. These materials are often added as cement substitutes and react with calcium hydroxide to reduce the coarse porosity of the concrete and to decrease the porosity at the paste-aggregate interfaces. Several conference proceedings exist documenting the positive benefits of these materials. References 10-12 are a sampling of some of the literature available.

Numerous references showed that even if concrete is produced to the most stringent of the codes, chloride will ingress into the concrete and corrosion of the steel reinforcement will initiate (13-15).  Figs. 1 and 2 from references 16 and 17 show that even for very low permeability concretes, chloride will eventually ingress into the concrete and initiate corrosion.  When chloride ingress is modeled as a function of w/(c+p) and pozzolan contents for various geometries and environmental exposures, times to corrosion initiation are typically 25 to 75 years, depending upon concrete cover, ambient temperature, and the severity of chloride exposure (18).  Thus, additional protection systems are necessary to meet extended design lives that are increasingly becoming specified.

Corrosion Inhibitors -- Corrosion inhibitors are chemical substances that reduce the corrosion of embedded metal without reducing the concentration of the corrosive agents.  This definition paraphrased from ISO 8044-89 makes the distinction between a corrosion inhibitor and other additions to concrete that improve corrosion resistance by reducing chloride ingress into the concrete.

Corrosion inhibitors can influence the anodic, cathodic or both reactions of the corrosion process.  Since the anodic and cathodic reactions must balance, a reduction in either one will result in a lowering of the corrosion rate.  Fig. 3 illustrates the effects of both types of inhibitors acting alone or in combination when the chloride concentration has not been changed.  When no inhibitors are present, the anodic ($A_1$) and cathodic curves ($C_1$) intersect at point W.  Severe pitting corrosion is occurring.  The addition of an anodic inhibitor (curve $A_2$) promotes the formation of $\gamma$-FeOOH (passive oxide), which raises the protection potential $E_p$, so that the anodic and cathodic curves now intersect at point X.  The corresponding corrosion rate, $i_{corr}$, is reduced by several orders of magnitude and the steel is passive.  Increasing quantities of anodic inhibitor will move curve $A_2$ to more positive $E_p$ values.

The addition of a cathodic inhibitor in the absence of an anodic inhibitor results in a new cathodic curve ($C_2$) as shown in Fig. 3.  The new intersection with the anodic curve ($A_1$) is at point Y.  Though the corrosion rate is reduced, pitting corrosion still occurs, because the potential remains more positive than $E_p$.  Therefore, a cathodic inhibitor would have to reduce cathodic reaction rates by several orders of magnitude to be effective by itself.

The case of combined anodic and cathodic inhibition is illustrated in Fig. 3 as the intersection of the anodic ($A_2$) and cathodic ($C_2$) curves at point Z.  The steel is passive as in the case of the anodic inhibitor alone (point X), but the passive corrosion rate is reduced further.

Commercially available inhibitors include calcium nitrite, sodium nitrite and morpholene derivatives, esters, dimethyl ethanol amine, amines and phosphates.  Several review and other papers discuss the performance of inhibitors in concrete (19,24).

The long-term performance benefits of calcium nitrite are well documented (16-20, 25-30).  Based upon these results, Table 1 was developed to indicate the level of chloride that a given addition of 30% solution of calcium nitrite protects against.

Furthermore, as noted in these papers the use of calcium nitrite or any other inhibitor is not a substitute for good quality concrete, and guidelines for reducing chloride ingress must be followed. Figs. 4 and 5 show the benefits achieved when calcium nitrite was present in the concretes shown in Fig. 1 and 2 (16,17).

Performance criteria for an amine and ester commercially available inhibitor are available (23). This inhibitor at a dosage of 5 $L/m^3$ was stated to protect to 2.4 $kg/m^3$ of chloride. A reduction in the chloride diffusion coefficient of 22 to 43%, depending on concrete quality was determined using accelerated test methods. The latest recommendations for this inhibitor are included in Life 365 (49), in which 2.8 $kg/m^3$ of chloride is the threshold value, and diffusion is reduced by 10%.

Shrinkage Reducing Admixtures -- Shrinkage reducing admixtures (SRAs) are added to concrete to reduce drying shrinkage (31). This results in a reduction in stresses that develop due to restraint under drying conditions. Numerous researchers have documented that a reduction in drying shrinkage results in a reduction of cracking under restrained drying conditions (32-35).

The reduction in shrinkage obtained in low permeability concretes was shown to be sufficient to substantially reduce cracking (36). Fig. 6 shows the shrinkage result improvements and Table 2 the improvement in cracking. Fig. 7 shows the improvement in an actual bridge in which the side without SRA cracked.

A further benefit in the use of some SRAs is a reduction in chloride ingress (37) and in water absorption (38). Fig. 8 shows how this can lead to an improvement in time to corrosion initiation. Thus, the use of some SRAs can result in reduced cracking and permeability in high performance concretes, which should result in improved corrosion performance.

Since concrete exposed to chlorides is often subjected to freezing and thawing, it is important that it be durable under those conditions. Data in Table 3 show that a good air void system and performance in ASTM C666 Method A is obtainable with low permeability concrete containing SRAs.

Fiber Reinforcement -- Though steel fibers can be used to control crack opening and improve concrete toughness, they are susceptible to corrosion near the surface resulting in staining. Naaman and Kosa indicated that under load there can be substantial reduction in cross sectional area of steel fibers in concrete exposed to corrosive conditions (39). Thus, fibers that are non-corrosive are desirable to reduce crack openings in corrosive conditions.

Recent developments have resulted in synthetic fibers that can significantly increase toughness (40-43). Fig. 9 shows the substantial improvement in ductility and resistance to crack opening with a new structural polyolefin fiber.

Work by Shah et al. indicates that if cracks can be kept under 100 micrometers, chloride ingress can be significantly reduced (32). Corrosion studies by Schiessl and Raubach (44) and Ohno et al. (45) show that reduced crack widths combined with lower water-to-cement ratios and increased cover result in lower corrosion rates. This is especially true when calcium nitrite corrosion inhibitor is used (46,47). Thus, reduction in cracking and crack widths should result in a substantial improvement in corrosion performance.

## LIFE-CYCLE COST ANALYSIS

The benefits of using low permeability concretes and corrosion inhibitors can be demonstrated by modeling the life-cycle cost performance. Several programs are available to perform this analysis, with the DuraModel™ (48) and Life 365 (49) being two examples used in North America. Common features are noted below:

In order to perform a life-cycle analysis several pieces of information are needed. These include:

1. environmental exposure and member geometry
2. estimate of the chloride diffusion coefficient
3. effect of the protection system on chloride ingress, chloride threshold value for corrosion initiation, and corrosion rate after corrosion initiates
4. initial costs of the protection system
5. repair costs
6. time between repairs

### Environment and Geometry

The exposure environment needs to be assessed to estimate the chloride exposure. Important parameters are temperature and degree of salt exposure. The salt exposure is different for submerged marine, splash tidal zones, airborne chlorides, and deicing salt applications. Submerged marine and tidal zone exposures quickly come to a fixed high surface concentration of chloride. In the case of deicing salts and air-born chlorides, the surface concentration increases more slowly over time. Geometry plays a key role as a pile in the ocean represents a two-dimensional case of chloride ingress, whereas a deck or wall is a one-dimensional exposure.

### Chloride Diffusion Coefficient

The ingress of chloride can be predicted from the diffusion coefficient for chloride in the concrete of interest. Due to the heterogeneous nature of concrete the exact rate of chloride diffusion cannot be calculated, and it will vary with differences in concreting materials. However, approximate values can be obtained with sufficient accuracy for estimating the ingress of chloride into concrete structures as a function of exposure conditions.

The diffusion of chloride in concrete follows Fick's second law of diffusion (3,13,14,50-55). This correlation can be used to calculate an apparent diffusion coefficient, $D_a$, if the chloride concentration at any time is known as a function of depth. A more rigorous approach would account for chloride binding and sorption effects (55). Work in our laboratory and others showed that if chloride profiles are used after longer exposure times of one to two years, then the apparent diffusion coefficient calculated is a good approximation of future chloride ingress (18,52-55). A method for adjusting $D_a$ for temperature is given in Reference 18.

The one-dimensional solution to Fick's second law with a constant surface chloride and semi-infinite slab is given by:

$$C(x,t)=C_o[1-\text{erf}(x/2(D_a t)^{0.5})] \qquad (1)$$

where  $C(x,t)$ is the chloride concentration at depth x and time t,
   $C_o$ is the surface concentration
   $D_a$ is the apparent diffusion coefficient and
   erf is the error function.

The solution for a square pile under similar constant surface chloride conditions is:

$$C(x,y,t)=C_o[1-\text{erf}(x/2(D_a t)^{0.5}) \times \text{erf}(y/2(D_a t)^{0.5})] \qquad (2)$$

where $C(x,y,t)$ is the chloride concentration at a distance x and y from the perpendicular sides at time t. Numerical methods can also be used and are necessary when $C_o$ or $D_a$ change as a function of time.

Cost Analysis

In order to perform a life-cycle cost analysis one needs to know the costs of initial corrosion protection as well as the Net Present Value (NPV) costs associated with repairs. The NPV is defined as:

$$NPV = \text{Cost} \times (1-D)^{-n} \qquad (3)$$

where Cost is the price for the repair if performed today, D is the discount rate or the interest rate less the inflation rate, and n is the number of years to repair. A typical value for the discount rate is 4% (D = 0.04). Thus the cost of a given protection system is the initial cost plus the NPV of the repairs.

Example of Life-Cycle Cost Analysis for a Bridge Deck -- A bridge deck in the Northern United States is exposed to deicing salts over time. For this example a typical average yearly temperature of 10 °C is chosen. The surface chloride concentration will increase over time at a rate of 0.6 kg/m$^3$/yr a constant maximum value of 15 kg/m$^3$ is reached. The rate of increase is less than that in a parking deck in the North due to the effects of rain. Since the surface chloride concentration is a function of time a numerical solution to Fick's Second Law is used.

Fig. 10 shows the chloride concentration at a depth of 65 mm for several types of concretes at 75 years. One with an effective diffusion coefficient of $1.3\times10^{-12}$ m$^2$/s represents a concrete with a water/cement ratio of 0.40. The lower curves represent concretes produced with 40% ground blast furnace slag or 7.5% silica fume ($0.78\times10^{-12}$ m$^2$/s) and the effects of a damp-proofing inhibitor in the concrete ($0.98\times10^{-12}$ and $0.58\times10^{-12}$ m$^2$/s).

Protection systems considered in Fig. 10 in addition to the use of lower permeability concrete are a 30% calcium nitrite solution at 10 and 15 L/m$^3$, and epoxy coated reinforcing bars alone and in combination with calcium nitrite. Repairs were considered to occur at 5 years after corrosion initiation and at 20 year interval after that. The repairs were set at $350/m$^2$ of surface and it was estimated that only 10% of the surface would need to be repaired. These numbers are fairly representative but should be adjusted for specific locations. Traffic delay costs, which can be considerable, were not included. Table 4 summarizes the initial, repair and total NPV of the various combinations, in Figure 10 as well as additional combinations. Clearly, low permeability concrete alone is not the best life-cycle option. Fig.11 ranks the 9 lowest costs systems against the base case, and further supports the benefits for corrosion protection systems even though initial costs are higher.

Membranes, stainless steel bars and cathodic protection were not included in the above analysis. At an initial cost of $43 to $130/m$^2$ cathodic protection is not cost effective for a new bridge deck. Membranes have a service life of about 20 years and an initial cost of about $30/m$^2$ so they are also not cost effective for corrosion protection, but could have other benefits where leaks to a lower level need to be avoided. Finally, 304 stainless steel comes in at about $39/m$^3$ in new construction which is much more costly than alternative protection mechanisms. The addition of an SRA and/or synthetic fibers would add $6 to $24/m$^2$. This would still be cost effective over the use of the alternative protection systems and would provide an increase in ductility and decrease in cracking.

## CONCLUSIONS AND RECOMMENDATIONS

Several systems exist for the protection of steel in concrete. To determine the merits of these different systems when used alone or in combination, it is necessary to be able to determine their expected service lives and to perform a life-cycle cost analysis.

An example was given for a life-cycle cost analysis for a bridge deck in a Northern U. S. exposure. As a result of this analysis it was shown that for a 75 year service life:

- Good quality and low permeability concrete while necessary, is not sufficient for providing the most cost effective system
- Corrosion inhibitors, in particular calcium nitrite, provide the most cost effective solutions
- New advancements in controlling cracking can further enhance concrete performance.

## REFERENCES

1. Somayaji, S., Keeling, D., and Heidersbach, R., "Corrosion of Reinforcing Steel in Concrete Exposed to Marine and Freshwater Environments", ACI SP 122-9, 1990, pp 139-172, Paul Klieger Symposium on Performance of Concrete, David Whiting editor.

2. Weyers, R.E. and Cady, P.D., "Deterioration of Concrete Bridge Decks from Corrosion of Reinforcing Steel", Concrete International, V. 9, No. 1, January 1987, pp. 15-20.

3. Bamforth, P. B. and Price, W.F., "Factors Influencing Chloride Ingress into Marine Structures", presented at Economic and Durable Construction Through Excellence, Dundee, Scotland, September 1993.

4. Wilkins, N. J. H., and Lawrence, P.F., "The Corrosion of Steel Reinforcement in Concrete Immersed in Sea water"; Corrosion of Reinforcement in Concrete Construction Ed. Crane E.P. Soc. of Chem. Ind. London (1983).

5. Transportation Research Board, Strategic Highway Research Program Research Plans, Final Report, NCHRP Project 20-20, TRB, Washington, DC (1986), pp. TRA 4-1 - TRA 4-60.

6. ACI Committee 318, "Building Code Requirements for Reinforced Concrete, (ACI 318-83)," American Concrete Institute, Detroit, 1982.

7 ACI Committee 357, "Guide for the Design and Construction of Fixed Offshore Concrete Structures", American Concrete Institute, Detroit, 1982.

8. Eurocode 2, Design of Concrete Structures, European pre-standard ENV 1992.

9. Norwegian Code N5 3474.

10. Proceedings of the Second CANMET/ACI International Conference on Fly Ash, Silica Fume, Slag, and Natural Pozzolans in Concrete, V. M. Malhotra Editor, American Concrete Institute, Detroit, ACI SP-91, V. I and II, 1986, pp 1601.

11. Proceedings of the Third CANMET/ACI International Conference on Fly Ash, Silica Fume, Slag, and Natural Pozzolans in Concrete, V. M. Malhotra Editor, American Concrete Institute, Detroit, ACI SP-114, V. I and II, 1989, pp 1706.

12. Proceedings of the Fourth CANMET/ACI International Conference on Fly Ash, Silica Fume, Slag, and Natural Pozzolans in Concrete, V. M. Malhotra Editor, American Concrete Institute, Detroit, ACI SP-132, V. I and II, 1992, pp 1671.

13. Browne, D., "Design Prediction of the Life for Reinforced Concrete in Marine and Other Chloride Environments", Durability of Building Materials, V. 1, (1982), pp 113-125.

14. Tuutti, K., Corrosion of Steel in Concrete, Swedish Cement and Concrete Research Institute, Stockholm, p. 469, (1982).

15. Berke, N. S., Scali, M. J., Regan, J. C., and Shen, D. F., "Long-Term Corrosion Resistance of Steel in Silica Fume and/or Fly Ash Concretes," Second CANMET/ACI Conference on Durability of Concretes, ed. V. M. Malhotra, CANMET, Canada, 1992, pp 899-924.

16. Berke, N. S. and Hicks,M. C., "Predicting Long-term Durability of Steel Reinforced Concrete with Calcium Nitrite Corrosion Inhibitor", 13th International Corrosion Conference, Melbourne, Australia, November 1996.

17. Berke, N. S., Hicks, M. C., Abdelrazig, B. I., et al., "A Belt and Braces Approach to Corrosion Protection, International Conference on Corrosion and Corrosion Protection of Steel in Concrete", University of Sheffield, July 1994, pp 893-904.

18. Berke, N. S. and Hicks, M. C., "Predicting Chloride profiles in Concrete", Corrosion, V. 50, No. 3, Corrosion, March 1994, pp 234-239.

19. Berke, N. S., and Weil, T. G., "World Wide Review of Corrosion Inhibitors in Concrete", Advances in Concrete Technology, CANMET, Ottawa, Canada, 1994, pp 891-914.

20. Nmai,C. K. and Kraus, P. D., "Comparative Evaluation of Corrosion-Inhibiting Chemical Admixtures for Reinforced Concrete", ACI SP 145-16, 1994, pp 245-262.

21. Berke, N. S., Hicks, M. C., Hoopes, R. J., et al., "Use of Laboratory Techniques to Evaluate Long-Term Durability of Steel Reinforced Concrete Exposed to Chloride Ingress", ACI SP 145-16, 1994, pp 299-328.

22. Maeder, U., "A New Class of Corrosion Inhibitors for Reinforced Concrete", Proceedings of the Third ANMET/ACI International Conference on Performance of Concrete in Marine Environment, V. M. Malhotra Ed., ACI SP 163, 1996, pp 215-232.

23. Johnson, D. A., Miltenberger, M. A. and Amey S. L., "Determining Chloride Diffusion Coefficients for Concrete Using Accelerated Test Methods", Third CANMET/ACI International Conference on Concrete in Marine Environment", St. Andrews by-the-Sea, New Brunswick, Canada, August 4-9, 1996, Supplementary Papers, pp 95-114.

24. Vogelsang, J. and Meyer, G., "Electrochemical Properties of Concrete Admixtures", Fourth International Symposium on Corrosion of Reinforcement in Concrete Construction, Cambridge, UK, July 1-4, 1996, C. L. Page, P. Bamforth and J. W. Figg, Eds., pp 579-588.

25. Berke, N. S., Hicks, M. C., and Tourney, P. G., "Evaluation of Concrete Corrosion Inhibitors, 12th International Corrosion Congress, Houston, TX, September 1993, pp 3271-3286.

26. Virmani, Y. P., "Effectiveness of Calcium Nitrite Admixture as a Corrosion Inhibitor", Public Roads, V. 54, No.1, June 1990, pp 171-182.

27. Berke, N.S., and Rosenberg, A.M., "Technical Review of Calcium Nitrite Corrosion Inhibitor in Concrete", Transportation Research Record 1211, Transportation Research Board, Washington, DC (1989), pp. 18-27.

28. Tomosawa, F., et al., "Experimental Study on the Effectiveness of Corrosion Inhibitor in Reinforced Concrete", RILEM Durability Symposium, 1990, pp 382-391.

29. Berke, N. S., Pfeifer, D. E., and Weil, T. G., "Protection Against Chloride-Induced Corrosion", Concrete International, V. 10, No. 12, 1988, pp. 45-55.

30. Montani, R., " Corrosion inhibitors - new options and possibilities", Concrete Repair Bulletin, p 10f, July-August 1996.

31. Transportation Research Circular No. 494, December 1999, "Durability of Concrete", Transportation Research Board, National Research Council, Washington DC.

32. Shah, S. P., Karaguler, M. E and Sarigaphuti, M., "Effects of Shrinkage-Reducing Admixtures on Restrained Shrinkage Cracking of Concrete", ACI Materials Journal, V. 89, No. 3, May-June 1992, pp 289-295.

33. Brooks, J. J. and and Jiang, X., "The Influence of Chemical Admixture and Restrained Drying Shrinkage of Concrete", Proceedings of the 5th CANMET/ACI Conference on Superplasticizers and other Chemical Admixtures in Concrete, V. M. Malhotra, editor, ACI SP-137, 1997, pp 249-265.

34. Berke, N. S., Dallaire, M. P. and Hicks, M. C., "New Developments in Shrinkage-Reducing Admixtures", SP 173-48, Chemical Admixtures, pp. 971-998.

35. Nmai, C. K., Tomika, R., Hondo, F. and Buffenbarger, J., "Shrinkage-Reducing Admixtures", Concrete International, V. 20, No. 4, 1998, pp 31-37.

36. Folliard, K. J. and Berke, N. S., "Properties of High-Performance Concrete Containing Shrinkage-Reducing Admixture", Cement and Concrete Research, V. 27, No. 9, 1997, pp 1357-1364.

37. Berke, N. S., Dallaire, M. P., Hicks, M. C. and Macdonald, A. C., "Holistic Approach to Durability of Steel Reinforced Concrete", Concrete in the Service of Mankind, Radical Concrete Technology, Proceedings of the International Conference, University of Dundee, Scotland, June 1996, pp 25-45.

38. Weiss, J. W., and Shah, S. P., "Restrained Shrinkage Cracking: The Role of Shrinkage Reducing Admixtures and Specimen Geometry", Pre-Proceedings of the RILEM International Conference on Early Age Cracking in Cementitious Materials (EAC'01), Haifa, Israel, March 12-14, 2001.

39. Naaman, A. E. and Kosa, K., "Corrosion of Fiber Reinforced Concrete and SIFCON: Project Summary", Proceedings of the National Science Foundation, Singapore, July 1993.

40. Morgan, D. R., Heere, R., McAskill, N. and Chan, C., "Comparative evaluation of system ductility of mesh and fiber reinforced shotcretes",

Eighth International Conference on Shotcrete for Underground Support VIII, S. Paulo, Brazil, April 1999, pp 216-239.

41. Trottier, J-F and Mahoney, D., "Innovative Synthetic Fibers", Concrete International, V. 23, No. 6, June 2001, pp 23-28.

42. Lin, Z., Kanda, T. and Li, V. C., "On interface property characterization and performance of fiber-reinforced cementitious composites", Concrete Science and Engineering, V.1, Sept.1999, pp 173-184.

43. Rieder, K.-A. and Berke, N. S., "Comparison of the fracture behavior of various fibers using the Wedge Splitting test method', presented at the fall 2000 ACI meeting in Toronto. To be published in ACI SP "Fiber Reinforced Concrete: Innovations for Value"

44. Schiessel, P. and Raupach, M., "Laboratory Studies and Calculations on the Influence of Crack Width on Chloride-Induced Corrosion of Steel in Concrete", ACI Materials Journal, V. 94, No.1, January-February 1997, pp. 56-62.

45. Ohno, Y., Praparntanatorn, S. and Suzuki, K., "Influence of cracking and water cement ratio on macrocell corrosion of steel in concrete", Corrosion of Reinforcement in Concrete Construction, Society of Chemical Industry, C.L. Page, P.B. Bamforth and J.W. Figg Editors, Cambridge, UK, July 1996, pp 24-32.

46. Schiessel, P., "Effectiveness and Harmlessness of Calcium Nitrite as a Corrosion Inhibitor", RILEM International Symposium on the Role of Admixtures in High Performance Concrete, Monterrey, Mexico, 1999.

47. Berke, N. S., Dallaire, M. P., and Hicks, M. C., "Corrosion of Steel in Cracked Concrete", Corrosion, V. 49, No.11, November 1993, pp. 934-943.

48. Berke, N. S., Macdonald, A. C. and, Hicks, M. C., "Use of the DuraModel for the Design of Cost-Effective Concrete Structures", ICCRRCS, Orlando, Florida, December 1998.

49. Thomas, M. D. A., and Bentz, E. C., "Life 365", Computer Program for Predicting Service Life and Life-Cycle Costs of Reinforced Concrete Exposed to Chlorides, University of Toronto, August 2000.

# 150  Berke et al.

50. Hansson, C. M. and Berke, N. S., "Chlorides in Concrete", MRS Symposium Proceedings, Pore Structure and Permeability of Cementitious Materials, L. R. Roberts and J. P. Skalny Eds., V. 137, November 1988, pp 253-270.

51. Short, N. R. and Page, C. L., "The Diffusion of Chloride Ions through Portland and Blended Cement Pastes", Silicates Industriels, V. 10, 1982, pp 237-240.

52. Goto, S. and Roy, D. M., "Diffusion of Ions Through Hardened Cement Pastes", Cement and Concrete Research, V. 11, No. 5/6, 1981, pp 751-757.

53. Garbozi, E. J., "Permeability, Diffusivity, and Microstructural Parameters: A Critical Review", Cement and Concrete Research, V. 20, No. 4, 1990, pp 591-601.

54. Dhir, R. K., Jones, M. R., Ahmed, H. E., et al., "Rapid Estimation of Chloride Diffusion Coefficient in Concrete", Magazine of Concrete Research, V. 42, No. 152, September 1990, pp 177-185.

55. Thomas, M. D. A. and Mathews, J. D., "Chloride Penetration and Reinforcement Corrosion in Fly Ash Concrete Exposed to a Marine Environment", ACI SP-163-15, pp 317-338, Proceedings of the Third CANMET/ACI International Conference on Concrete in Marine Environments, Canada, 1996.

Table 1
Calcium nitrite dosage rates vs. chloride protection

| Calcium Nitrite (30% sol.) $L/m^3$ | Chloride Ion $kg/m^3$ |
|---|---|
| 10 | 3.6 |
| 15 | 5.9 |
| 20 | 7.7 |
| 25 | 8.9 |
| 30 | 9.5 |

Table 2

Summary of free shrinkage measurements and
restrained shrinkage ring cracking

| Mixture Description | 28-day Drying Shrinkage (%) | 120-day Drying Shrinkage (%) | Average Time to Ring Cracking (days) | Average Crack Area per Ring $(mm^2)$ |
|---|---|---|---|---|
| Control Con | 0.049 | 0.070 | 44 | 5.9 |
| Concrete with 1.5% SRA (% on cement) | 0.032 | 0.050 | No cracks after 120 days | 0 |
| Silica Fume Concrete | 0.051 | 0.077 | 38 | 25.3 |
| Silica Fume Concrete With 1.5% SRA (% on cement) | 0.024 | 0.044 | 95 | 3.1 |

Table 3

Properties of the plastic and hardened concrete containing SRA
Concrete made with Type I cement, at W/C=0.40 and CF=390 kg/m$^3$

| | Reference | 5 L/m$^3$ SRA |
|---|---|---|
| Adva flow (mL/100 kg) | 228 | 228 |
| Darex II (mL/100 kg) | 19.6 | 6.5 |
| Slump (mm) | 127 | 170 |
| Air (%) | 6.3 | 7.5 |
| Initial set (hours) | 4.34 | 5.06 |
| Final set (hours) | 6.25 | 6.37 |
| 28 days f'c (Mpa) | 39.4 | 41.5 |
| Hardened air (%) | 5.8 | 6.3 |
| Specific surface(mm$^2$/mm$^3$) | 34.6 | 28.1 |
| Avg chord length (mm) | 0.1143 | 0.1422 |
| Spacing factor (mm) | 0.132 | 0.155 |
| RDME | 99 | 100 |

Table 4
75 Year Analysis for a Bridge Deck

| Case Design | Initial cost ($/m²) | Initiation (years) | First Repair (years) | Number of Repairs | NPV Repair ($/m²) | NPV Total ($/m²) |
|---|---|---|---|---|---|---|
| Base Case | 0 | 26 | 31 | 3 | 17.3 | 17.3 |
| 10 L/m³ CN | 3.7 | 50 | 55 | 2 | 5.9 | 9.6 |
| 15 L/m³ CN | 5.6 | 76 | 81 | 0 | 0 | 5.6 |
| 40% Slag | 0 | 36 | 41 | 2 | 10.2 | 10.2 |
| 7.5% Silica Fume (SF) | 3 | 36 | 41 | 2 | 10.2 | 13.2 |
| 5 L/m³ BO/A | 3.5 | 48 | 53 | 2 | 6.4 | 9.9 |
| ECR | 19.5 | 26 | 46 | 2 | 8.4 | 27.9 |
| 10 L/m³ CN + 40% Slag | 3.7 | 74 | 79 | 0 | 0 | 3.7 |
| 10 L/m³ CN+7.5% Silica Fume | 6.7 | 74 | 79 | 0 | 0 | 6.7 |
| 5 L/m³ BO/A + 40% Slag | 3.5 | 65 | 70 | 1 | 2.3 | 5.8 |
| 5 L/m³ BO/A + ECR | 23.0 | 48 | 68 | 1 | 2.4 | 25.4 |
| 10 L/m³ CN + ECR | 23.2 | 50 | 70 | 1 | 2.3 | 25.5 |
| 10 L/m³ CN+ECR+40% Slag | 23.2 | 74 | 94 | 0 | 0 | 23.2 |
| 5 L/m³ BO/A+ECR+40% Slag | 23.0 | 65 | 85 | 0 | 0 | 23.0 |

CN = 30 % solution of calcium nitrite
BO/A = Butyl Oleate/Amine
ECR = Epoxy-coated reinforcing steel (top and bottom mats)

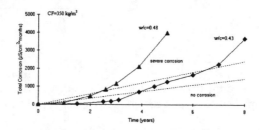

Fig. 1 - Total corrosion of lollipops* - control samples

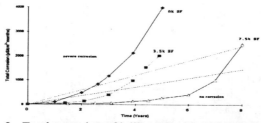

Fig. 2 - Total corrosion of lollipops* with various amounts
of silica fume; w/c=0.48, CF=350 kg/m³
*75mm × 150 mm concrete cylinders with an embedded #3 rebar

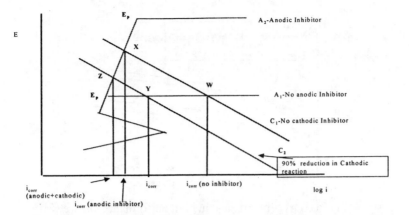

Fig. 3 – Comparison of cathodic and anodic inhibitors
W >> Y

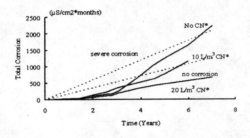

Fig. 4 - Total corrosion of lollipops* w/c=0.43,
CN − 30% solution of calcium nitrite, CF = 350 kg/m³

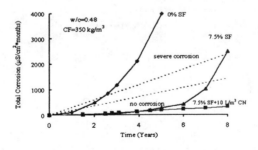

Fig. 5 - Total corrosion of lollipops*

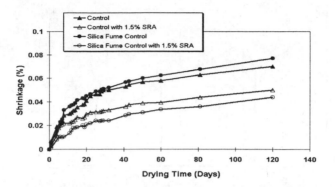

Fig. 6 – Drying shrinkage in high performance concrete
with and without SRA (ASTM C 157)

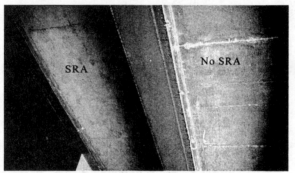

Fig. 7 – Underside of bridge deck

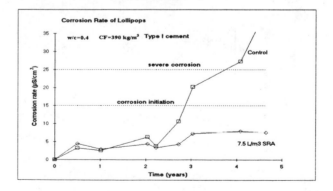

Fig.8 - Improvement in time to corrosion initiation in good
quality concrete, containing SRA

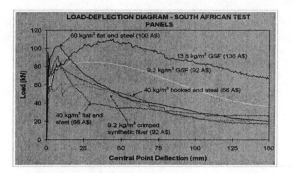

Fig. 9 – Use of fibers in a shotcrete application

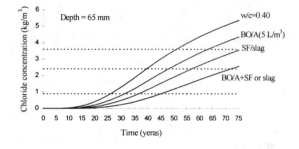

Figure 10 - Estimated chloride concentrations for a bridge deck.
Chloride build-up = 0.6 kg/m$^3$/year

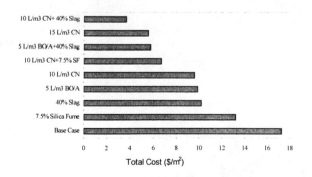

Figure 11 - Total costs ($/m$^2$), deck for a 75 year design bridge

156 Berke et al.

# Study of the Behavior of Concrete with Shrinkage Reducing Admixtures Subjected to Long-Term Drying

## by R. Gettu, J. Roncero, and M. A. Martín

Synopsis: Shrinkage is a critical characteristic of concrete that can lead to undesired cracking, thereby limiting the serviceability of concrete structures. The ability to design concretes with significantly lower shrinkage is, therefore, of great practical importance. Along these lines, new shrinkage reducing chemical admixtures have been developed in order to reduce the extent of the shrinkage strains. The present work analyses the effectiveness of the incorporation of three such admixtures using long-term drying shrinkage tests. The results indicate a remarkable reduction of the shrinkage for two admixtures based on polypropylene glycol formulations while no significant improvement was obtained in the case of a wax-based admixture. Additionally, the influence of these shrinkage reducing admixtures on other basic properties of the concrete, such as workability and 28-day compressive strength, has been quantified. A plasticizing effect, which can be exploited for reducing the superplasticizer dosage or the water/cement ratio, was observed in the case of the two glycol-based admixtures. A slight decrease of the compressive strength was measured in all the cases due to the incorporation of the admixtures.

Keywords: chemical admixtures; concrete; deformations; drying; shrinkage

Dr. Ravindra Gettu, ACI member, is the Director of the Structural Technology Laboratory of the Universitat Politècnica de Catalunya (UPC). His research interests include concrete technology, high-performance concretes, fiber reinforced concrete and fracture mechanics. He has co-authored more than 100 technical publications in journals and conference proceedings.

Dr. Joana Roncero is Senior Researcher at the Construction Engineering Department of the UPC. Her research interests include chemical admixtures for concrete, rheology and analytical techniques for microstructural characterization.

Miguel Á. Martín is Senior Technician at the Structural Technology Laboratory of the UPC.

## INTRODUCTION

In structural design, shrinkage is an important consideration for serviceability and crack width limits. The magnitude of the shrinkage deformations depends on the concrete composition, especially the content and properties of the cement, and the water/cement ratio (w/c), as well as the environmental conditions. Moreover, the use of water reducing chemical admixtures or plasticizers has permitted the reduction of the w/c without sacrificing workability, contributing to lower shrinkage under drying conditions.

A new class of chemical admixtures, denominated as shrinkage reducing admixtures (SRAs), emerged in the 1990s. These products have been designed with the specific aim of reducing the capillary stresses within the pore structure that are responsible for the shrinkage in concrete when subjected to drying. These admixtures are generally organic polymers that reduce the surface tension of the water leading to lower capillary stresses when the pore water is removed due to drying. Their chemical composition seems to be based on alcohol alkylen oxide adducts, polypropylene glycol or glycol ether derivatives.

Work reported in the literature (1-6) clearly demonstrate the decrease in the shrinkage of concrete due to the incorporation of SRAs. Furthermore, their results have demonstrated that, in addition to decreasing the drying shrinkage, SRAs can reduce the volume of macropores in the hardened paste and the permeability, as well as the crack widths in restrained concrete under drying conditions. Other benefits that have been reported include the increase in the fluidity of the fresh concrete and the decrease in the tensile creep. Nevertheless, some negative influences have also been attributed to the SRAs, such as a slight decrease in compressive and tensile strengths, and in the modulus of elasticity.

In the present work, the unrestrained or free shrinkage deformation of concretes incorporating two SRAs based on polypropylene glycol formulations is evaluated over a drying period of seven months. Additionally, the effectiveness of another product based on a wax formulation, which may be treated as an SRA, is studied. The effect of this admixture is expected to be different from that described earlier since it acts by preventing the loss of water from the pores during early ages (instead of decreasing the surface tension of the pore water). Companion tests on two base (reference) concretes without SRA have also been performed.

## MATERIALS USED

Cement of type CEM I 52.5 R, crushed limestone sands with the grain size ranges of 0-2 mm and 0-5 mm, crushed limestone gravels with the size ranges of 5-12 mm and 12-20 mm were used in the concretes. The properties of the superplasticizers (denoted as SN and SC) and the SRAs (denoted as SRA1, SRA2 and SRA3) that have been used are shown in Table 1.

The concretes were designed to have high workability (or slump) and a characteristic compressive strength of 35 MPa. The admixture dosages correspond to the ranges indicated by the respective suppliers. In the case of SRA1, two different dosages were used: 1.5 and 2% by weight of cement. Since the incorporation of the SRA led to an increase in the workability of the concrete, the superplasticizer content was adjusted to yield the desired slump level, which was 17-20 cm immediately after mixing and at least 10 cm after 15 minutes. The compositions of the six concretes, along with their basic properties are given in Table 2. Note that all superplasticizer dosages are given as dry superplasticizer/cement ratios (sp/c), and their water content has been accounted for in the w/c. Regarding the SRAs, only SRA3 contains some water and this has also been taken into account. Also, the water added to the mix includes that considered to be absorbed by the aggregates, which are taken to be saturated, surface-dry within the concrete.

The results of compression tests, at the age of 28 days, of standard 150×300 mm cylinders of each concrete are given in Table 2. These specimens, three in each case, were demolded after 24 hours and cured in a fog room up to the age of testing. The average results indicate a slight decrease in the strength with the incorporation of an SRA. For SRA1, the dosages of 1.5% and 2% result in strength reductions of 5% and 12% on average, respectively. On the other hand, the strength decreased by 4% and 7% for SRA2 and SRA3 on average, respectively. In all the comparisons, the reference is taken as the concrete with the same superplasticizer and without SRA. The observed strength decrease confirms the conclusions of other researchers (2-4).

## SHRINKAGE TEST RESULTS

Shrinkage deformations were evaluated using standard cylinders of 150×300 mm, with a 100 mm long strain gage embedded at the center of each specimen for measuring the axial strain. The strain measurements were begun immediately after casting. The specimens were demolded at the age of 24 hours and then submerged in lime-saturated water at 20°C until the age of 28 days. The age of 24 hours, just after demolding, was taken as the reference point (corresponding to "zero" strains) for the shrinkage measurements. Accordingly, all strains that occur during the first 24 hours (i.e., settlement, plastic shrinkage, thermal contraction, chemical shrinkage) are attributed to the fresh or setting concrete and are not included in the values presented as shrinkage. After the age of 28 days, the specimens were subjected to a temperature of 20°C and relative humidity of 50% in a climatic chamber.

The strain evolution for all the specimens during the curing period and seven months of subsequent drying are shown in Figure 1, where negative values indicate swelling (or expansion) during the curing. Note that each curve gives the average trend of 3 specimens. As expected, all six concretes experience some expansion under water (30-50 $\mu\varepsilon$), during the 28-day curing period. The strain measured during curing can be considered as the difference between the swelling due to being submerged in water and the autogenous shrinkage (which is taken as the sum of the chemical shrinkage, thermal shrinkage and shrinkage due to self-desiccation). During the drying period, the strains measured in the concrete correspond to the total shrinkage or the sum of the autogenous and drying shrinkage, where the contribution of the former is expected to be negligible since most autogenous shrinkage occurs earlier (7).

When the two reference concretes (denoted as CREF-SN and CREF-SC) are compared, it is observed in Figure 1 that the concrete with the superplasticizer SC exhibits a strain that is about 17% higher than the concrete with superplasticizer SN, after 7 months of drying. This confirms the conclusion of a previous work (8) that the incorporation of a polycarboxylate based superplasticizer could lead to slightly higher shrinkage than a naphthalene-based product.

The total shrinkage in the concretes with SRA1 and SRA2 is significantly lower than in the corresponding base concretes. Both admixtures show similar trends after 7 months of drying, with a reduction in the shrinkage strain of about 56% and 50%, respectively, when compared with the reference concrete.

On the other hand, the responses of the concretes CSRA1(1.5%)-SN and CSRA1(2%)-SN, incorporating the two different dosages of SRA1, are similar.

In both cases, after 7 months of drying, the strain is about 56% of that of the base concrete (CREF-SN).

However, the incorporation of SRA3 does not yield similar reductions in shrinkage strains, with the strain in CSRA3-SC after 7 months of drying being only about 15% lower than that of the corresponding reference CREF-SC.

The evolutions of the drying shrinkage can be more clearly seen in the semi-logarithmic plots of Figure 2, where each curve is the average of the results obtained in 3 specimens. Note that the curves begin from zero since only the strain increments during drying are shown. At the end of the test period, the decrease in drying shrinkage strains in the concretes with 1.5% and 2% of SRA-1 are about 50% and 56%, respectively, when compared to that in CREF-SN. During the same period, the drying shrinkage in CSRA2-SC is also about 50% of that in CREF-SC. However, in the case of the concrete with SRA3, the decrease in drying shrinkage is only 13%, when compared with the concrete CREF-SC.

DISCUSSION

It is observed from the test results that the incorporation of an appropriate SRA can affect the shrinkage behavior of the concrete dramatically. The use of the two polypropylene glycol-based SRAs (i.e., SRA1 and SRA2) leads to a significant reduction in the drying shrinkage of the concrete. In both cases, reductions of 50-56% were obtained in the shrinkage strain after 7 months of drying. The glycol based SRAs seem to act mainly by reducing the surface tension of the water in the pores of the hardened concrete and, consequently, reducing the capillary stresses when the water evaporates. Although the loss in weight of the specimens was not measured over the test duration, other studies have concluded that these SRAs do not decrease the rate of water loss (8). Therefore, it appears that the decrease in shrinkage cannot be attributed to any retardation of moisture loss from the concrete. On the other hand, the wax-based SRA3 is supposed to prevent the evaporation of water from the pores. It appears that this is not as effective since the decrease in the drying shrinkage is only about 13%.

Regarding the effect of the SRA dosage, the results obtained with the two dosages of SRA1 indicate that the benefit due to the increase in the dosage from 1.5 to 2% is not significant. This may suggest the existence of a saturation or optimum SRA dosage, similar to that observed in the case of superplasticizers (9).

In the fresh concrete, both SRA1 and SRA2 produce a considerable plasticizing effect, as seen in the slump values in Table 2. This can be exploited for reducing the superplasticizer dosage or, better still, the water/cement ratio (2, 4). In the present study, the w/c has been maintained relatively constant in all the concretes in order to conserve the reference, and the superplasticizer dosage was reduced to maintain the workability, which will lead to lower material costs, in practice.

On the other hand, a reduction of the w/c will lead to a higher class of concrete (i.e., higher strength and durability) and will result in further decreases in shrinkage. This could compensate the reduction in the 28-day compressive strength (5) that is observed in all the concretes with SRAs (Table 2).

## CONCLUSIONS

Drying shrinkage tests have been performed on 35 MPa strength concretes in order to study the effect of three shrinkage reducing admixtures. Additionally, the possible influence of a higher dosage is evaluated with one of the SRAs. The concretes were subjected to drying conditions (50% R.H., 20°C) during 7 months after a curing period of 28 days.

On the basis of the results obtained, the following conclusions can be made:

- A significant reduction of the drying shrinkage of concrete is observed when a polypropylene glycol based SRA is incorporated. A decrease of about 50% in the shrinkage strain has been obtained.

- The incorporation of a wax-based SRA led only to a decrease in shrinkage of 13% with respect to the reference concrete, after a drying period of 7 months. This admixture is, therefore, not as effective as those based on polypropylene glycol formulations.

- Regarding the effect of different SRA dosages, there is not much difference in the shrinkage evolution between the concretes with SRA/cement dosages of 1.5% and 2%.

- There is a slight decrease in 28-day compressive strength due to the incorporation of the SRAs studied. In the concretes incorporating glycol based SRAs, this can be compensated by reducing the water/cement ratio without compromising the workability since these admixtures also exhibits a plasticizing effect. The plasticizing effect was not observed in the case of the wax-based admixture.

Acknowledgments: Partial financial support for the present work has been provided by Technology Transfer projects funded by Bettor MBT and Grace, and by the Spanish CICYT grant PB98-0298, and the CICYT and European Commission grant FEDER 2FD97-0324-C02-02 to the UPC. Cementos Molins, Bettor MBT and Grace donated the materials used in this study.

## REFERENCES

1.  M. Shoya, S. Sugita and T. Sugiwara, Improvement of drying shrinkage and shrinkage cracking of concrete by special surfactant, *Admixtures for Concrete: Improvement of properties*, Ed. E. Vázquez, Chapman and Hall, London, pp. 484-495 (1990).

2.  S.P. Shah, M.E. Karaguler and M. Sarigaphuti, *ACI Mater. J.*, V. 89, pp. 289-295 (1992).

3.  J.J. Brooks and X. Jiang, The influence of chemical admixtures on restrained drying shrinkage of concrete, *Superplasticizers and Other Chemical Admixtures in Concrete* (Proc. Fifth CANMET/ACI Intnl. Conf., Rome), SP-173, Ed. V.M. Malhotra, ACI, Detroit, pp. 249-265 (1997).

4.  K.J. Folliard and N.S. Berke, *Cem. Concr. Res.*, V. 27, pp. 1357-1364 (1997).

5.  W.J. Weiss, B.B. Borichevsky and S.P. Shah, The influence of a shrinkage reducing admixture on the early-age shrinkage behavior of high performance concrete, *Utilization of High Strength/High Performance Concrete* (5th Intnl. Symp., Sandefjord, Norway), Ed. I. Holand and J. Sellevold, Norwegian Concrete Association, V. 2, pp. 1339-1350 (1999).

6.  J. Mora, M.A. Martín, R. Gettu and A. Aguado, Study of plastic shrinkage cracking, and the influence of fibers and a shrinkage reducing admixture, *Proc. Fifth CANMET/ACI Intl. Conf. on Durability of Concrete* (Barcelona), Supplementary papers, Ed. V.M. Malhotra, ACI, Detroit, pp. 469-483 (2000).

7.  O.M. Jensen and P.F. Hansen, Autogenous deformation and change of the relative humidity in silica fume-modified cement paste, *ACI Mater. J.,* Vol. 93, No. 6, pp. 539-543 (1996).

8. J. Roncero, *Effect of superplasticizers on the behavior of concrete in the fresh and hardened states: Implications for High Performance Concrete*, Doctoral Thesis, Universitat Politècnica de Catalunya, Barcelona, Spain (2000).

9. L. Agulló, B. Toralles-Carbonari, R. Gettu and A. Aguado, Fluidity of Cement Pastes with Mineral Admixtures and Superplasticizer – A study based on the Marsh cone test, *Mater. Struct.*, V. 32, pp. 479-485 (1999).

Table 1 - Properties of the chemical admixtures used

| Designation | Type | Formulation | Density $(g/cm^3)$ | Water content (%) | Range of dosage, given by supplier |
|---|---|---|---|---|---|
| SN | Superplasticizer | Naphthalene based | 1.18 | 65 | 0.18-0.7%* |
| SC | Superplasticizer | Ether polycarboxylate based | 1.05 | 80 | 0.3-1.5%* |
| SRA1 | Shrinkage reducing admixture | Polypropylene glycol based | 0.93 | 0 | 1.0-2.5%* |
| SRA2 | Shrinkage reducing admixture | Polypropylene glycol based | 0.99 | 0 | 1.0-2.5%* |
| SRA3 | Shrinkage reducing admixture | Wax based | 0.90 | 60 | 5 kg/m$^3$ |

\* By weight of cement

Table 2 - Composition and properties of the concretes

| Concrete | CREF-SN | CREF-SC | CSRA1(1.5%)-SN | CSRA1(2%)-SN | CSRA2-SC | CSRA3-SC |
|---|---|---|---|---|---|---|
| Superplasticizer | SN | SC | SN | SN | SC | SC |
| SRA | None | None | SRA1 | SRA1 | SRA2 | SRA3 |
| **Composition** (per m$^3$) | | | | | | |
| Cement I 52.5 R | | | 325 kg | | | |
| Water added* | | | 180 lit. | | | |
| 0-2 mm sand | | | 250 kg | | | |
| 0-5 mm sand | | | 740 kg | | | |
| 5-12 mm gravel | | | 200 kg | | | |
| 12-20 mm gravel | | | 725 kg | | | |
| Superplasticizer | 5.85 lit. | 2.09 lit. | 3.37 lit. | 2.81 lit. | 1.86 lit. | 2.09 lit. |
| SRA | -- | -- | 5.36 lit. | 7.14 lit. | 7.50 lit. | 5.56 lit. |
| **Ratios** (by weight) | | | | | | |
| water/cement | 0.45 | 0.45 | 0.45 | 0.45 | 0.45 | 0.45 |
| sp/c** | 0.69% | 0.135% | 0.40% | 0.33% | 0.12% | 0.135% |
| SRA/cement | -- | -- | 1.5% | 2.0% | 2.29% | 1.54% |
| Slump immediately after mixing (cm) | 17 | 18 | 17 | 17 | 20 | 17 |
| Slump 15 minutes after mixing (cm) | 17 | 10 | 12 | 15 | 15 | 10 |
| Mean 28-day compressive strength in MPa (and coefficient of variation), from 3 specimens | 45.0 (±2.0%) | 45.2 (±4.9%) | 42.8 (±2.0%) | 39.8 (±2.0%) | 43.2 (±2.9%) | 42.2 (±0.5%) |

*Accounting for the water contained in the admixtures, and the humidity and
coefficient of absorption of the aggregates
** dry superplasticizer/cement ratio

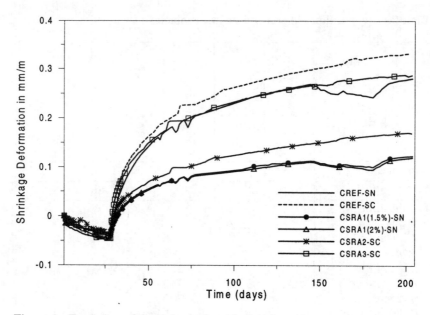

Figure 1 - Evolution of the deformations during the tests

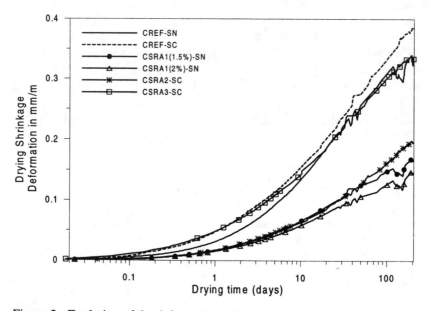

Figure 2 - Evolution of the deformations during the drying period

# Influence of Superplasticizer on the Volume Stability of Hydrating Cement Pastes at an Early Age

by B. Bissonnette, J. Marchand, C. Martel, and M. Pigeon

Synopsis: The influence of superplasticizer on the chemical (total) and autogeneous (external) shrinkage of hydrating cement pastes was investigated. Three different commercial CSA Type 10 cements were tested. Test variables also included type of superplasticizer (melamine-based and naphtalene-based) and dosage in admixture (three different dosages). All neat paste mixtures were prepared at a water/cement ratio of 0.35. Chemical shrinkage measurements were carried out using the classical dilatometric method initially developed by Le Chatelier. Autogeneous shrinkage measurements were performed according to the immersion method. All tests were performed in a temperature-controlled bath kept at 20°C. Test results indicate that the dosage in admixture influences the kinetics and magnitude of both chemical shrinkage and autogeneous shrinkage, especially during the first 24 hours. Beyond that period, the overall effects of dosage were observed to be less pronounced. Data also emphasize the potential importance of the type of superplasticizer upon early volume changes. Though the investigated cements are known to sometimes exhibit quite different early-age behaviors in the field, no significant differences were observed as far as chemical shrinkage and autogeneous shrinkage are concerned.

Keywords: cement; early-age; hydration; Le Chatelier contraction; self-desiccation; shrinkage; superplasticizer

**B. Bissonnette** is Research Associate at the Civil Engineering Department of Laval University, Quebec, Canada. His research interests are in the areas of concrete repairs and durability, deformations and testing.

**J. Marchand** is Associate Professor of Civil Engineering at Laval University, Quebec, Canada. His research interests are in the areas of numerical modeling, durability of concrete and performance of dry concrete products.

**C. Martel** is a M.Sc. student in Civil Engineering at Laval University, Quebec, Canada.

**M. Pigeon** is Vice-Dean for Research at Laval University, Quebec, Canada. His research interests are in the areas of durability of concrete and performance of concrete repairs.

## INTRODUCTION

The improvement of mixture design methods, the widespread use of supplementary cementing materials, and the development of new chemical admixtures have largely contributed to increase the overall performance of hydrated cement systems. However, despite their superior mechanical properties and their improved durability, high-performance cement-based materials often appear to be more sensitive to early-age cracking (1-3).

Given the importance of the problem, significant effort has been made to understand the basic mechanisms that control the volume instability of hydrating cement systems at an early age. Among other things, special emphasis was placed on the improvement of existing experimental methods and the development of new test procedures to investigate the behavior of neat cement paste, mortar and concrete mixtures during the first few days of hydration (1, 3-5).

The systematic application of these new procedures has shed new light on the parameters that control the autogeneous deformations of these systems. For instance, there is now considerable evidence that the mineralogical composition of cement and its particle size distribution have a significant influence on the volume instability of cement systems during the first few days of hydration (6, 7). Parameters like water/cement ratio and type of supplementary cementing materials have also been found to be central to the early-age behavior of hydrating cement-based materials (1, 8, 9).

Curiously, the addition of chemical admixtures has only been the subject of a limited number of studies. A few years ago, Paulini (10) studied the influence of various chemical admixtures on the chemical (or total) shrinkage of neat cement paste mixtures. According to the author, accelerators (such as calcium chloride) tend to markedly accelerate the initial rate of shrinkage of the material. Mixtures prepared with a water-reducing admixture, a retarding agent or a superplasticizer usually tend to swell during the first hours that follow the contact of water with the cement grains. The addition of superplasticizer was also found to delay setting and affect the kinetics of shrinkage.

In an early investigation of the problem, Edmeades et al. (11) found the autogeneous (or external) shrinkage of cement pastes containing 0.1% of calcium lignosulphonate to be more important than that of reference cement pastes, particularly after 10 hours of hydration. The admixture was also found to delay setting.

Tazawa and Miyazawa (12) investigated the influence of the type and dosage of superplasticizing agents and the effect of shrinkage-reducing admixtures on the behavior of neat cement pastes prepared at a fixed water/binder ratio of 0.30. According to their results, superplasticizing agents do not have any marked effect on autogeneous shrinkage. However, shrinkage-reducing admixtures tend to reduce the autogeneous shrinkage of neat cement pastes. The authors found that both alcohol alkylene oxide and glycol ether types of admixtures could contribute to reduce linear autogeneous shrinkage.

The lack of information concerning the influence of superplasticizing agents on the volume stability of hydrating cement systems is rather unfortunate since this type of chemical admixture is nowadays currently used in the production of concrete, particularly for the fabrication of high-performance mixtures. In order to bring more information on the effects of superplasticizers, eighteen neat cement paste mixtures were prepared and tested. Test variables included source of cement, type of superplasticizing agent and admixture dosages. Chemical (total) and external shrinkage measurements were performed for all mixtures.

It should also be emphasized that the scope of the present work was limited to autogeneous deformations of neat cement paste mixtures. Although temperature (particularly temperature gradients) can be, in certain cases, the main cause of the premature failure of some concrete structures, the present research was solely restricted to the behavior of cement systems kept under isothermal conditions at 20° C.

## MATERIALS

Three different commercially available CSA Type 10 (ASTM Type I equivalent with calcareous fillers) cements (labeled *C1*, *C2* and *C3*, respectively) were used in the preparation of the neat cement paste mixtures. Pertinent information on each cement is given in Table 1.

Mixtures were prepared with two different types of superplasticizing agent. A first series of nine neat cement paste mixtures was produced with a melamine-based superplasticizer (labeled *M*). A naphtalene-based superplasticizing agent was used in preparation of the remaining nine mixtures (labeled *N*).

For each combination of cement and admixture, three different dosages of superplasticizing agent were tested. The reference dosage (labeled 100%) corresponds to the saturation point in superplasticizer as determined by Marsh cone experiments (13). Typical flow curves are given in Figure 1. The two other dosages (labeled 50% and 150%, respectively) were adjusted by cutting by half or by multiplying by 1.5 the reference dosage.

All mixtures were prepared at a water/cement ratio of 0.35 using deionized water. Mixtures were batched in a high-speed mortar mixer placed under vacuum (at 10 mbar) to prevent, as much as possible, the formation of air bubbles during mixing.

## EXPERIMENTAL PROCEDURES

The kinetics of chemical shrinkage was determined for all mixtures. The chemical shrinkage test (inspired by Le Chatelier's pioneer work (14)) consisted in immersing a given mass of paste in an erlenmeyer flask. The amount of material was adjusted in order to have a 7.5 mm-thick layer of paste in the flask. Previous experience has shown that the test is rather sensitive to the thickness of the paste layer (1, 15). At the end of the bleeding period[1], the flask was filled with water and surmounted by a graduated capillary tube (Figure 2a). Afterwards, the volume of water gradually consumed by the hydration reaction was regularly measured. During the entire duration of the test, the flask was immersed in a thermo-regulated bath kept at 20°C. It should be emphasized that this test measures the total volume change of the hydrating material (i.e. the sum of the external shrinkage and the internal gas volume created by self-desiccation during the hydration process).

The kinetics of external (or autogeneous) shrinkage was also determined for all mixtures. The external shrinkage test consisted in filling an elastic plastic envelope (latex membrane) with a given mass of cement paste. The sample (approximately 150 mm-long and with a diameter of 50 mm) was then immersed in water. Its change of mass, always measured immersed in water, was periodically monitored during the first few days of hydration (15, 16). By applying Archimedes' principle, the change of mass of the material in water can be directly related to its change in volume. The main advantage of this

---

[1] The bleeding period extended from 2 to 5 hours according to the type of mixture.

method is that measurements begin right after the preparation of the mixture. During this test, the temperature is maintained at 20°C (Figure 2b). This procedure can only be used to determine the apparent (or external) volume change of the material.

After 24 hours and six days of hydration, the evaporable water content of the paste samples was measured by determining the loss in mass between test conditions and drying at 110° C. The non-evaporable water content of all mixtures was also measured. Loss of weight between drying at 110°C and 1000°C was considered to represent the non-evaporable water content. All measurements were conducted on specimens used for the external shrinkage tests (i.e. kept in sealed conditions). For each test condition, three samples were tested.

## TEST RESULTS

### Chemical Shrinkage Experiments

Chemical shrinkage test data are in Figures 3 to 8. Each figure summarizes the results of all six mixtures made of one of the three investigated cements. Test data obtained after 24 hours and 6 days of hydration are also summarized in Tables 3 to 5. All chemical shrinkage test results are expressed in mL/100 g of cement. In the figures, the volume changes are plotted as a function of time (elapsed since the initial contact of water with the cement grains). Each curve corresponds to the average result of 4 specimens. The repeatability was found to be very good (±1.5%).

As previously emphasized, the initial recording was performed once bleeding had stopped. The volume change experienced in the meantime was assumed to be equal to that measured for the companion external shrinkage specimens, which were monitored almost from the time of casting into the latex membranes. In the figures, the chemical shrinkage curves are offset accordingly.

As can be seen in Figures 3 to 8, the kinetics of chemical shrinkage tends to decrease monotonically during the experiment. As discussed by Barcelo et al. (17), chemical shrinkage can be linearly related to the degree of hydration of cement (as measured by isothermal calorimetry). In that respect, the reduction in the kinetics of volume change can be associated with a slower rate of reaction of the binder.

Irrespective of the mixture characteristics, the chemical shrinkage recorded after 6 days varies between 5 and 6 ml/100 g of cement (or from approximately 8 to 10.0% of the total volume of the mixture). These results are consistent with the theoretical values found in the literature[1] (17, 18), which fluctuate around 6 ml/100 g of cement for a fully hydrated cement paste mixture.

To assess the validity and reproducibility of these experiments, two additional test series were carried out on specimens made with the same mixture composition, but from separate batches. The results are presented in Figure 9, along with the corresponding external shrinkage data. As can be seen, the reproducibility of both types of tests is good.

External (autogeneous) Shrinkage Experiments

The external shrinkage test results are also presented in Figures 3 to 8. Again, each curve represents the average result computed from 4 specimens. The repeatability was again found to be good (±11.0%). As can be seen on the figures, in the first few hours (i.e. from 12 to 32 hours after the initial contact of water with the cement particles), chemical shrinkage measurements essentially coincide with the autogeneous (external) deformations, thus indicating that the material still behaves as a suspension. The good correlation between the two series of results during the first hours of hydration is consistent with the previous observations of numerous authors (1, 8, 19, 20).

These results should however be considered with caution. In a careful analysis of the early-age deformations of hydrating cement systems, Barcelo et al. (17) could establish that, beyond a critical point (called the percolation threshold), the rate of external shrinkage was in fact slightly (approximately 5 to 10%) slower than the rate of chemical shrinkage. The reduced rate of deformation was attributed to the fact that, once the percolation threshold is reached, the movement of particles in the suspension is impeded in at least one direction. As emphasized by the authors, this percolation threshold usually occurs early in the hydration process and should therefore be distinguished from the setting point of the material.

The moment at which the autogeneous shrinkage curve begins to significantly depart from the chemical shrinkage deformation corresponds to the formation of a self-supporting skeleton. From that moment on, the hydrating paste cannot deform freely to accommodate the volume reduction, and significant internal stresses are being generated in the material. In her comprehensive investigation of the chemical shrinkage and self-desiccation of cement-based materials, Boivin (15) has shown that the point where the autogeneous shrinkage curve deviates from the chemical shrinkage deformation occurs between the initial and the final setting times of the cement paste (i.e. well after the time required to reach the percolation threshold).

---

[1] Assuming that it takes 0.23 g of water to fully hydrate 1 g of cement, and that the Le Chatelier contraction is equal to 0.254 the volume of non-evaporable water.

As can be seen in Figures 3 to 8, the occurrence of the «kink» point in the external shrinkage curve (point where it separates from the chemical shrinkage curve) tends to vary significantly according to the cement, the type and dosage in superplasticizing agent. The «kink» point in the external shrinkage curve is always observed at relatively low degrees of hydration. In addition, the discrepancy between the chemical shrinkage and the external shrinkage is ultimately quite significant, the latter value being limited to a little less than 1.25 ml/100 g of cement after 6 days of hydration.

## DISCUSSION

### Influence of Cement Composition

The ordinary portland cements used in this investigation, which originate from three different plants located in the Eastern Canada region, are known to exhibit rather different behaviors in the field, especially with regard to rheology and setting times. Although the measurements of early volume changes in isothermal conditions do not necessarily reveal drastic differences, it is obvious that the volume instability of the neat cement paste mixtures is slightly influenced by the source of cement.

The comparison of the chemical shrinkage data obtained for the three cements, for each given combination of type of superplasticizer and admixture dosage, shows that chemical shrinkage values measured for mixtures made of cement $C2$ were systematically higher than those obtained for samples prepared with cements $C1$ and $C3$ (see Figures 3 to 8). As can be seen in Table 1, cement $C2$ has the highest $C_3S$ content among the three binders, while their respective $C_3A$ contents do not differ much.

Evidence of the strong influence of $C_3S$ hydration on the volume instability of hydrating cement systems at early-age has been reported by many investigators (21-22). This phenomenon is linked to the high reactivity of $C_3S$ and to the marked reduction of the absolute volume of the system upon hydration. These observations are also in good agreement with degree of hydration data reported in Tables 3 to 5. As can be seen, values measured for the mixtures made of cement $C2$ are systematically higher than those determined for the samples prepared with the two other cements. The higher reactivity of cement $C2$ appears to be solely linked to its mineralogical composition since the three binders were all ground to the same fineness (around 375 $cm^2/g$ in all three cases).

Test results also indicate that the source of cement had a different influence on the external shrinkage of the hydrating cement pastes. Irrespective of the type and amount of superplasticizer, the external shrinkage is observed to increase with binder *C1*, *C2* and *C3*, respectively (see Figures 3 to 8 and Tables 3 to 5). Although the differences between the various mixtures are not very marked, they are nevertheless very consistent from one test series to another.

Values obtained for the mixtures made of cement *C3* are particularly interesting considering the fact that chemical shrinkage results observed for these pastes were usually lower than those determined for the samples prepared with cement *C2*. Since the use of cement *C3* did not apparently increase the consumption of water, the higher external shrinkage values can probably be explained by the effect of this cement on the pore network morphology of the paste mixtures. The validity of this assumption needs to be investigated.

Influence of Superplasticizer

As can be seen in Figures 3 to 8 and in Tables 3 to 5, the use of the naphthalene-based superplasticizer has contributed to systematically increase the kinetics of shrinkage of the various mixtures. This phenomenon is presumably related to the greater efficiency of this admixture to deflocculate cement particles. The better distribution of cement particles accelerates hydration and accentuates chemical shrinkage. This assumption is in good agreement with the non-evaporable water content data that indicate that the use of the naphthalene-based superplasticizer has generally resulted in slightly higher degree of hydration values (see Table 4).

It should also be emphasized that the use of this type of superplasticizer has also led to higher chemical and external shrinkage values after 6 days of hydration. A similar trend was observed in a previous investigation devoted to the influence of plasticizing agents upon drying shrinkage (24).

It can be seen that the effect of the type of superplasticizer on both chemical shrinkage and external shrinkage is roughly the same during the first days of hydration. However, as hydration progresses, the difference between the two series of mixtures vanishes in the case of chemical shrinkage, while it remains roughly the same for the external shrinkage experiments. These data emphasize the fact that, beyond the «kink» point, the external shrinkage deformation is not solely influenced by the kinetics of hydration but is also affected by a series of other parameters such as the pore structure of the material.

For both types of superplasticizer investigated, the dosage appears to have a significant influence on the early-age deformations. For the chemical shrinkage experiments, the effect of dosage appears to vary according to the stage of the hydration process. During the initial stage (i.e. before the «kink» point), the kinetics of chemical shrinkage appears to be inversely proportional to the dosage in superplasticizer. This effect is particularly important when dosage is brought from 50% to 100%. The retarding effect is much less significant when the dosage is further increased from 100% to 150%. Accordingly, an increase in the amount of superplasticizer also offsets the location of the «kink» point.

Beyond the «kink» point, the phenomenon is reversed and the pastes prepared with higher admixture dosages suddenly exhibit larger rates of deformation. The chemical shrinkage values after 6 days of hydration also systematically increase with the dosage in superplasticizer. As can be seen in Figures 3 to 8 and in Tables 3 to 5, final chemical shrinkage values (after 6 days of hydration) tend to increase with the dosage in superplasticizer.

The dosage in superplasticizer also affects the behavior of the pastes observed in external shrinkage experiments. Unsurprisingly, prior to the «kink» point, the effect of the admixture dosage is similar to what has been observed in the chemical shrinkage tests. Afterwards, an increase in the admixture dosage accelerates the kinetics of external shrinkage. Higher dosages also lead to higher shrinkage values at the end of the test period.

Globally, test results tend to indicate that the dosage and type of superplasticizer affect the time at which the «kink» point occurs as well as the kinetics of both types of shrinkage beyond the «kink» point (i.e. after setting). These effects are probably linked, at least in part, to the influence of superplasticizer addition on the chemistry of the system. However, it also very likely that the rate of deformation after the «kink»point is sensitive to the arrangements of the cement particles during setting.

## CONCLUSION

Chemical and external shrinkage tests conducted simultaneously have led to a better understanding of the early-deformation of cement pastes as a function of some composition parameters. These complementary experimental techniques allow a characterization of the material behavior at both the micro- (particles) and meso- (material) levels. The main conclusions that can be drawn from the present investigation are summarized as follows:

- The use of three different CSA Type 10 cements did not result in very significant differences as far as the early-age volumetric behavior is concerned, though their behavior in the field is known to be sometimes quite different (especially binder $C1$).

- As compared to the melamine-based superplasticizer, the addition of naphthalene-based admixture was found to systematically increase chemical shrinkage and external shrinkage.

- Both the melamine and naphthalene-base superplasticizers were observed to influence the early-volume change kinetics and magnitude as their rate of addition was increased. The increase in admixture addition slowed down very early volume changes, up to the «kink» point, but then resulted in higher shrinkage rates and larger ultimate shrinkage values. In the case of chemical shrinkage, superplasticizer dosages beyond saturation were needed to cause a significant increase.

In order to better analyze the scope and significance of such results, it will be necessary to relate them to the effective degree of hydration. Also, the temperature during curing operations might have very important consequences and it will need to be addressed as well.

From a practical standpoint, the results reported in this paper call for some attention in the formulation of moderate to low water-binder ratio materials, especially in the selection and dosage of the superplasticizing agent.

## ACKNOWLEDGMENTS

This project was funded by the Natural Sciences and Engineering Research Council of Canada (NSERC) and by the Fonds pour la Formation de Chercheurs et l'Aide à la Recherche (FCAR) of the government of the province of Quebec.

## REFERENCES

1. Bissonnette, B., Marchand, J., Delagrave, A., Barcelo, L. (2001) *Early-age behavior of cement-based materials*, Materials Science of Concrete - VI, American Ceramic Society, (in press).

2. Burrows, R.W. (1998) *The visible and invisible cracking of concrete*, Monograph N° 11, American Concrete Institute, 71 p.

3. Bjøntegaard, Ø. (1999) *Thermal dilation and autogenous deformation as driving forces to self-induced stresses in high performance concrete*, Ph.D. Thesis, 1999:121, NTNU Trondheim (Norway), 256 p.

4. Hammer, T.A. (1999) *Test methods for linear measurement of autogenous shrinkage before setting*, Autogeneous Shrinkage of Concrete, E & FN Spon, London, U.K., pp. 143-154.

5. Barcelo, L., Boivin, S., Rigaud, S., Acker, P., Clavaud, B., Boulay, C. (1999) *Linear vs volumetric autogeneous shrinkage measurement :*

*Material behavior or experimental artefact*, Proceedings of the Second International Research Seminar on self-desiccation and its importance in concrete technology, Edited by B. Persson and G. Fagerlund, Lund, pp. 109-126.

6. Bentz, D.P., Garboczi, E.J., Haecker, C.J., Jensen, O.J. (1999) *Effects of cement particle size distribution on performance properties of Portland cement-based materials*, Cement and Concrete Research, Vol. 29, N° 10, pp. 1663-1671.

7. Bentz, D., Jensen, O.M., Hansen, K.K., Olesen, J.F., Stang, H., Haecker, C.J. (1999) *Influence of cement particle size distribution on early age autogeneous strains and stresses in cement-based materials*, Journal of the American Ceramic Society, Vol. 84, N° 1, pp. 129-135.

8. Justnes, H., Van Gemert, A., Verboven, F., Sellevold, E. (1996) *Total and external chemical shrinkage of low w/c ratio cement pastes*, Advances in Cement Research, Vol 8, N° 31, pp. 121-126.

9. Justnes, H., Ardoullie, B., Hendrix, E., Sellevold, E.J., Van Gemert, D. (1998) *The chemical shrinkage of pozzolanic reaction products*, ACI Special Publication SP 179-11, pp. 191-205.

10. Paulini, P. (1990) *Reaction mechanisms of concrete admixtures*, Cement and Concrete Research, Vol. 20, N° 6, pp. 910-918.

11. Edmeades, R.M., James, A.N., Wheeler, J. (1966) *The formation of ettringite on aluminium surfaces*, Journal of Applied Chemistry, Vol. 16, pp. 361-368.

12. Tazawa, E., Miyazawa, S. (1995) *Influence of cement and admixture on autogenous shrinkage of cement paste*, Cement and Concrete Research, Vol. 25, N° 2, pp. 281-287.

13. Aïtcin, P.C. (1998) *High-performance concrete*, E & FN Spon, London. U.K., 590 p.

14. Le Chatelier, H. (1900) *Sur les changements de volume qui accompagnent le durcissement des ciments*, Bulletin de la Société pour l'Encouragement de l'Industrie Nationale, 5ᵉ Série, Tome 5, pp. 54-57, (in French).

15. Boivin, S. (1999) *Retrait au jeune age du béton: Dévelopement d'une méthode expérimentale et contribution à l'analyse physique du retrait endogène*, Doctoral Thesis, École Nationale des Ponts et Chaussées, Paris (France), 249 p. (in French).

16. Justnes, H., Van Gemert, A., Verboven, F., Sellevold, E., Van Gemert, D. (1997) *Influence of measuring method on bleeding and chemical shrinkage*

*values of cement pastes*, 10th International Congress on the Chemistry of Cement, Vol. II, 8 p.

17. Barcelo, L., Boivin, S., Acker, P., Toupìn, J., Clavaud, B. (2001) *Early-age shrinkage of concrete: Back to physical mechanisms*, Concrete Science & Engineering, Vol. 3, N° 10, pp. 85-91.

18. Buil, M., (1979) *Contribution à l'étude du retrait de la pâte de ciment durcissante*, Doctoral Thesis, École Nationale des Ponts et Chaussées, Paris (France), 67 p. (in French).

19. Gagné, R., Aouad, I., Shen, J., Poulin, C. (1999) *Development of a new experimental technique for the study of the autogeneous shrinkage of cement paste*, Materials and Structures, Vol. 32, pp. 635-642.

20. Charron, J.-P., Marchand, J., Bissonnette, B. (2001) *Early-age deformations of hydrating cement systems: Comparison of linear and volumetric shrinkage measurements*, RILEM International Conference on Early-Age Cracking in Cementitious Systems, p. 245-257.

21. Powers, T.C. (1935), Absorption of water by Portland cement paste during the hardening process, Industrial and Engineering Chemistry, Vol. 27, N° 7, pp. 790-794.

22. Tazawa, E., Miyazawa, S. (1997), Influence of constituents and composition on autogenous shrinkage of cementitious materials, Magazine of Concrete Research, Vol. 49, N° 178, pp. 15-22.

23. Justnes, H., Sellevold, E.J., Reyniers, B., Van Loo, D., Van Gemert, A., Verboven, F., Van Gemert, D. (1999), *The influence of cement characteristics on chemical shrinkage*, in Autogeneous Shrinkage of Concrete, E & FN Spon, London, U.K., pp. 71-80.

24. Brooks, J.J. (1989) *Influence of mix proportions, plasticizers and superplasticizers on creep and drying shrinkage of concrete*, Magazine of Concrete Research, Vol. 41 N° 148, pp. 145-153.

Table 1 – Composition of cement

| Chemical (%) | C1 | C2 | C3 |
|---|---|---|---|
| $SiO_2$ | 20.50 | 20.10 | 20.31 |
| $Al_2O_3$ | 4.21 | 4.14 | 4.17 |
| $Fe_2O_3$ | 2.99 | 2.74 | 2.87 |
| CaO | 62.10 | 62.00 | 62.04 |
| MgO | 2.80 | 1.93 | 2.41 |
| $SO_3$ | 3.20 | 3.05 | 3.12 |
| Alkalis | 0.80 | 0.83 | 0.81 |
| LOI | 2.10 | 3.01 | 0.31 |
| **Minerological (%)** | | | |
| $C_3$-S | 55.6 | 58.7 | 57.3 |
| $C_2$-S | 16.6 | 13.1 | 14.7 |
| $C_3$-A | 6.3 | 6.1 | 6.1 |
| $C_4$-A-F | 9.1 | 7.8 | 8.5 |

Table 2 – Mixture characteristics

| Mixture ID | Cement (kg/m$^3$) | Water* (kg/m$^3$) | N SP (mL/kg C) | M SP (mL/kg C) |
|---|---|---|---|---|
| *C1N50* | 1495 | 523 | 6.0 | |
| *C1N100* | 1491 | 521 | 12.0 | |
| *C1N150* | 1487 | 520 | 18.0 | |
| *C1M50* | 1495 | 524 | | 4.2 |
| *C1M100* | 1492 | 523 | | 8.3 |
| *C1M150* | 1490 | 521 | | 12.5 |
| *C2N50* | 1494 | 523 | 6.2 | |
| *C2N100* | 1490 | 521 | 12.3 | |
| *C2N150* | 1485 | 520 | 18.5 | |
| *C2M50* | 1495 | 524 | | 4.3 |
| *C2M100* | 1492 | 522 | | 8.6 |
| *C2M150* | 1489 | 522 | | 12.9 |
| *C3N50* | 1494 | 523 | 6.2 | |
| *C3N100* | 1490 | 521 | 12.5 | |
| *C3N150* | 1485 | 520 | 18.7 | |
| *C3M50* | 1495 | 523 | | 4.6 |
| *C3M100* | 1492 | 522 | | 9.2 |
| *C3M150* | 1488 | 521 | | 13.8 |

*Note: The total amount of water in the mixture included the added water and
also the water fraction in the liquid-phase superplasticizer

Table 3 - Shrinkage test results (*C1*)

| Mixture | Time elapsed since W-C contact (days) | Degree of hydratation (sealed cond$^n$) | Chemical shrinkage | External shrinkage |
|---|---|---|---|---|
| | | | (mL per 100 g of cement) | |
| *C1N50* | 1 | 0.245 | 2.41 | 0.85 |
| | 6 | 0.695 | 5.13 | 0.96 |
| *C1N100* | 1 | 0.239 | 2.40 | 0.95 |
| | 6 | 0.589 | 5.00 | 1.04 |
| *C1N150* | 1 | 0.235 | 2.50 | 1.04 |
| | 6 | 0.572 | 5.97 | 1.17 |
| *C1M50* | 1 | 0.232 | 2.29 | 0.81 |
| | 6 | 0.655 | 4.88 | 0.91 |
| *C1M100* | 1 | 0.231 | 2.33 | 0.88 |
| | 6 | 0.556 | 4.85 | 0.98 |
| *C1M150* | 1 | 0.225 | 2.35 | 0.98 |
| | 6 | 0.552 | 5.61 | 1.12 |

Table 4 - Shrinkage test results (*C2*)

| Mixture | Time elapsed since W-C contact (days) | Degree of hydratation (sealed cond$^n$) | Chemical shrinkage | External shrinkage |
|---|---|---|---|---|
| | | | (mL per 100 g of cement) | |
| *C2N50* | 1 | 0.282 | 2.55 | 0.91 |
| | 6 | 0.777 | 5.44 | 1.02 |
| *C2N100* | 1 | 0.260 | 2.59 | 1.02 |
| | 6 | 0.615 | 5.40 | 1.12 |
| *C2N150* | 1 | 0.255 | 2.76 | 1.14 |
| | 6 | 0.763 | 6.57 | 1.30 |
| *C2M50* | 1 | 0.233 | 2.29 | 0.84 |
| | 6 | 0.763 | 5.22 | 0.95 |
| *C2M100* | 1 | 0.223 | 2.33 | 0.92 |
| | 6 | 0.646 | 5.54 | 1.02 |
| *C2M150* | 1 | 0.222 | 2.35 | 1.03 |
| | 6 | 0.593 | 6.48 | 1.18 |

Table 5 - Shrinkage test results (*C3*)

| Mixture | Time elapsed since W-C contact (days) | Degree of hydratation (sealed cond$^n$) | Chemical shrinkage | External shrinkage |
|---|---|---|---|---|
| | | | (mL per 100 g of cement) | |
| *C3N50* | 1 | 0.239 | 2.34 | 0.93 |
| | 6 | 0.667 | 4.97 | 1.04 |
| *C3N100* | 1 | 0.236 | 2.37 | 1.03 |
| | 6 | 0.565 | 4.95 | 1.14 |
| *C3N150* | 1 | 0.233 | 2.40 | 1.16 |
| | 6 | 0.570 | 5.72 | 1.33 |
| *C3M50* | 1 | 0.251 | 2.38 | 0.87 |
| | 6 | 0.580 | 5.07 | 0.98 |
| *C3M100* | 1 | 0.238 | 2.50 | 0.97 |
| | 6 | 0.654 | 5.06 | 1.07 |
| *C3M150* | 1 | 0.230 | 2.45 | 1.08 |
| | 6 | 0.572 | 5.67 | 1.25 |

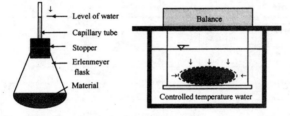

Figure 1 – Schematic representation of *(a)* chemical shrinkage test setup and *(b)* external shrinkage test setup

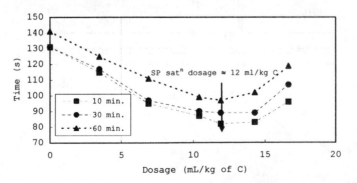

Figure 2 – Typical flow curves obtained during a Marsh cone experiment

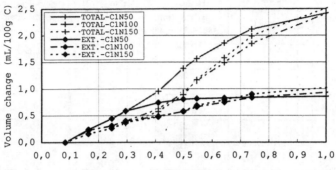

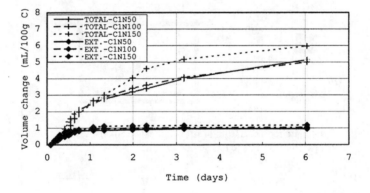

Figure 3 - Shrinkage test results (*C1* and naphtalene-based superplasticizer)

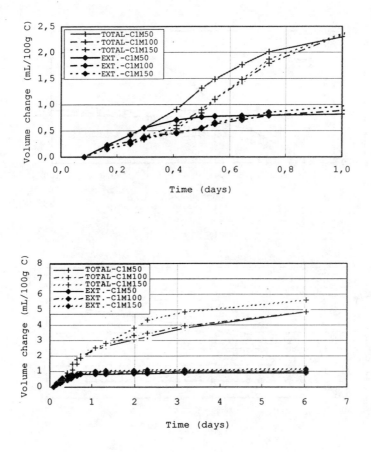

Figure 4 -  Shrinkage test results (*C1* and melamine-based superplasticizer)

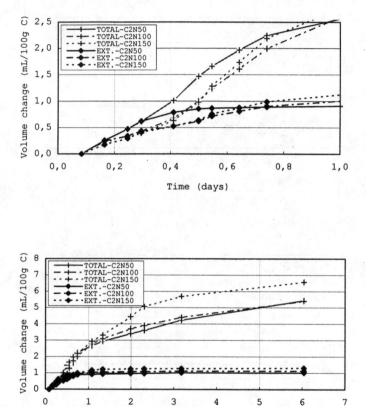

Figure 5 -  Shrinkage test results (*C2* and naphtalene-based superplasticizer)

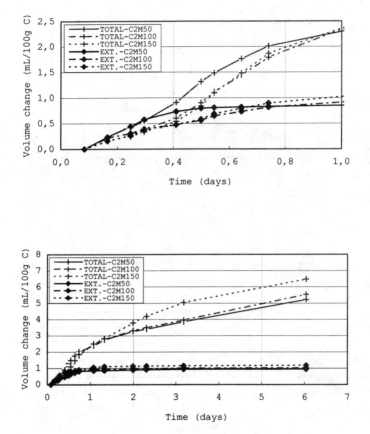

Figure 6 - Shrinkage test results (*C2* and melamine-based superplasticizer)

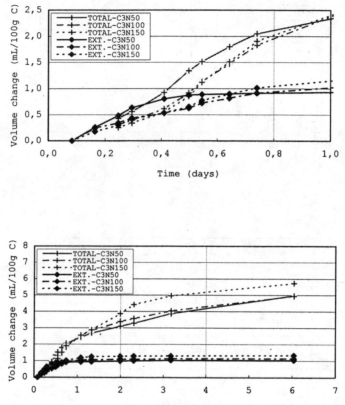

Figure 7 -  Shrinkage test results (*C3* and naphtalene-based superplasticizer)

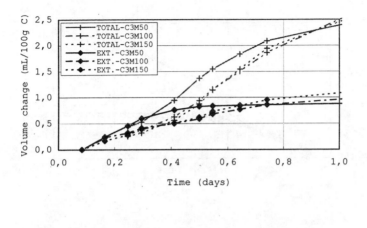

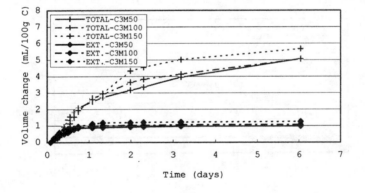

Figure 8 - Shrinkage test results (*C3* and melamine-based superplasticizer)

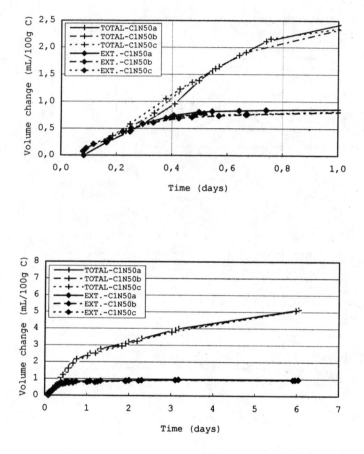

Figure 9 - Repeated shrinkage tests (*C1* and naphtalene-based superplasticizer)

# SP 206–12

# Grip-Specimen Interaction in Uniaxial Restrained Test

## by S. A. Altoubat and D. A. Lange

Synopsis: Restrained tests are used to evaluate the risk of early age cracking and the cracking sensitivity of concrete mixtures. One test that has become common in recent years is the active uniaxial restrained test in which the length change due to shrinkage is recovered by applying external load to maintain the concrete sample at constant length. The length change is measured by linear variable differential transformer (LVDT), which is used as the control signal in this test. In such tests, the dog-bone geometry is used to grip the ends. To ensure a fully restrained test, the LVDT response to the loads and to shrinkage should reflect the deformation in the concrete sample. Therefore, the grip-specimen interaction should not interfere with the measurement of deformation, and this depends on the instrumentation and how the LVDT is attached to the concrete specimen. Some experiments in the literature have the LVDT attached to the steel grips, a practice vulnerable to possible error due to the interaction between the grip and the concrete. This study considered two methods of attaching the LVDT. First, the LVDT is attached to the steel grips; second, the LVDT is attached to the concrete within the zone of reduced cross-section. The results indicate that attaching the LVDT to the grips results in errant measurement of the shrinkage stress, creep, and elastic strains due to the grip-specimen interaction. The consequences will be false interpretation of fully restrained shrinkage and creep characteristics because the grip-specimen interaction leads to a partially restrained test. The study suggests mounting the LVDT to the concrete sample away from the grips to achieve a fully restrained test. Results for two concrete mixtures with w/c ratio of 0.51 and 0.56 are discussed for both methods of attaching LVDTs.

Keywords: end effects; restrained shrinkage; shrinkage; tensile creep; tensile test

*ACI Member Salah A. Altoubat is a research associate at W. R. Grace in Cambridge, Massachusetts. He received his Ph.D in civil engineering from the University of Illinois at Urbana-Champaign in 2000, and his MS in structural engineering from the Jordan University of Science and Technology, Jordan in 1990. His current research interest includes early age behavior of concrete, creep, shrinkage and cracking.*

*ACI Member David A. Lange is an Associate Professor of Civil Engineering at the University of Illinois at Urbana-Champaign. He received his Ph.D. from Northwestern University. He is a member of ACI Committees E802, Teaching Methods and Materials; 544, Fiber Reinforced Concrete; 549, Thin Reinforced Cement Products, and serves as chair of Committee 236, Materials Science of Concrete. His research interests include early age properties of concrete, microstructure of porous materials, water transport in repair and masonry materials, and industrial applications of high performance cement based materials.*

## INTRODUCTION

Early-age shrinkage and cracking of concrete has been a focal research in recent years due to the advent of high strength and high performance concrete with low water/binder (w/b) ratios, which are more prone to cracking. Early age shrinkage cracking is a key issue of long term durability and serviceability. Shrinkage of concrete is important because it is the main driving force for cracking, but the relaxation properties and the extent of restraint which will determine whether the shrinkage will lead to cracking are also important in the assessment of the consequences of shrinkage.

In view of the variety of factors that must be considered, quantitative cracking tests under restrained conditions are essential. Uniaxial restrained shrinkage tests have been developed and used in the literature to assess the early age shrinkage cracking (1-8). These tests can be divided into two main types according to the mode of restraint: passive restraint and active restraint. The passive restraint test fixes the end grips of the concrete sample by external rigid frame. The test used grips with tapered geometry to reduce stress concentration that may lead to premature cracking, and can be partially instrumented to measure the restraining force (1). The stresses developed in the passive test depend on the rigidity of the concrete and the restraining frame, and therefore the data obtained are not sufficiently fundamental. This shortcoming has been avoided in the development of the active restraint test. In the active test, one grip is fixed and the other is free to move. It is returned to its original position periodically, after some shrinkage has occurred. This is achieved by a special arrangement at the moving grip, whereby a screw or hydraulic mechanism is activated to bring the grip back to its original position, and a load cell measures the induced load. The active tests are either partially automated (3,4) or fully automated closed-loop systems (5,6). In both cases, the original position and the movement of the grip are either determined by a strain gage or by an LVDT.

A survey of the testing rigs indicated that the LVDT used to control the movement of the grip to maintain original length of the concrete sample is in most cases attached to the moving grip (e.g. 2,3,4,5). Consequently, the measured deformation may not exclusively reflect the shrinkage of the concrete sample between the grips because the LVDT measurement also incorporates whatever is happening within the grips. This could include the deformation associated with material damage or slip at the contact surfaces between the concrete and the grip. Consequently, a fully restrained test would be falsely assumed, and the data generated in these tests regarding shrinkage, creep and cracking would be erroneous if interpreted as for a fully restrained test. Recently, a fully closed-loop active restrained test has been developed at the University of Illinois (6,7), and the issue of the grip/specimen interaction has been investigated to ensure a fully restrained test. This paper sheds light on this important issue.

## EXPERIMENTAL

### Restrained Test Device

A fully automated restrained shrinkage test was developed to study the restrained shrinkage behavior and the relaxation properties for early age concrete [6,7]. The principle of the tests was based on the concept of Bloom and Bentur (4), which was integrated into a closed-loop system by Kovler (5). The developed system tests two identical "dog-bone" samples; one specimen is restrained and the load developed by drying shrinkage is measured, and the other specimen is unrestrained and the shrinkage deformation is measured. The dimensions of the specimen were selected to accommodate a maximum aggregate size of 25 mm. Each specimen is 1000 mm long and 76.2x76.2 mm in cross-section. The experiment is controlled by a closed loop system capable of highly accurate measurements and smooth loading.

A vertical layout of the experimental set-up was designed in which the test specimens were mounted vertically in a Universal Testing Machine. The bottom grip of the restrained specimen was fixed to the base of the machine, whereas the top grip was movable and was connected to the machine through a load cell. A swivel-joint was installed between the grip and the load cell to minimize eccentric loading. The free shrinkage specimen was vertically mounted on the base of the machine. The specimen cross-section is gradually enlarged to fit into the end grips. This design configuration minimizes stress concentration at contact surfaces. A general view of the experimental device is shown in Figure 1.

The restrained condition was simulated by maintaining the total deformation of the restrained sample within a threshold value of 5 μm, which is defined as the permissible change in the gage length of the specimen before restoration to the original length. The computer- controlled test checked shrinkage deformation continuously, and when the threshold was exceeded, an increase in tensile load was applied by the actuator to restore the concrete specimen to its original length. In this way, a restrained condition was achieved and the stress generated by shrinkage mechanism was measurable.

Comparison of the free shrinkage results with the shrinkage of the restrained specimen enabled discrimination of creep strain from shrinkage strain. Figure 2 shows how creep strain can be calculated from the restrained and free shrinkage test. The free shrinkage was measured from the free shrinkage specimen and the restrained shrinkage was based on the recovery cycles by which the specimen was loaded to restore its original length. Thus, each recovery cycle consisted of shrinkage and creep strain recovered by instantaneous elastic strain that was induced by incremental tensile load applied by the actuator. The sum of the elastic strain at any time is equal to the combined shrinkage and creep strains. Knowing the free shrinkage component, the creep strain can be quantified. The computer controlled recovery cycle in this test can be used to perform additional tests such as creep and relaxation, by programming the system to follow a different pattern. A variety of mechanical properties of concrete at early age such as components of strain, shrinkage stress, moduli of elasticity and creep coefficient can be determined by this experiment.

## Instrumentation and LVDT Attachment

The longitudinal shrinkage was measured by a linear variable differential transformer (LVDT). Each measurement was an average value of 100 readings per second of the LVDT. Such a procedure permitted very high accuracy and reproducibility of linear displacement measurement of less than ± 0.1 μm.

Two ways of measuring the deformation were investigated. First, the LVDT was attached to the steel grips, and the gage length was the total length of the sample between the grips. In this case, the LVDT measures all deformations between the grips including the interaction at the contact surfaces. Second, the LVDT was attached to the concrete sample through a metal stud hooked to the concrete within the reduced cross-section. The gage length in this case is in the middle of the concrete sample away from the end grips. Several tests were performed for each configuration, and results from both methods will be presented and discussed. Schematic presentation of the LVDT attachment is shown in Figure 3.

## Materials and Test Program

Two normal concrete mixtures with w/c ratio of 0.51 and 0.56 were tested under drying conditions. Materials used were Type I portland cement, crushed limestone aggregates with maximum size of 25 mm, and natural sand. The gradation of coarse and fine aggregates satisfied ASTM C33 requirements, and the fine aggregates had a fineness modulus of 2.2. The normal concrete mixtures had a paste volume fraction of 0.35.  Proportions of the concrete mixtures are presented in Table 1.

Two series of tests were conducted for the two mixtures considered in this study. In the first series, the deformation measuring devices (LVDTs) were attached to the surface of the grips, while in the second series, the LVDTs were attached to the concrete samples a way from the end grips as shown in Figure 3. Several replicate tests were performed in each series.

In each test, two linear specimens were cast; one used for free shrinkage and the second for restrained shrinkage. Concrete specimens were cast, covered with plastic sheets and stored in a humidity chamber for 18 hours before installation in the machine. At this age, it was possible to handle the specimens and to instrument the test in the vertical layout of the experiment. Specimens were left unrestrained for 1-2 hours after exposure to minimize the effect of thermal shock that may cause premature failure as described by Kovler (9). The specimens were then exposed to drying at relative humidity (RH) of 50 %, and a temperature of 23 degrees C.

## RESULTS AND DISCUSSIONS

Typical test results for the two ways of LVDT attachment are presented. The reproducibility of the test, shrinkage stress evolution, free shrinkage, tensile creep, creep coefficient, elastic modulus, and stress-elastic strain are presented. The effect of the method of LVDT attachment in the reliability and accuracy of the data generated from the test is also discussed.

To evaluate reproducibility of the experiment, replicate tests were performed for both methods of attachments. Figures 4 and 5 present the stress development of replicate tests when the LVDTs were attached to the grip surface and to the concrete, respectively. Clearly, both methods generated reproducible data and both methods seemed reliable for such a test. However, the magnitude of the stress developed for each method differed for the same concrete mixture as shown in Figure 6. This suggested that the two methods represented different restraint conditions. Further analysis of the data of the free shrinkage, creep and stress-strain helped address this issue.

The free shrinkage strain measured from the free shrinkage specimen is shown in Figure 7 for the two methods of LVDT attachment. The results indicate that the free shrinkage is similar for both methods, which suggests that the stress development should not differ. But the fact that the stress development was different created concern about the degree of restraint and the grip-specimen interaction. The interaction between the grip and the specimen and its effect on the degree of restraint will exist only if there is a load applied to the grips as in the case of restrained test. Therefore, similar free shrinkage strain measurements are expected, and the results indicated a reasonable consistency and similarity between the two methods.

From the measured load and the strain recovered on each recovery cycle of the restrained test, a stress – elastic strain can be established. The stress is calculated from the cumulative loads measured in the restrained specimen, and the strain is cumulative strain obtained from the compensation cycles. The stress-strain relation obtained for the two methods is shown in Figures 8 and 9 for the two concrete mixtures tested in this study. It is clear that the calculated stress when the LVDT is attached to the steel grips is much lower than when the LVDT is attached to the concrete. Despite the lower stresses however, the calculated strain "presumably elastic strain" is much higher, which cannot be true for the same concrete.

The LVDT attachment method also affects measurement of the elastic modulus. The calculated secant modulus is presented in Figure 10 for the concrete mixture with w/c ratio of 0.56. The calculated values of the secant modulus when the LVDT was attached to the grips did not agree with literature values at this age. The concrete (w/c = 0.56) was expected to reach a modulus of elasticity of about 20 MPa after 24 hours, however the calculated values from the test results ranged between 8 and 12 MPa. The attachment of the LVDT on the steel grips resulted in errant measurement of the elastic strain due to grip-specimen interaction whereby part of the strain recovered in the compensation cycle was recovered in the form of slip and not as elastic strain in the concrete. As a result, the specimen was partially restrained and a false low value of elastic modulus was obtained.

When the LVDT was attached to the concrete, the shrinkage stress measured in the experiment was higher as shown in Figures 6, 8 and 9, which indicated a fully restrained condition. Furthermore, the calculated modulus of elasticity was in reasonable agreement with normal values for concrete.

## Interpretation of the Interaction

Past studies that used LVDTs mounted on the grips typically considered the gage length of the specimen to be the free length of concrete in between the grips (e.g. Figure 3 shows the free length as 673 mm). However, the concrete volume within the gripped ends is also subject to stress. In fact, the state of stress within the gripped ends is very complex with some regions in compression and other regions in tension. The concrete volume within the gripped ends certainly undergoes elastic and creep deformation, and contributes to the overall measured deformation if the LVDTs are mounted on the grips. Furthermore, there is potential for slip to occur between the grips and the concrete. Since such tests assume that the gage length is the free length between the grips, the measured deformation is always higher than the true value for the assumed gage length. The excess deformation is evident in Figures 8 and 9, and the magnitude of the excess deformation increases as tensile load increases.

If the LVDTs are mounted on the grips, the specimen would behave as one that is not fully restrained. The shrinkage stresses in the test would reflect a partially restrained condition, and the time of first cracking would be delayed. The resolved creep strain would not accurately reflect the true relaxation properties because creep depends on the level of stress which would be higher in the fully restrained case. Figure 11 shows the creep coefficient for the two methods of measurement, and the creep coefficient was underestimated when the LVDTs were mounted to the grips. The excess deformation would be falsely interpreted as elastic strain in the concrete, and will lead to an underestimated modulus of elasticity as shown in Figure 10.

This study has shown that the error associated with mounting the LVDTs on the grips is of a significant magnitude. While there may be little impact in measurement of free shrinkage, the error will affect all properties that depend on the restrained specimen measurements. In such a case the relaxation properties and cracking of concrete characterized by this test would falsely overestimate performance of the concrete in the field.

## SUMMARY AND CONCLUSIONS

The interaction between the end grips and the concrete samples in dog-bone-shaped uniaxial restrained tests is substantial. The common practice of mounting LVDTs on the steel grips for measuring specimen length change is a procedure that introduces error if fully restrained conditions are desired. Tests use cross-head measurements may not also be appropriate. When the LVDT is attached to the grips, excess deformation is included in the measurement. The excess deformation is due to elastic and creep strains in the concrete volume located within the grip ends, and slip between the grips and the concrete material. The excess deformation becomes more critical in active closed-loop restrained shrinkage tests where the LVDT signal is the controlling parameter. The consequences will be false interpretation of fully restrained shrinkage and creep characteristics because the grip-specimen interaction leads to a partially restrained test. The true degree of restraint is difficult to estimate because the state of stress within the end regions is a complex combination of compression and tension. The data generated about shrinkage stress, creep, and shrinkage cracking may not accurately characterize the true restrained shrinkage behavior.

This study has identified a source of error introduced by the practice of mounting LVDTs to the steel grip ends in active uniaxial restrained shrinkage tests. The error is avoided by mounting the LVDTs on the free length of the concrete specimen so as to avoid all influence of deformation that may occur in the concrete volume within the gripped ends.

## ACKNOWLEDGMENT

This research project was supported by the Federal Aviation Administration (FAA) Center of Excellence (COE) at the University of Illinois and by the National Science Foundation (CAREER Award # CMS-9623467).

## REFERENCES

1. RILEM TC 119-TCE: Avoidance of thermal cracking in concrete at early ages-recommendations, Materials and Structures, Vol. 30, 1997, pp. 451-461.

2. Springenschmid, R., Breitenbucher, R., and Mangold, M., " Development of the cracking frame and the temperature-stress testing machine", in Thermal Cracking in Concrete at Early Ages, R. Springenschmid (editor), Proc. RILEM Symp., E&FN SPON, 1994, pp. 137-144.

3. Paillere, A. M., Buil, M., and Serrano J. J., " Effect of fiber addition on the autogenous shrinkage of silica fume concrete", ACI Materials Journal, Vol. 86, no. 2, 1989, pp. 139-144

4. Bloom, R., and Bentur, A., " Free and restrained shrinkage of normal and high strength concretes", ACI Materials Journal, Vol. 92, No. 2, 1995, pp. 211-217.

5. Kovler, K., " Testing systems for determining the mechanical behavior of early age concrete under restrained and free uniaxial shrinkage", Materials and Structures, Vol. 27, No. 170, 1994, pp. 324-330.

6. Altoubat S. A., "Early age stresses and creep-shrinkage interaction of restrained concrete," Ph.D thesis in the Department of Civil Engineering at the Univ. of Illinois at Urbana-Champaign, 2000.

7.  Altoubat, S. A. and Lange, D. A., " Creep, shrinkage and cracking of restrained concrete at early age," ACI Materials Journal, Vol. 98, No. 4, 2001, pp. 323-331.

8.  Pigeon, M., Toma, G., Delagrave, A., Bissonnette, B., Marchand, J., and Cprince, J., " Equipment for the analysis of the behavior of concrete under restrained shrinkage at early age", Magazine of Concrete Research, Vol. 52, No. 4, 2000, pp. 297-302.

9.  Kovler K., "Shock of evaporative cooling of concrete in hot dry climates," Concrete International, No. 10, 1995, pp. 65-69.

Table 1 Proportions of concrete mixtures

| Constituents | NC-0.51 | NC-0.56 |
|---|---|---|
| Coarse Agg. Kg/m$^3$ | 925.8 | 925.8 |
| Fine Agg. kg/m$^3$ | 741.8 | 741.8 |
| Cement kg/m$^3$ | 421.4 | 397.6 |
| Water kg/m$^3$ | 214.9 | 222.6 |
| W/C | 0.51 | 0.56 |

Figure 1 General view of the experimental setup

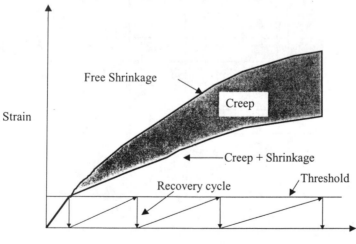

Figure 2 Separation of creep strain

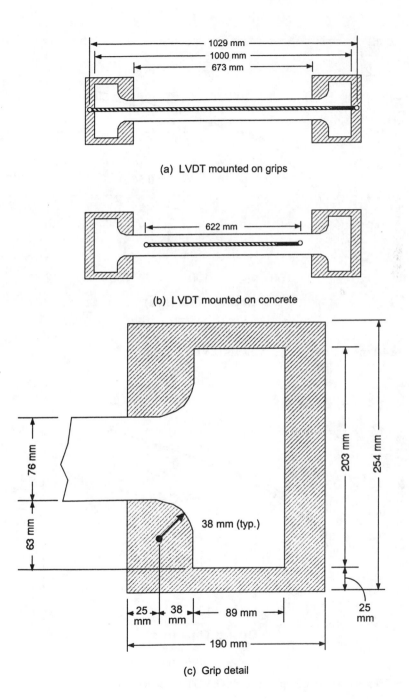

(a) LVDT mounted on grips

(b) LVDT mounted on concrete

(c) Grip detail

Figure 3.  Two LVDT mounting methods and grip details

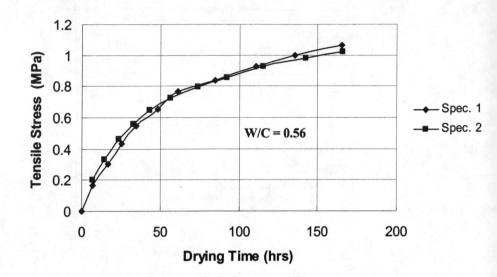

Figure 4 Tensile stress of replicate tests (LVDT on grips)

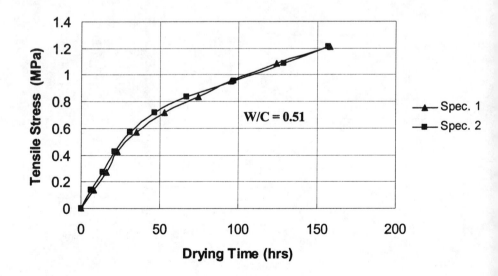

Figure 5 Tensile stress of replicate tests (LVDT on concrete)

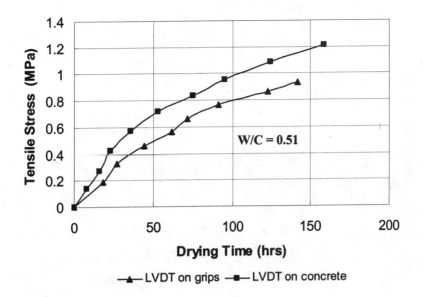

Figure 6 Shrinkage stress for the two methods of LVDT attachment

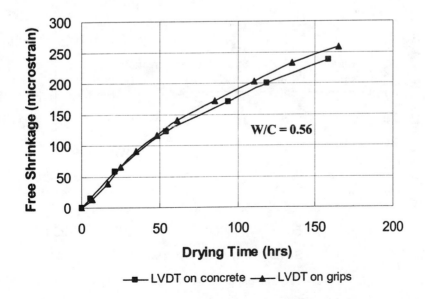

Figure 7 Free shrinkage for the two methods of LVDT attachment

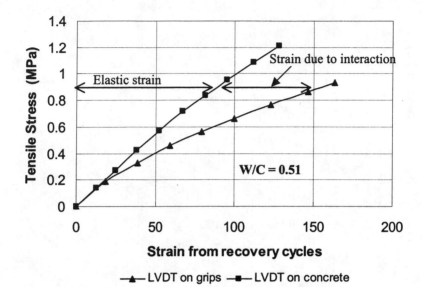

Figure 8 Stress-strain diagram for the two methods of LVDT attachment (w/c = 0.51)

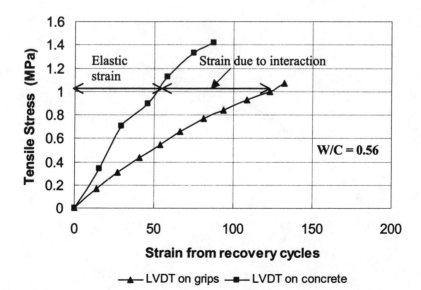

Figure 9 Stress-strain diagram for the two methods of LVDT attachment (w/c = 0.56)

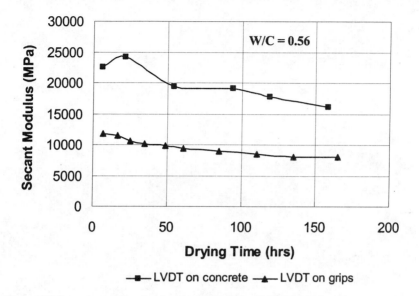

Figure 10 Secant modulus for the two methods of LVDT attachment

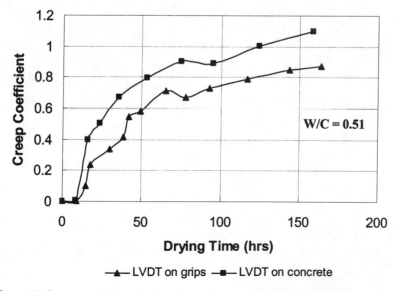

Figure 11 Creep coefficient for the two methods of LVDT attachment

# SP 206-13

# Using Acoustic Emission to Monitor Damage Development in Mortars Restrained from Volumetric Changes

## by T. Chariton and W. J. Weiss

<u>Synopsis:</u>
Early-age cracking can occur in cementitious materials when volumetric changes caused by temperature or moisture fluctuations are prevented by the surrounding structure. This paper describes a preliminary study in which early-age damage development was monitored in restrained cementitious mortar specimens using acoustic emission. A steel testing frame was used to provide passive restraint to uni-axial specimens. Free shrinkage characteristics were measured using geometrically similar specimens. Acoustic sensors were mounted on the surface of the mortar and acoustic activity was recorded continuously. The experiment revealed that the acoustic activity in the free and restrained specimens was initially similar, however the restrained specimens generated an increase in acoustic activity at later ages. This presumably occurs as a result of the increase in the residual stress to strength ratio. The age of visible cracking was observed to correlate well with a discrete, sudden increase in acoustic activity. Acoustic energy was used to indicate the change in the properties of the acoustic events as the concrete grows closer to the age of cracking. The location of damage was determined in the specimens using a linear approach that corresponded well with visual observation.

<u>Keywords:</u> acoustic emission; cracking; damage; early-age; fracture; microcracking; moisture profile, restrained shrinkage; shrinkage

**Todd Chariton** was an undergraduate student in the School of Civil Engineering at Purdue University at the time this study was completed. He currently anticipates graduation in December of 2001.

ACI Member **Jason Weiss** is an assistant professor in Civil Engineering at Purdue University. He earned his BAE from Penn State and MS and Ph.D. from Northwestern. He is secretary of ACI committee 123 and 231 and is a member ACI 209, 446, and 522. His research interests include durability, early-age materials, fracture mechanics, and non-destructive testing.

## INTRODUCTION

It is known that when concrete dries it shrinks. If concrete is prevented from shrinking freely, residual tensile stresses develop which, if high enough, can result in cracking. This can be especially problematic at early-ages while the concrete is gaining strength. Historically, models have been proposed to predict the age of cracking based on comparing the time-dependent tensile strength of concrete with the residual stress level that develops as a result of restraint as shown in Figure 1. In this approach stress develops as a function of the shrinkage strain, elastic modulus, and stress relaxation (1). Cracking is assumed to occur when the residual stress that develops exceeds the tensile strength of concrete (i.e., the intersection of stress and strength). While this method provides a valuable phenomenological description of shrinkage cracking, the accurate prediction of cracking can be more complicated as described in the following sections.

## QUANTIFYING MICROCRACKING USING ACOUSTIC EMISSION

Recently many researchers have studied how material properties influence the age of cracking. Researchers have focused almost exclusively on either assessing the level of residual stress that develops as a result of restraint (2,3,4,5,6) or on developing models to better understand the mechanisms of early-age cracking (1,7,9). This information has been used to develop models to predict whether or not cracking should be expected for each mixture and the severity of cracking that may be expected (7,8,9).

This paper describes an approach to quantify cracking that occurs when concrete is restrained from shrinking freely. Figure 2a illustrates a case in which the residual stress that develops never exceeds the time-dependent strength (or fracture resistance) of the concrete. Consequently, no cracking would be expected. Alternatively, Figure 2b illustrates the case in which the stresses that develop exceed the strength and a large visible crack develops. While these options are consistent with a strength-based approach and are a reasonable first approximation, they do not explicitly consider the development of micro-cracking. As a result, in this approach 'failure' is inherently assumed to correspond to the development of visible surface cracking. However, this does not consider that internal micro-cracking may have developed due to the restraint as shown in Figure 2c. It is the central idea of this paper that unseen microcracks can develop in concrete in response to restraint. Although invisible to the eye, these cracks can accelerate deterioration, thus ultimately reducing long-term performance.

This idea that unseen damage is occurring in concrete is not new. Previous research (10,11,12) showed that micro-cracking in concrete may occur at the interface of the aggregate and matrix even at extremely low stress levels. As the load is increased, micro-cracking continues to develop at a slow rate until the stress reaches approximately 40-50% of the maximum stress, after which time a dramatic increase in interfacial bond cracking occurs. This effect is responsible for producing the well-known non-linear nature of the stress-strain response of concrete. As the stress approaches 60-70% of the maximum stress cracks begin to form in the matrix and at stresses near 85-95% of the peak stress these cracks begin to coalesce forming a localized crack. While these trends were observed for concrete tested in compression, similar results were obtained for concrete tested in tension (13).

In restrained shrinkage, the extent of micro-cracking that develops as a result of restraint may be significant. Figure 1b illustrates the ratio of residual stress and strength to illustrate that the concrete can be at relatively high stress levels for a long period of time. At these high stress levels micro-cracking may occur. While this can explain micro-crack development in the restrained specimen, it can not be assumed that no cracking occurs in the free shrinkage specimen. Bazant and Chern (14) and Granger et al. (15) have both described how the development of cracking at the surface of the concrete could occur due to the high level of stresses that develop which will act to reduce the overall free shrinkage that is measured. This implies that a combination of restraint and moisture gradients would be responsible for cracking in both specimens.

## EXPERIMENTAL PROGRAM

In this research program tests were conducted using two mortar mixtures. Both mortars consisted of 45% fine aggregate by volume with a water-to-cement ratio (w/c) of 0.5. Type I cement was used in the first series of specimens (L-I-45) while Type III cement from the same source was used in the second series (L-III-45). Both cements were obtained from the same source and the primary difference between these cements was their fineness. A Blaine fineness of 360 $m^2$/kg was measured for the Type I cement and a fineness of 535 $m^2$/kg was measured for the Type III cement. (Note both cements had essentially the same chemical make-up with $C_3S$ of 56-60%, $C_2S$ of 11-13%, $C_3A$ of 8-9%, and a $Na_2O$ equivalent alkali content of 0.5-0.6%). The mortar was mixed using a standard Hobart mixer. The mixture was placed in the forms and vibrated. The specimens were sealed with an acrylic sheet and stored under wet burlap at room temperature 23°C +/- 1°C for 24 hours at which time they were demolded, prepared (as described in the following paragraph), and moved to an environmental chamber which was maintained at 23°C +/- 1°C and 50% +/- 2% RH for the remainder of the study.

Two different specimen geometries were used during this study as illustrated in Figures 3 and 4. The first specimen geometry was used for free shrinkage measurements determined in accordance with ASTM C-157 with a 25 mm x 25 mm cross section, a 250 mm gage length, and a 275 mm overall length (1 in x 1 in x 11 in). Steel gage studs were placed in the mortar bars at the time of casting. To simplify the drying condition, two sides and two ends of the specimen were sealed with aluminum tape at the time of demolding. This enabled preferential drying to occur from the unsealed (top and bottom) faces as shown in Figure 3. The restrained specimens were also prepared at the time of demolding by sealing the sides and the ends to prevent moisture loss. This results in the free and restrained specimens having identical drying surface to volume ratios.

In the restrained system, length change was prevented through the use of a steel frame. The mortar was cast in a barbell shape. As the mortar dried it would attempt to shrink causing the steel to compress. It should be noted that the frame is not perfectly rigid, therefore the stress in the steel resulting from the shrinkage of the specimen will cause it to undergo some displacement. Through tensile testing the elastic modulus of the steel was the determined to be 210 GPa. Based on simple calculations it can be shown that the steel is roughly 10x as stiff as the mortar.

The acoustic emission system used for these experiments was a Vallen Model AMS4. The broad-band sensors were attached to the concrete surface using grease and an elastic band was used to apply a slight pressure to the sensor to maintain its contact with the concrete. The transducers were used to convert the detected waves into electrical signals, which were amplified, processed, and recorded by the Vallen acquisition 32 software and analysis was performed using the Visual-AE software.

In addition to the shrinkage tests, split-cylinder testing was performed at various ages for each series to determine an approximate tensile strength using the approach outlined in ASTM C-496. It should be noted that the cylinders initially cast as a 100 mm x 200 mm cylinder and subsequently cut to minimize any thickness effects to a size (100 mm x 38 mm) using a wet saw with a diamond blade before testing. Loading was applied to the cylinder through wooden strips at a rate of 1 MPa/min.

## DISCUSSION OF MECHANICAL RESULTS

The two unrestrained (free shrinkage) specimens that were not equipped with acoustic sensors were monitored for free shrinkage and weight loss. Measurements were taken twice a day for the first two days after demolding, and once a day after an age of three days. The resulting plots of free shrinkage and weight loss over time are shown in Figure 5. As expected the shrinkage rate is high at early ages and begins to level off at later ages. This can be attributed to the fact that at early ages the surface of the specimen dries rapidly and a very steep moisture gradient is established. As the specimen is exposed to drying for a longer time the diffusion of water from within the specimen is more difficult and takes an increased amount of time. As most of the water is lost at later ages the free shrinkage approaches the asymptotic limit of the ultimate shrinkage value. Comparing the plots of L-I-45 and L-III-45, it can be observed that these curves deviate over time. It is also apparent however, that the L-III mixture experiences a significantly lower weight loss than the L-I mixture. Both of these observations can be attributed to the fact that the L-III cement is finer (i.e., the cement particles are much smaller). As water is added to the system the outer surface of each cement grain is in contact with pore water resulting in a more rapid hydration per volume of cement. As a result, more water has chemically reacted (hydrated) in the L-III specimens than in the L-I specimen at the same age. This water is then not available to evaporate; consequently, the specimen has a substantially lower weight loss. In addition, it can be argued that the finer cement (L-III) may have a finer pore structure resulting in the generation of higher capillary stresses and a greater shrinkage (16).

In order to determine the strength development of the L-I and L-III mixtures, split-cylinder tests were performed at 1, 3, 7, 14, and 28 days. The results of the strength tests are provided in Figure 6. From the plot, it is apparent that the L-I gains strength more slowly up to an age of 7 days, whereas the L-III has approximately 90% of its ultimate strength at one day. After 7 days, both mixtures achieve approximately the same split tensile strength, and this strength does not substantially increase after this time. The difference in rate of strength gain can again be explained by the fact that the L-III cement particles are smaller than those of the L-I cement. Again, the greater surface area for the L-III mixture results in increased rate of hydration that is consistent with a higher initial strength.

While one would commonly think that the higher strength material would be more desirable, it can be noted that cracking was observed at an earlier age for the L-III restrained specimen. Initially this may appear to be counter-intuitive since the shrinkage is similar at early ages. However the mixtures made using the finer cement (L-III) will have a more rapid increase in elastic modulus and a lower creep relaxation both of which would result in less stress relaxation for the same free shrinkage (1,9,16). This implies that strength and free shrinkage are not the only criterion that should be considered in the development of specifications to limit early-age cracking.

## DISCUSSION OF ACOUSTIC EMISSION MEASURMENTS

Figures 7-12 show some of the acoustic emission measurements that were obtained from tests in the L-I-45 and L-III-45 test series. The number of acoustic events (with an amplitude greater than 30 dB) is plotted versus specimen age in Figures 7 and 8 for the L-I and L-III specimens respectively. In both plots the acoustic events are seen to rise from 1 day when the specimens were demolded, the sensors were placed on the specimen, and drying began. It can be observed that numerous acoustic events occur in both the restrained and unrestrained specimens for each mixture. The similarity between the plots for the restrained and unrestrained specimens indicates that micro-cracking is occurring in both specimens. While it is easy to understand that micro-cracking could occur in the restrained specimen, the micro-cracks have also been proposed to occur in the free shrinkage specimen occur as a result of moisture gradients (14,15). As stated previously, the water at the surface evaporates more readily than the water from the concrete core. Therefore, the surface is subject to a higher degree of shrinkage, which is restrained by the 'core' concrete. In the restrained specimen, micro-cracks continue to develop until a significant jump in the number of acoustic events is observed (approximately 2.7 days for L-III-45 and 4 days for L-I-45). It is speculated that this discrete increase in acoustic activity occurs as the micro-cracks coalesce to form a larger, visible crack in the specimen. The dashed lines indicate the age at which cracking was observed on these specimens. It should be noted for the L-III specimen that cracking occurred between 2.5 days and 2.9 days, when the specimen was not being monitored to determine the precise time of cracking. Cracking was observed immediately after the discrete increase in acoustic activity in the L-I series although it should be mentioned that the size of the crack at these early ages is very small and a microscope is typically needed to locate the crack. Slightly after the time that the crack has formed, the restrained specimen is now essentially able to shrink freely, and the acoustic activity curve continues along a slope that is nearly identical to that of the unrestrained specimen.

While monitoring the development of acoustic events provides some useful information, it can only provide limited information. It does not provide information about the physical nature of the event. Researchers [17,18,19] have previously reported that by characterizing the shape of the acoustic waveform, further information can be learned about the crack. To illustrate this, the concept of acoustic energy was employed. Acoustic energy is defined simply as the absolute value of the area under the acoustic waveform. More recent research by Landis and Whittaker [20] has linked the concept of acoustic energy with fracture energy in a specimen. While the acoustic energy is typically only a fraction of the mechanized fracture energy, a linear relationship between the two has been reported. Figures 9 and 10 show the plot of the acoustic energy versus specimen age for the restrained and unrestrained specimens for both mixtures. Similar to the graph of acoustic events versus specimen age, it can be seen that there is a discrete rise in acoustic energy just before the time that cracking is visually observed. It can also be observed from these figures that the acoustic energy of restrained specimens diverges from the acoustic energy of the unrestrained specimens well before visible cracking occurs. It is currently thought that this can be attributed to the fact that the residual stress that is developing in the specimen is approaching the tensile strength. It has been argued that as the specimen dries and shrinks, there is an increase in the number of micro-cracks. In addition it can be argued that the cracks that are present continue to grow deeper into the specimen. It can be seen that the acoustic energy continues to show an increase after the visible cracking is observed. This may be attributed to two possible effects, either the crack has not propagated through the complete thickness of the specimen or there are "crack-bridging" effects that transfer stress at the crack face.

In addition to cumulative plots of acoustic events and energy, spatial resolution of the acoustic activity was also performed. Since the specimen was linear, the use of two sensors (one on either end of the specimen) allowed the location of acoustic events to be determined by comparing the arrival time of an event at both sensors. A plot of the location of acoustic events at various ages, and a corresponding picture of the location of the observed crack is provided in Figures 11 and 12. When viewing Figure 11 it is important to recall that the L-I-45 specimen was observed to crack at an age of approximately 4 days. Figure 11 shows the location of the acoustic events during three time ranges (1-3 days, 3-5 days, and 5-7 days). It can be seen that at early and late ages there is relatively uniform acoustic activity along the length of the specimen, however between the ages of 3 and 5 days there is a large increase in the number of events at a location that is almost identical to that of the observed crack. It is apparent that the majority of the acoustic events occur at the location where the crack is observed which would be consistent with the idea of crack localization or crack band formation. Similarly, the L-III-45 specimen showed a uniform distribution of acoustic activity up to 2.5 days (Fig. 12). After 2.5 days the acoustic activity increases sharply at the location where the crack became visible. Again, after visible cracking is observed the events are relatively evenly distributed throughout the specimen with no discernable concentration of events.

## SUMMARY AND CONCLUSIONS

In conclusion, this paper has illustrated that acoustic emission can provide information on damage that develops as a result of restraint. It was illustrated that cementitious mortars demonstrate substantial acoustic activity during drying. This was observed for both the unrestrained (free) or restrained specimens. This can be attributed to micro-cracking which is consistent with previous explanations by Bazant and co-workers (14) and Acker and co-workers (15). It can however be seen that restrained specimens show a sudden increase in acoustic events which occurs slightly before cracks are noticed visually. More substantial micro-cracks occur due to restraint as the residual stress increases. These stresses are distinguished from other micro-cracks with events having a higher energy. To illustrate the difference in acoustic waves generated during cracking, the acoustic energy was plotted for both restrained and unrestrained specimens. It can be seen that these curves deviate at an age that is substantially earlier than the age at which cracking is observed on the surface. This would be consistent with the explanation of micro-cracking at higher load levels. In addition, the use of acoustic energy is consistent with the idea that energy associated with the acoustic waveform can be related to specimen distress.

## ACKNOWLEDGEMENTS

The authors gratefully acknowledge support received from the Purdue Research Foundation and the Center for Advanced Cement-Based Materials (project C-1). Furthermore, the second author expresses his continuing appreciation to Professor Shah for his instruction, counseling, and mentoring, which have been instrumental in guiding him in the development of his professional career.

## REFERENCES

1. Weiss, W. J., (1999) "Prediction of Early-Age Shrinkage Cracking in Concrete Elements," Ph.D. Dissertation, Evanston, IL

2. Altoubat, S. A., and Lange, D. A., (1997) "Early-Age Creep and Shrinkage of Fiber Reinforced Concrete for Airfield Pavement," Aircraft Pavement Technology, ed. F. V. Hermann, pp. 229-243

3. Bloom, R., and Bentur, A. (1995) "Free and Restrained Shrinkage for Normal and High Strength Concretes", ACI Materials Journal, Vol. 92, No. 2, pp. 211-217

4. Kovler, K. (1994) "Testing System For Determining The Mechanical Behavior Of Early Age Concrete Under Restrained And Free Uniaxial Shrinkage", Materials and Structures, RILEM, London, U.K., 27(170), 324-330

5. Sellevold, E., (1999) "Report 2.4: Mechanical Properties of Young Concrete: Evaluation of Test Methods for Tensile Strength and Modulus of Elasticity. Determination of Model Parameters," NOR-IPACS Report, SINTEF, 45 pp.

6. Toma, G., Pigeon, M., Marchand, J., Bissonnette, Bercelo, L., (1999) "Early-Age Autogenous Restrained Shrinkage: Stress Build Up and Relaxation," Self-Desiccation and Its Importance in Concrete Technology, eds. Persson, B., and Fagerlund G., pp. 61-72

7. Gryzbowski, M., and Shah, S. P. (1989) "Model to Predict Cracking in Fiber Reinforced Concrete due to Restrained Shrinkage," Magazine of Concrete Research, Vol. 41, No. 148, 125-135

8. Weiss, W. J., Yang, W., and Shah, S. P., (1998) "Shrinkage Cracking of Restrained Concrete Slabs", ASCE Journal of Engineering Mechanics, Vol. 124, No. 7, pp. 765-774

9. Weiss, W. J., Yang, W., and Shah, S. P., (2000) "Influence of Specimen Size/Geometry on Shrinkage Cracking of Rings", ASCE Journal of Engineering Mechanics, Vol. 126, No. 1, pp. 93-101

10. Hsu, T.T.C, Slate, F. O., Struman, G. M., and Winter, G., (1963), "Microcracking of Plain Concrete and the Shape of the Stress Strain Curve", Journal of the American Concrete Institute, Vol. 60, pp. 209-224

11. Shah, S. P., and Chandra, S., (1968) "Critical Stress, Volume Changes, and Microcracking of Concrete," Journal of the American Concrete Institute, 65, 770-781

12. Darwin, D., and Attigobe, E. K., (1983) "Load Induced Cracks in Cement Paste," Proceedings of the 4th Engineering Mechanics Specialty Conference, Purdue University

13. Li, Z., Kulkarni, K., and Shah, S.P., (1993) "New Test Method for Determining Post-Peak Response of Concrete Specimens Under Uniaxial Tension, Experimental Mechanics, 33, pp. 181-188

14. Bazant, Z. P., and Chern, J., C., (1985), "Concrete at Variable Humidity: Constitutive Law and Mechanisms", Materials and Structures, RILEM, Paris, Vol. 18, Jan., pp. 1-20

15. Granger, L., Torrenti, J. M., and Acker, P., (1997) "Thoughts About Drying Shrinkage: Experimental Results and Quantification of Structural Drying Creep," Materials and Structures, Vol. 30, pp. 588-598

16. Shah, S. P., and Weiss, W. J., (2000) "High Strength Concrete: Strength, Permeability, and Cracking," Proceedings of the PCI/FHWA International Symposium on High Performance Concrete, Orlando Florida, © 2000, pp. 331-340

17. Ouyang, C., Landis, E., and Shah, S. P., (1991) "Damage Assessment in Concrete Using Quantitative Acoustic Emission," ASCE Journal of Engineering Mechanics, Vol. 117, No. 11, pp. 2681-2698

18. Ohtsu, M., (1996) "The History and Development of Acoustic Emission in Concrete Engineering," Magazine of Concrete Research, Vol. 48, No. 177, 321-330

19. Yoon, D.-J., Weiss, W. J., and Shah, S. P., (2000) "Assessing Corrosion Damage in Reinforced Concrete Beams Using Acoustic Emission." J. of Engrg. Mechanics Div., ASCE, Vol. 126, No. 3, pp. 273-283

20. Landis, E. N., and Whittaker, D. B., (2000) "Acoustic Emission as a Measure of Fracture Energy", Fourteenth Engineering Mechanics Conference, ASCE, May 21-24 Austin Texas

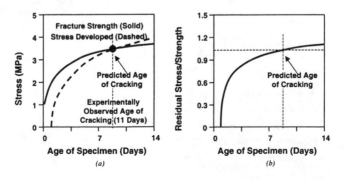

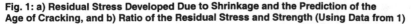

Fig. 1: a) Residual Stress Developed Due to Shrinkage and the Prediction of the Age of Cracking, and b) Ratio of the Residual Stress and Strength (Using Data from 1)

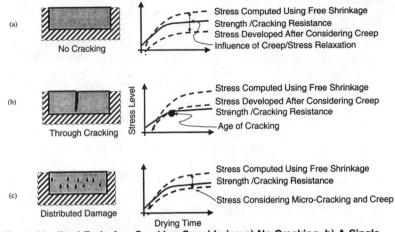

Fig. 2: Idealized Early-Age Cracking Considering a) No Cracking, b) A Single Crack, and c) Distributed Damage Development without Localized Cracking

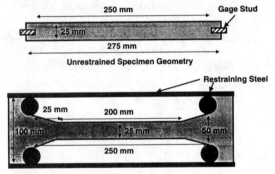

Fig. 3: Specimen Geometry

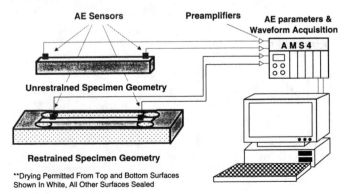

Fig. 4: Acoustic Emission Experimental Set-up

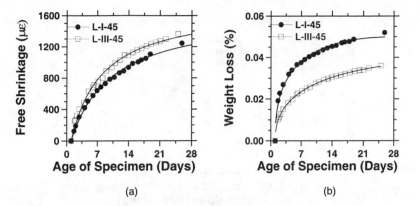

(a)                                    (b)

Fig. 5: a) Free Shrinkage Versus Time and b) Weight Loss Versus Time

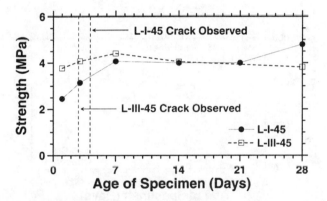

Fig. 6: Split-Tensile Strength Development Versus Time

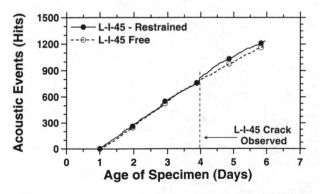

Fig. 7: L-I-45 Cumulative Acoustic Events Development as a Function of Time

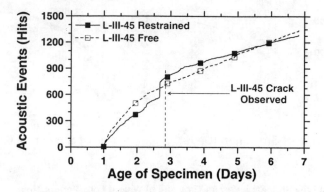

Fig. 8: L-III-45 Cumulative Acoustic Events Development as a Function of Time

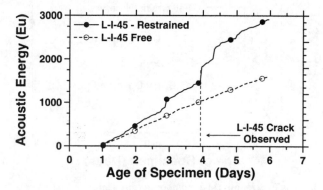

Fig. 9: L-I-45 Acoustic Energy Development as a Function of Time

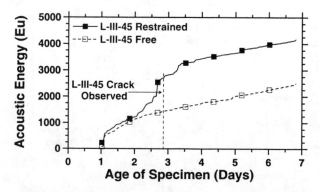

Fig. 10: L-III-45 Acoustic Energy Development as a Function of Time

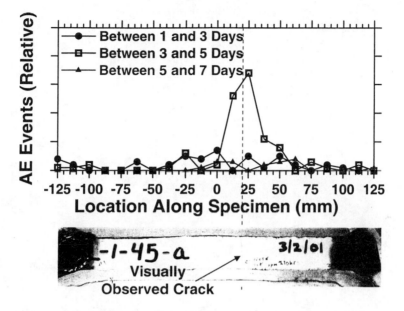

Fig. 11: L-I-45 Spatial Resolution of AE Events Along the Restrained Specimen At Different Times

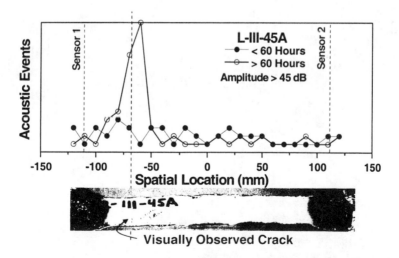

Fig. 12: L-III-45 Spatial Resolution of AE Events Along the Restrained Specimen At Different Times

# Modeling of Restrained Shrinkage Cracking in Concrete Materials

## by S. A. Mane, T. K. Desai, D. Kingsbury, and B. Mobasher

**Synopsis:** An experimental study was conducted to evaluate the restrained shrinkage cracking in plain and fiber reinforced concrete. The experiment utilizes a constant humidity chamber holding the restrained shrinkage specimens. The chamber is subjected to constant flow of air around the specimens. The strain in the restraining steel and the crack width in the concrete samples were monitored continuously. The experimentally obtained results are affected by geometry of the specimen, the humidity and shrinkage conditions, and the restraint offered by stiffness of the steel ring. In addition, concrete properties such as the stiffness, shrinkage and creep affect the response.

In order to better understand the restrained shrinkage of concrete under the proposed test method and eliminate the influence of test conditions, an analytical approach was developed. The model incorporates key influential parameters of shrinkage, creep, aging, and microcracking, in the stress analysis of a restrained concrete section. The theoretical model was used to calibrate and interpret the experimental test results.

Keywords: cementitious composites; cracking; creep; fibers; restrained shrinkage

**S. A. Mane** and **T. K. Desai** are Graduate Research Assistant at the Department of Civil and Environmental Engineering at Arizona State University.

**D. Kingsbury** is Manager of Integrated Mechanical Testing Laboratories at the College of Engineering and Applied Sciences, Arizona State University.

ACI member **Barzin Mobasher**, Ph.D. is an associate professor of civil and environmental engineering at Arizona State University. He is a member of ACI Committee 544, Fiber Reinforced concrete, ACI 446, Fracture Mechanics. His research activities include fiber reinforced concrete, toughening mechanisms, and modeling of durability.

## INTRODUCTION

Experimental and theoretical procedures for measurement of cracking potential of concrete mixtures were developed. Cementitious materials shrink due to loss of moisture from capillary and gel pore microstructure when subjected to a dry environment. Chemical shrinkage, caused by a reduction of the volume of the products in comparison to the reactants, also occurs due to hydration of cement. It is also well known that the cementitious materials have very poor properties in resisting tensile stress. If concrete is restrained from shrinkage, tensile stresses generate and ultimately lead to cracking. The extent of shrinkage depends on many factors including the material properties, the temperature, relative humidity, the age of concrete when it is subjected to drying, and the geometry of the structure (1).

One possible method to reduce the adverse effect of cracking is the addition of fibers. Quantitative methods for assessing performance of fibers to resist cracks in cementitious composites are lacking. In addition, use of high performance concrete mixtures has become quite popular requiring the need to develop performance based specification criteria. This necessitates the development of analytical understanding of the results of this and other shrinkage tests.

A restrained shrinkage test configuration has been recently adopted to measure the susceptibility of concrete mixtures to cracking. This procedure is referred to as AASHTO PP34-99 (2) and is similar to other restrained tests developed by other researchers (3). In this procedure the experimentally obtained results are affected by geometry of the specimen, properties of concrete, the humidity and shrinkage conditions, the restraint offered by stiffness of the steel ring, the effect of microcracking, and finally creep parameters of concrete.

Recently various theoretical approaches for modeling the restrained shrinkage cracking of concrete have been developed (4). These models try to address the interaction between materials property and the shrinkage characteristics of concrete.   In order to better understand the restrained shrinkage of concrete under the proposed test method and eliminate the influence of test conditions, an analytical approach was developed.  The model incorporates many of the key influential parameters of shrinkage, creep, aging, and microcracking, with the stress analysis of a restrained concrete section. The proposed method allows continuous monitoring of the stress in the concrete section, and the extent of cracking in the concrete.

## RESEARCH METHODOLOGY

To assess cracking due to restrained shrinkage, a ring type test specimen similar to the AASHTO specification PP34-99 was used.  This test simulates restrained shrinkage cracking. The ring consists of a 2.625 in. (66.675 mm) thick annulus of concrete cast around a rigid steel ring 0.5 in (11.2 mm) in thickness with a diameter of 11.4 inches (289.56 mm) and a height of 5.25 inches (133.35 mm). One concrete ring for each restrained shrinkage test was prepared. Two strain gages were mounted on inside surface of steel ring at the mid height level and 90 degrees apart.  The response of the strain gages were averaged and reported as the strain in the steel.  In addition a control steel ring was instrumented with strain gages and its response was measured as a baseline measurement to account for the drift in the response.

### Mix Design Materials

Concrete mixtures with a water-to-cementitious solids ratio (w/c) of 0.40 – 0.55 and having a slump of 2-2.5"(50-65 mm) were designed.  Three different mixtures were considered including flyash blended cement, fiber reinforced, and a control concrete mixture. Table 1 presents the summary of the mix design information.  For all the concrete mixes, superplasticizer was used at a dosage of 200 ml per 100 kg of total cementitious materials. The superplasticizer was mixed in water prior to the start of the mixing procedure. The coarse aggregate used was 20 mm crushed gravel with a sand/total aggregate ratio of 0.76. The mix quantities listed in table 1 are based on the saturated surface dry (SSD) methodology.

Two types of AR glass fibers were obtained from VETROTEX, Cem-FIL SAINT-GOBAIN.  High dispersion (HD) AR Glass fibers were used in chopped strand form.  These fibers are formulated for mixing with concrete, mortar, and other cement-based mixes where a uniform dispersion is needed. The dosage of glass fiber in this category was limited to 0.6 kg/m$^3$ and aimed at early age strength and plastic shrinkage crack control.  High Performance (HP) AR glass fibers were also used.  The dosage of glass fiber in this category was limited to 5.0 Kg/m$^3$ and aimed at long-term strength and ductility.  A total of four number of specimens were prepared, one with control concrete, one with HP fibers, one with HD fibers, and one with 25% class F flyash which was studied extensively in an earlier study. (5)

In order to achieve a desired slump and a cohesive concrete mix, the following mixing procedure was adopted.  The dry coarse aggregates and sand were introduced in the mixer and blended for 90 seconds with one half of the mixing water.  Then cement, flyash or fibers (depending on the case), and remaining water were added to the mixer and blended for 3 additional minutes. Prior to casting lubricant oil was applied on the steel surfaces to minimize the frictional resistance.  All the specimens were filled in two layers with proper compaction in between the layers. A vibration table was used to help with the consolidation of the fresh mixture in the molds. After 24 hrs, specimens were demolded and placed in a curing room at 90% (RH) and $23^0$ C (70 $^0$ F) for 3 days of curing.    The specimens were placed in the shrinkage chamber immediately after the 3 days of curing.  The relative humidity within the chamber was maintained at $35\pm1\%$ RH.  Drying was allowed from outer, circumferential and  top surface of the specimen whereas the bottom rested on the wooden plate.

In addition to the shrinkage specimens, a cylinder of 2.35 in (60 mm) diameter and 5.5 in (140 mm) high were prepared for direct tension test.  A circular saw cut of 8 mm is made along the circumference of the cylinder with a water-cooled diamond blade.  Results of the tension test results were used in the theoretical modeling of the shrinkage test.

**Experimental Setup**

Fig. 1 shows the environmental chamber containing two racks. Four specimens were placed in the chamber with two specimens at each level. The temperature of the environmental chamber was maintained at $40.1^0$ C ($100\pm5^0$ F).  In addition a fan was used to continuously circulate the air through the closed chamber.  The speed of the fan may be controlled to simulate the effect of wind conditions.  However, in this test procedure, a constant slow speed was used to ensure proper mixing of the air and temperature distribution throughout the chamber.  The strain gages were connected by cable to a computer based data acquisition board.    Data were collected for up to 90 days using a continuous recording of the strain gages attached to the inner surface of the steel. A LABVEIW program was written to measure the strain gage readings. The response from the strain gage readings was recorded for every minute for the duration of the test.

In order to measure the crack width, a special microscope set up was used. This set up is shown in Fig. 2 and is able to scan the entire surface of the specimen. The surface of the specimen was examined for new cracks and the measurements of the widths of already existing cracks was performed every 24 hrs during the first few days after cracking. The images of crack propagation were collected using a traveling microscope with a 10X objective. These images were converted to a digital image using a frame grabber and stored in the computer for further analysis.

Fig. 3a and 3b shows the experimental setup of direct tension test on cylinders. The cylinder were held at each end by a special designed grips which hold the specimen and is connected to actuator by 0.25 in (6.25 mm) round pins. An extensometer was placed across the notch to measure the displacement in the direction of load. Fig. 3b gives close view of experimental setup with the grips holding the specimen and extensometer mounted near the notch. The test was conducted under the extensometer control to measure the post peak response.

## EXPERIMENTAL RESULTS AND DISCUSSION

In the initial period, the specimen has sufficient moisture stored in the capillary pores, parts of which is being used for hydration process. Loss of this moisture does not contribute to significant shrinkage. During the first few days of shrinkage deformation, reduction in the relative humidity throughout the cross section decreases the moisture content, resulting in contraction of the specimen around the steel ring and generation of clamping stresses. These stresses may initiate and propagate cracks. In the present study a crack was defined as observable if it could be detected by means of the microscope. Formation of partial cracks through the width and thickness of the specimens were accounted for in the measurements. Cracking took place in all the specimens, however at different times and under different crack growth conditions.

Fig. 4a and 4b represent the response of shrinkage tests recorded by the strain gages. Results of the control concrete are compared with the flyash concrete mixtures in fig. 4a. The zero time (in days) corresponds to the time when the specimens were first placed in the environmental chamber. The shrinkage results in concrete experiencing tensile stresses, which are offset by the compressive stresses in steel. The strains build up until cracking of concrete releases the stress, causing a gradual recovery of the stresses as a function of time. This gradual recovery is dependant on the creep properties of concrete. As shown in fig. 4a it is observed that after almost 7 days of shrinkage, the deformations recorded by strain gages indicated existence of cracking. These were not however detectable or observable by the microscope. This stage of deformation was referred to as the microcracking phase. The control specimen had the first crack at the age of 13 days and it measured at 0.008 mm wide. This specimen developed a through crack at the age of 25 days and at that stage the crack width was measured as 0.235 mm.

The fiber reinforced concrete composites had more resistance to the crack propagation than control and flyash concrete. However, due to the low volume fraction of the fibers (in the range of 0.2%), one cannot expect to see a significant impact due to the use of fibers. Fig. 4b shows the strain on the steel as a function of time for glass fiber reinforced concrete specimens. Note that initial increase in the strain is because of the high temperature when the specimens are placed in the environmental chamber. Gradually as the concrete shrinks under given condition, the strain gage mounted on the steel ring show compression. With the initiation of the crack in the specimen, the concrete relaxes. It is evident from the plot that the specimen with HD12 glass fibers had first visible crack at the age of 10 days. Whereas the sample with HP12 glass fibers had no measurable crack.

After cracking, the uncracked portion of the specimen will continue to shrink, resulting in further widening of the crack. The load carrying capacity of the cracked section decreases with the widening of the crack and the stress in both concrete and steel will decrease. The post peak response of the concrete is quite important in the modeling of this phase of response. In the case of plain concrete or concrete with flyash, the only source of resistance to cracking opening is due to aggregate interlock. Furthermore, the only resistance against shrinkage is the friction between the concrete and steel ring.

Fig. 5 represents the crack width as a function of time. As shown in figure, the specimen with 25% flyash has less resistance to crack propagation. The first crack formed at the age of 10 days and it became a through crack at 14 days. The crack widths were measured as 0.0152 mm and 0.9374 mm wide respectively. This early cracking may be due to lack of hydration of flyash concrete in three days of curing, and also the changes to the pore structure of the hydrated microstructure. The HD fiber has the first crack at the age of 10 days measured as 0.0068mm wide. This crack had propagated along half of the height of the specimen and was not a through crack, whereas the HP fiber concrete mixtures did not have any cracks measured by microscope at the age of 30 days. The glass fiber concrete specimens have more resistance to the crack propagation than fly ash concrete mix since they showed no observable, or small scale cracking at ages of up to 30 days.

It was observed that once the crack became visible, it propagated quickly through the whole thickness of the concrete ring. In addition, in the plain and flyash concrete specimens only a single crack was developed throughout the observation period. The single crack hindered the formation of any other crack in the specimen as its width was continuously increased. The final values of the crack widths on the outer surface, exposed to drying, and on the inner surface sealed off by the steel ring, were found to be almost equal. These two observations indicate that the dimensions of specimen were sufficient so as to obtain a uniform stress distribution conditions in the cross section of the concrete ring.

## THEORETICAL MODEL, RESULTS AND DISCUSSION

The experimental results are affected by the geometry of the specimen, the properties of concrete, the humidity and shrinkage conditions, the restraint offered by the stiffness of the steel ring, the effect of microcracking, and finally creep parameters of concrete. In order to better understand the restrained shrinkage of concrete under the proposed test method and eliminate the influence of test conditions, an analytical approach was developed. The model incorporates many of the key parameters of shrinkage, creep, aging, and microcracking, with the stress analysis of a restrained concrete section. The proposed theoretical model is used to predict the strain-time history and the cracking progress in a cementitious composite under imposed strains due to shrinkage. The material modeling can be applied to plain as well as fiber reinforced concrete.

The humidity profile through the thickness of the section was assumed to follow the Fick's law of diffusion with $h_s$ and $h_0$ representing the humidity at the surface and the interior section of the specimen. The humidity at node i, h(i) is represented by distant x from the surface of the specimen and *erf(x)* represents the error function (6). This relationship is shown in Fig. 6 for a specimen experiencing a relative humidity of 70% at the inside and 20% on the outside. A cubic function was used to relate the shrinkage strain as a function of the humidity profile throughout the thickness as shown in Equation 1.

$$h(i) = h_s - (h_s - h_0) * erf(x)$$
$$\epsilon_{sh}(i) = \epsilon_{sh}^u * (1 - h(i)^3)$$

(1)

The two equations presented above are independent of time. In order to produce the relationship as a function of time, several simplifications wee implemented. Although it is possible to predict the ultimate shrinkage of a concrete mixture based on its constituents such as models by ACI, CEB FIP (7), or Bazant-Panula,(8), in the present study a fixed value of $\varepsilon^u_{sh}$ was used. Clearly, this is an approximation to the current mixes, and it is expected that the ultimate shrinkage strain for each mixture based on the free shrinkage response be used. This strain distribution was decomposed to be a function of two variables, the average shrinkage strain and the average curvature across the cross section defined as:

$$\epsilon_{sh}(t,x) = \bar{\epsilon}_{sh}(t) + \bar{K}_{sh}(t)\, x_i$$

$$\bar{\epsilon}_{sh}(t) = \frac{1}{n}\sum_{i=1}^{n}\epsilon_{sh}(t,i) \tag{2}$$

$$\bar{K}_{sh}(t) = \frac{1}{n}\sum_{i=2}^{n}\left[(\epsilon_{sh}(t,i)-\epsilon_{sh}(t,i-1))/(x_i - x_{i-1})\right]$$

The shrinkage strain profile represented by equations 2 is a linearized representation of the strain distribution presented in equation 1. The relationship of ultimate shrinkage strain and curvature as a function of time were then assumed to be based on the ACI equations [5] represented in equation 3.

$$\bar{\epsilon}_{sh}(t) = \bar{\epsilon}_{sh} * \frac{t}{35+t}$$

$$\tag{3}$$

$$\bar{K}_{sh}(t) = \bar{K}_{sh}\frac{t}{35+t}$$

To obtain an estimate of the stresses in concrete and steel based on the generalized elastic solutions, a case of shrink fit was considered. The solution for stress analysis of a ring of internal and external diameter "a" and "b", subjected to internal and external pressures, $p_i$ and $p_e$ was considered (9). The radial and tangential stresses in addition to radial displacement as a function of position "r" are obtained as:

$$\sigma_{rr} = \frac{a^2 b^2 (p_e\text{-}p_i)}{(b^2\text{-}a^2)r^2} + \frac{(a^2 p_i\text{-}b^2 p_e)}{(b^2\text{-}a^2)}$$

$$\sigma_{\theta\theta} = \frac{-a^2 b^2 (p_e\text{-}p_i)}{(b^2\text{-}a^2)r^2} + \frac{(a^2 p_i\text{-}b^2 p_e)}{(b^2\text{-}a^2)} \tag{4}$$

$$u_r = \frac{-(1+\upsilon)\, a^2 b^2\, (p_e\text{-}p_i)}{(b^2\text{-}a^2)\, E\, r} + \frac{(1\text{-}\upsilon)\, (a^2 p_i\text{-}b^2 p_e)r}{(b^2\text{-}a^2)\, E}$$

The problem is solved with steel ring subjected to external pressure and concrete ring subjected to internal pressure. Under the assumption that the pressure between the two disks will increase until the concrete cracks or steel yields, one can set up a solution based on the failure of concrete or the steel. Von-Mises stress conditions (equation 5) were used as the failure criteria. For the geometrical and material properties considered in the present case, the cracking in the concrete is the mode of failure.

Fig. 7a and 7b shows the geometry of a section and the stress distribution in concrete as a function of the position along the thickness. The stress is normalized with respect to the interface pressure, "p". Note that in this formulation, the strain distribution in the concrete is assumed to be opposite the strain distribution caused by the shrinkage, causing higher stresses at the surface. One can observe that there is a gradual decay in the tangential stress in the concrete as we move from inside surface towards outside surface. Using the criteria that the Von-Mises stress is equal to the tensile strength, the stress in the interface, "p", as a function of concrete tensile strength $f'_t$ and also the stain in steel ring is computed as functions of geometry and material properties:

$$\sigma_{mises} = \sqrt{\sigma_{rr}^2 + \sigma_{\theta\theta}^2 - \sigma_r \sigma_\theta} = f'_t \qquad (5)$$

$$p = 4 \frac{b^2 - c^2}{\sqrt{b^4 + 3c^4}} f'_t \qquad\qquad \epsilon_s = \frac{2b^2(b^2 - c^2)f'_t}{E(-b^2 + a^2)\sqrt{3c^4 + b^4}} \qquad (6)$$

One can use the tensile strength of concrete in terms of $f'_t$ as the failure criteria as the simplest form of failure surface definition. Alternatively, one could use a fracture based criteria for the growth of the crack. Results of this analysis, as indicated in figure 7.b, reflect that there may be a difference in the uniformity of the magnitude of stress throughout the thickness of the specimen by as much 10-20%. To simplify the problem, the interaction of stress distribution due to the shrinkage was ignored and a strain profile in accordance to the shrinkage profile was assumed. The concrete was sectioned into several layers, each acting under a uniaxial stress condition. Steel was assumed to be linear elastic.

The specimen is discretized as several layers of concrete and a layer of steel. The material properties in terms of elastic and strength of each layer are defined. Strain history at the top and bottom of each layer were initialized next. In addition, the state of cracking in each layer was also initialized. A 1-D model for concrete and steel was assumed. For each stack layer the applied shrinkage strain at the top and bottom of the layer were computed using the average strain, the average curvature, and the position along the depth. The creep strain was also added to this shrinkage level. An incremental method was used such that the stress at the top and bottom of the layer were updated using the layer strain and the creep compliance of the specimen computed using the ACI equation. Equation 7a represents the continuity condition such that the shrinkage and creep will result in elastic and inelastic strains. Parameter $i$ represents the time step. The current state of strain in concrete was imposed using Equation 7a indicating that the shrinkage strain subjected to creep would be offset by the elastically generated stress. The summation of elastic and inelastic strains is referred to as $\Delta\varepsilon_e^i$ as the "effective mechanical stress" as shown in equation 7b. The stress is updated from the previous stage of stress as defined in equation 7c, where parameter $K^i$ is the effective stiffness of the sample, and is a function of total strain obtained from the stress strain response.

$$\Delta\varepsilon^i_{total} = (\Delta\epsilon_{el} + \Delta\epsilon_{in})^i + \Delta\varepsilon^i_{cr} + \Delta\varepsilon^i_{sh} = 0$$

$$\Delta\epsilon^i_e = (\Delta\epsilon_{el} + \Delta\epsilon_{in})^i \qquad\qquad \text{(7 a, b, c)}$$

$$\sigma^i_k = K^i \Delta\varepsilon^i_e + \sigma^{i-1}_k$$

Since the concrete is restrained by means of the steel ring, the shrinkage of the concrete creates elastic stresses, which must be in equilibrium with the steel. As time increases, the free shrinkage strain, as $\Delta\varepsilon_{sh}$ is imposed on the specimen. This free shrinkage strain is used to calculate the restrained shrinkage as $\Delta\epsilon^i{}_{sh}$. The shrinkage of concrete is therefore less than the free shrinkage. The strain in the specimen is redistributed assuming that the steel provides a restraint according to the ratio of the stiffness of steel and concrete. This relationship is shown in equation 8a where the amount of offset is determined by means of the relative stiffness of steel and concrete. The stress in concrete is subjected to relaxation as determined by the creep compliance reported in equations 8b and 8c. Equation 8c is the creep compliance of concrete as defined by (10). It was assumed that the creep of concrete in compression (as defined by this equation), can be applicable to the present tensile mode of loading. As time is taken into account, the creep of concrete would offset the shrinkage stresses and result in relaxation of the elastic stresses. The combined effect of shrinkage and relaxation in the stress due to creep of concrete would be used to update the current state of stress as shown in Equation 7c.

$$\Delta\epsilon^i_{sh} = \frac{A_s E_s - A_c E_m}{A_s E_s} * \Delta\epsilon_{sh}$$

$$\Delta\epsilon_{cr} = C \Delta\sigma^{i-1}_{cr}(t) \qquad\qquad \text{(8 a, b, c)}$$

$$C = C_u \frac{t^{0.6}}{10 + t^{0.6}}$$

A failure criterion based on the maximum stress theory was used. Once the failure stress denoted by $f'_t$ was reached, a decaying stress softening curve associated with a tangential function was used. The form of the strain softening law proposed by Horii (11) was used. The value of the stress $\sigma$, and post peak deformation $w$, are obtained from the damage parameter $\omega$. The ascending portion region is affected by the amount of damage reached at the peak:

$$\frac{\sigma}{f'_t} = \sqrt{\frac{tan(\pi\omega_0/2)}{tan(\pi\omega/2)}} \qquad \frac{w}{w_0} = \frac{\sigma}{f'_t}\left(\frac{log(sec\,\pi\omega/2)}{log(sec\,\pi\omega_0/2)}\right) - 1 \qquad (9\ a,\ b)$$

where, $\omega_0$ is the damage accumulated at the peak stress. The value of $\omega_0$ is obtained based on the curve fit from the experimentally obtained data. The deformation at peak is $w_0$, and obtained as: $w_0 = \varepsilon_p\,H$, where H is the gage length of the specimen, and $\varepsilon_p$ is the strain at peak stress.   The direct tension test runs under the extensometer control was conducted to obtain the tensile stress strain response of concrete. Fig. 8 shows experimental curve of stress vs. elongation across a circular specimen.

For each layer of concrete as shown in fig. 9 the procedure of uniaxial stress and strain calculation was applied, before the time step was increased. Once the stress distribution in concrete was calculated, results were integrated through the section to calculate the force. The equilibrium of forces in concrete and steel were imposed next resulting in obtaining the stress in steel as shown in equation 10.   The elastic recovery of steel was also accounted for in the model.

$$\varepsilon_s(t) = \frac{F_c(t)}{d_s E_s} \qquad where \quad F_c(t) = l\int_b^c \sigma_c^i(t)\,dx \qquad (10)$$

Where $d_s$, $E_s$, and $\varepsilon_p$ represent the thickness, Young's modulus and strain level in the steel.   The measured stresses in steel were compared to the experimentally collected data. At the end of each time increment the relaxation strain in steel was obtained from the equilibrium conditions.

$$\varepsilon_S^R = \varepsilon_c - \varepsilon_{cr}^R \qquad (11)$$

Finally the forces in steel and concrete sections would be calculated based on the initial values at the beginning of each time increment and the amount of relaxation.

$$\Delta F_c^i(t) = \epsilon_{cr}^R l d_c E_c$$
$$F_c^{i+1} = F_c^i + \Delta F_c^i(t) \qquad (12\ a,\ b,\ c)$$
$$F_s^{i+1} = F_s^i - \Delta F_c^i(t)$$

Fig. 10a represents the strain distribution through the thickness of concrete and steel for as a function of time. The various stages of loading are shown as numbers 1-21 indicating the number of incremental steps. Note that the strain gradually increases throughout the thickness for the concrete and steel (stages 1 through 4).   Fig. 10b represents the stress distribution in concrete and steel as a function of time for the imposed strain history shown in fig. 10a. The assumption of linearity of the strain distribution throughout the thickness is clearly evident and within the region of steps 1-4 the concrete remains linear as well.   Note that as the strain increases, the specimen stress- strain passes through the plateau stress level and the material enters the strain-softening zone.   This is clearly shown in the step #7 to step #8 where the tensile cracking of the concrete takes place.   As the loading stages increase, the formulation clearly shows the propagation of the crack throughout the depth of the specimen.   The crack extends to approximately 1/3 the depth at stage #8, and as the loading increases through stages 8-11, it propagates through the entire thickness.   Beyond stage 11, the response of concrete remains relatively flat throughout the thickness as concrete enters the post peak zone.

Fig. 11 represents the theoretical curve of strain in steel as a function of time.   It is noted that for the experimental curve, the initial increase in the strain readings is due to the heating of the steel ring placed in the chamber.   This represents a shift in the experimental data with respect to the calculated values.   By accounting for the initial rise of temperature in experimental curve, both experimental and the theoretical curves match.   Note that the compressive stress builds up in steel gradually, and as the concrete cracks due to tensile failure mode, there is a gradual relaxation and recovery of strain in steel. The first crack formation in the theoretical model is after 8 days, which is quite similar to the experimental results. Beyond the cracking of concrete, its creep properties in the post peak region contribute to the increase in the strain, causing further reduction of steel strain.

Fig. 12 compares the strain in steel for theoretical and experimental results for the concrete and concrete containing 25% flyash. Note that the response of quite similar to the previous case except that due to adjustment of scales, the curves seem relatively flat.   It is also noted that the age at which the cracking takes place is much earlier for flyash concrete. Fig. 13 compares the strain in steel for the case of fiber reinforced concrete.   Note that the strain in the steel does not decay quite as fast as the previous example.   It is possible to input the exact stress strain response of the material in the present model, and develop an algorithm to predict the stress strain response subjected to shrinkage.   These approaches are however not applied in the present formulation.

## CONCLUSION

An experimental and theoretical program is presented to address the testing and analysis of restrained shrinkage cracking in concrete.   Results indicate that the proposed test method is capable to measure the shrinkage and long term performance of concrete materials modified with mineral admixtures and also fiber reinforced concrete.

## REFERENCES

1   Grzybowski, M., "Determination of Crack Arresting Properties of Fiber Reinforced Cementitious Composites," Publication TRITA-BRO-8908, Royal Institute of Technology, Sweden, June 1989.

2   AASHTO Specification, "Standard Practice for Estimating The Cracking Tendency of Concrete," AASTHO Designation: PP34-99, December 1997.

3   Grzybowski, M.; and Shah, S. P., "A Model to Predict Cracking in Fiber Reinforced Concrete Due to Restrained Shrinkage," *ACI MATER J* 87 (2):138-148 Mar-Apr 1990.

4   W. Jason Weiss, Wei Yang, and S. P. Shah, "Influence of Specimen Size/Geometry on Shrinkage Cracking of Rings, *ASCE. J. of Engineering Mechanics*, V.126, No.1, pp. 93-101, 2000.

5   Mane, S. A.; Tixier, R.; and Mobasher, B., "Development and Application of High Performance Concrete with Blended Coal Flyash", Final Report, Salt River Project, Phoenix, AZ, pp. 70, 2001.

6   Tuma, J.J., Engineering Mathematics Handbook, McGraw-Hill, 1970.

7   Nawy, E. G., 1992, "Cracking of Concrete: ACI and CEB Approaches," Proceedings CANMET International Symposium on Advances in Concrete Technology, Athens 2$^{nd}$ ed., CANMET, ed. V. M. Malhotra, Ottawa, pp. 203-242.

8   Bazant, Z. P.; and Panula, L., "Practical Prediction of Time-Dependent Deformation of Concrete. Part 1: Shrinkage. Part2: Creep," *Materiaux et Construction*, Vol. 11, No. 65, Sept.-Oct. 1978, pp. 307-328.

8   Timoshenko,S.,Goodyear, J.N.,Theory of Elasticity,McGraw-Hill,1951.

10  Branson, D. E., "Deformation of Concrete Structures," McGraw- Hill New York, 1977; and "Compression Steel Effects on Long-Term Deflections," Proceedings, *ACI Journal*, Vol. 68, ACI, Detroit, pp. 555-559, 1971.

11  Karihaloo, B. L. "Fracture Mechanics & Structural Concrete," Longman scientific and Technical Publishers. 1995, p. 330.

Table 1: Portland cement concrete mix design

| Mix ID # | % FA | W/B ratio | Dry weight of materials, kg | | | | | Super-plasticizer, ml | Fiber, g |
|---|---|---|---|---|---|---|---|---|---|
| | | | Cement | Fly-ash | Sand | CA | Water | | |
| Con | - | 0.40 | 9.01 | - | 7.79 | 10.24 | 3.61 | 18 | - |
| 25FA | 25 | 0.40 | 7.26 | 1.82 | 7.84 | 10.31 | 3.63 | 18 | - |
| HD12 | 10 | 0.55 | 6.63 | 0.66 | 7.88 | 10.36 | 4.01 | 18 | 8 |
| HP12 | 10 | 0.55 | 6.63 | 0.66 | 7.88 | 10.36 | 4.01 | 18 | 67 |

Fig. 1: Experimental setup of shrinkage test and recording of strain gage readings

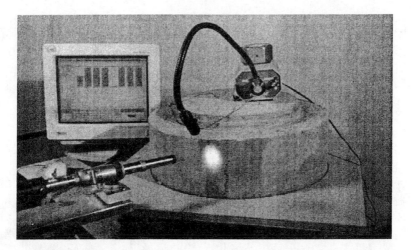

Fig. 2: Inspection of cracks using camera with microscope lens.

Fig. 3a and 3b: Setup of direct tension test on a cylinder

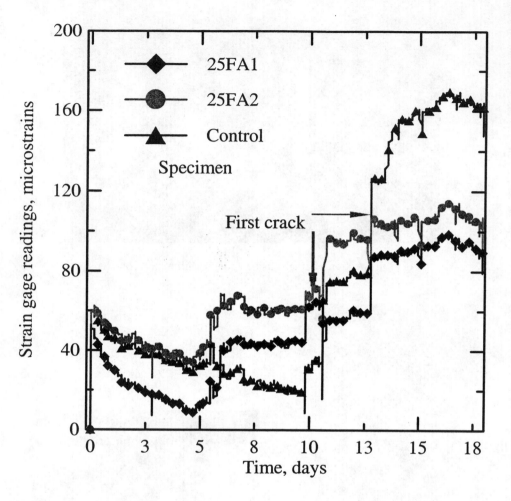

Fig. 4a: Strain gage readings vs time for control and fly ash specimen

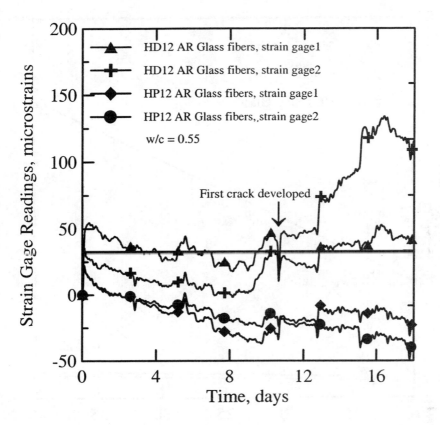

Fig. 4b: Strain gage readings vs time for HD12 and HP12 AR glass fiber reinforced specimen

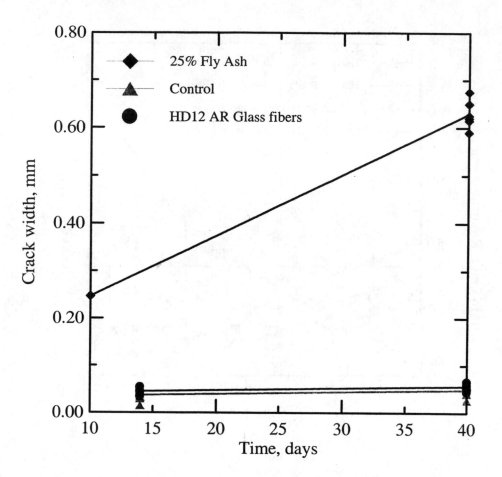

Fig. 5: Crack measurement with respect to time in shrinkage test

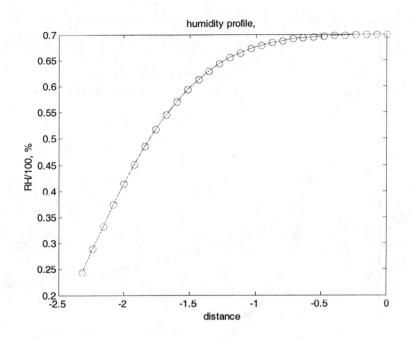

Fig. 6: Assumed humidity profile through the thickness of the section

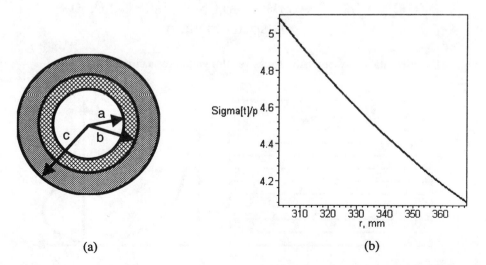

| (a) | (b) |

Fig. 7a: The geometry of the shrink fit problem. Fig. 7b: Transverse stress, $\sigma_{\theta\theta}$ through the concrete thickness as a function of the interfacial pressure between the two materials.

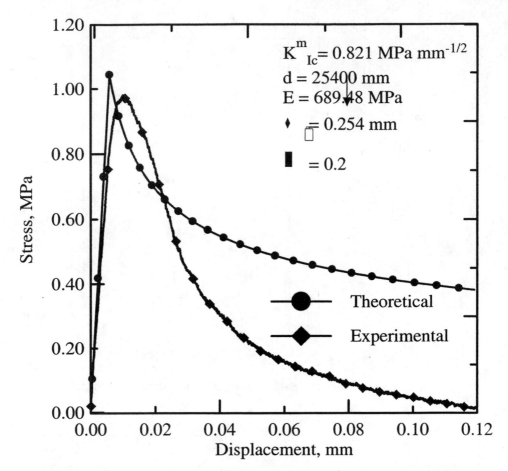

Fig. 8: Experimental curve of tensile stress vs. elongation for a circular specimen

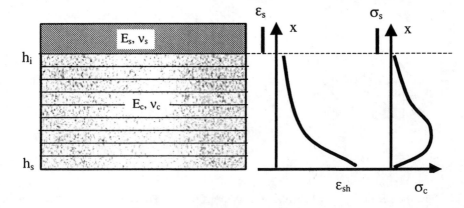

Fig. 9: Schematics of the model used

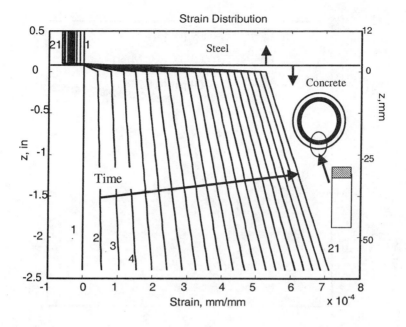

Fig. 10a: Strain distribution in specimen for various stages of loading as a function of time

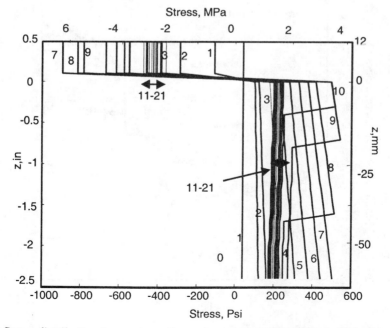

Fig. 10b: Stress distribution in specimen for various stages of loading as a function of time

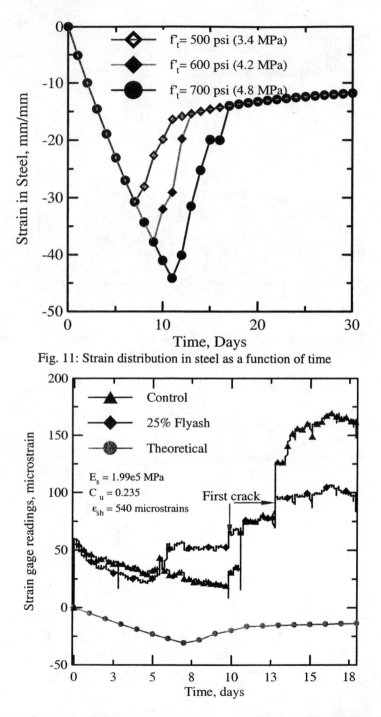

Fig. 11: Strain distribution in steel as a function of time

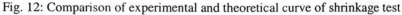

Fig. 12: Comparison of experimental and theoretical curve of shrinkage test

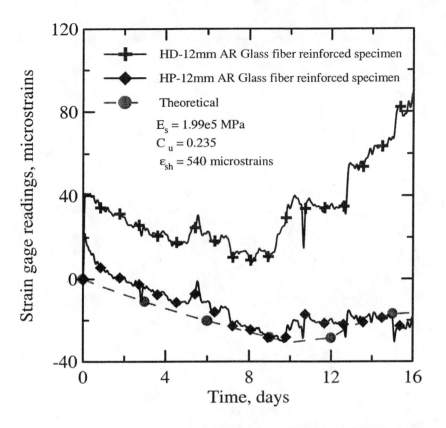

Fig. 13: Comparison of experimental and theoretical curve of shrinkage test

# Field Study of the Early-Age Behavior of Jointed Plain Concrete Pavements

## by H.-L. Chen, T. H. Schell, and J. G. Sweet

**Synopsis:** Field engineers have observed that Jointed Plain Concrete Pavements (JPCP) exhibit irregular joint cracking patterns. Upon inspection, it was seen that most of the joints remain uncracked at early ages, while many of those joints that do crack experience excessively large crack widths. This results in a quicker localized deterioration of these joints, and ultimately a shorter life span of the highway. This phenomenon, as well as many of the early-age mechanical properties of concrete, was investigated in this study. This paper describes the study of the early-age JPCP joint cracking. On-site investigations were conducted during the construction of three major JPCP highways. The pavements were monitored for slab temperature profile histories, ambient temperature histories, transverse joint crack development and overall behaviors. The concrete temperature histories were obtained at selected locations for each investigation using embedded thermocouples and a non-contact infrared thermometer. Crack growth histories were obtained for each site by measuring the crack widths at each joint. A time- and location-dependent analysis was developed which gives an acceptable representation of the observed cracking pattern. The analysis is based on such factors as temperature effect, varying properties according to concrete age and the frictional forces between the JPCP and its underlying layers. The results of this study can be used to help control the locations of joint cracking and crack widths of early-age concrete pavements.

**Keywords:** concrete pavement; crack width; early-age properties; field study; friction; highway pavement; plain concrete; temperature

Dr. Hung-Liang (Roger) Chen, Professor, Department of Civil and Environmental Engineering at West Virginia University (WVU). He received his undergraduate civil engineering education from National Taipei Institute of Technology, Taiwan and obtained his M.S. and Ph.D. in structural engineering from Northwestern University, Evanston, IL, in 1985 and 1989, respectively.

Troy H. Schell was a Graduate Research Assistant at WVU. He received his B.S. in civil engineering from WVU in 1997 and his M.S. in Civil Engineering from WVU in 2001. He is a structural engineer at Century Engineering in Elkins, WV.

Joseph G. Sweet is currently a Graduate Research Assistant at WVU. He received his B.S. in civil engineering from West Virginia University in May, 2000.

## INTRODUCTION

Newly constructed rigid concrete highway pavements have been found to experience peculiar joint movements, (1). The highways in question are constructed of jointed, plain concrete pavement (JPCP) slabs of 25.4 cm to 30.5 cm (10 in to 12 in) thickness, which contain no wire mesh or reinforcement. The slabs are saw-cut shortly after placement (usually within the first 12 to 24 hours) in order to control the surface texture and location of cracking, which occurs due to thermal and shrinkage stresses in the concrete. The joint spacing for all slabs in this study was either 4.6 m or 6.1 m (15 ft or 20 ft), depending on the site. Steel dowel bars are generally used to transmit shear forces while the pavement is subjected to service loads; these bars are usually coated with a bond-breaking substance to minimize the resistance to deformation at the joints caused by bonding with the concrete, (1,2). A schematic drawing of the saw cut and dowel bar is shown in Figure 1. Separation/filter fabric is sometimes placed beneath the base course to prevent contamination of the free-draining base by fine aggregates from the sub-grade. All of the pavements that will be discussed utilize a slip-form paving technique.

Uniform crack openings were originally anticipated at each joint. However, based on field observations, non-uniform joint cracking at early-age is common, as some joints open much wider than others do, resulting in the premature localized deterioration of the pavement at those joints, and ultimately a shorter life span for the highway, (2-4). Material properties of concrete at early-age have been reported, (1-5), and this information is needed for the analysis of cracking behavior of fresh concrete pavement, (1,2). In this study, a hypothesis was generated with respect to the general cracking behavior of these pavements. The ambient temperature and concrete temperature histories were found to be important factors. The effect of the sub-grade on the movement of the slab also has a direct relation to the stresses generated in the concrete, since both thermal contraction and shrinkage of the concrete pavement are resisted by the frictional force developed between the pavement and the underlying courses, (4-7). This creates an internal tensile stress in the concrete during contraction, causing the concrete slab to crack at a pre-designated area (a saw-cut joint), which in turn releases the stress at that joint.

## SITE INVESTIGATIONS

There were three JPCP highways that were instrumented and monitored in this study: (i) I-79 Southbound lanes near Marianna, Pa, (ii) I-80 Eastbound Lanes near Danville, Pa, and (iii) Corridor H near Elkins, WV. For all site investigations, temperature data was collected at selected depths and locations so that temperature profiles could be developed. At those locations, temperatures were taken at the surface and at predetermined depths, as shown in Fig. 2. These depths will be referred to as surface, top (5.1 cm, or 2 in, below the surface), middle (at a depth of one-half of the slab thickness), and bottom (2.5 cm, or 1 in, above slab-base course interface). All of these thermocouple groups, containing three thermocouples each, were located approximately halfway between the nearest saw cut joints. A hand-held, non-contact digital infrared thermal meter (thermal gun) was used to measure the surface temperature. Crack width measurements were made on all site investigations using crack comparators.

The first JPCP construction site to be instrumented and monitored was a section of I-79 southbound lanes near Marianna, PA. This pavement section was approximately 783.9 m (2,570 ft) in length. The concrete slab thickness, h, was 27.9 cm (11 in). It was placed on a 10.2cm (4 in) thick lean base course on a graded 15.2 to 20.3 cm (6 to 8 in) aggregate sub-base course. The mix designs for all sites can be seen in Table 1. The cross-sectional design can be seen in Figure 3(a). The concrete slab was cured with white curing compound applied after the surface texturing was performed. The transverse joints were all saw cut from about 12 to 22 hours after the construction start time. The cuts were made on a 1:6 skew at a 6.1 m (20 ft) spacing; they were 0.48cm (3/16 in) wide and approximately h/3 deep. The slab was monitored continuously for 43 hours after casting. Five thermocouple groups were placed at different locations along the slab length. Slab temperatures were taken at approximately every hour and the crack width measurements by location and time were checked.

The second investigation was performed on a section of I-80 eastbound lanes near the Danville, PA exit. The construction design was different from that of the I-79 construction design. The variations in mix design [see Table 1(b)], base course materials, various course depths and concrete curing techniques were all considered. The base course material used was Asphalt Treated Permeable Base Course (ATPBC), similar to that used in West Virginia (Free Draining Base course). The cross-sectional design can be seen in Figure 3(b). The slab's cross-section was 7.32 m (24 ft) wide with a minimum 33-cm

(13-in) thickness placed on a minimum 6.4 cm (2.5-in) ATPBC on the existing highway's rubblized concrete. The investigated section was 866.2 m (2840 ft) long. The transverse joints were cut to the same specifications as the I-79 slab; the timing of the saw cutting was also similar to that of the I-79 slab. Six groups of thermocouples were inserted into this slab. The thermal infrared gun was used to take the surface temperatures at the group locations and on the base course prior to concrete placement. Two thermocouples were used to observe the ambient temperatures, one at either end of the pavement section.

The third investigation performed was on a 553.6-m (1,815-ft) section of Corridor H near Elkins, WV. The slab section was 27.9 cm (11 in) deep and 7.3 m (24 ft) wide. The highway design was different from those of the other sites, but it was placed using a similar technique. The mix design can be seen in Table 1(c). The base course material was Free Draining Base, a mixture of 2% asphalt and 98% aggregate (#57 crushed limestone) that is placed and compacted on site. In this case, a layer of synthetic, separation/filter fabric was used beneath the base course. The cross-sectional design for the instrumented pavement can be seen in Figure 3(c). The concrete slab sections placed were cured with white curing compound as required by WVDOT specifications. Saw cutting began about 12.5 hours after concrete placement began and was completed around 8 hours later. The joints were cut perpendicular to the slab unlike the skewed cuts made in the Pennsylvania projects. The cuts were to be 0.38 cm (1/8 in) wide and cut to a depth of h/4 by design. The actual first cut depth on the Corridor H slab, however, was only 5.1 to 6.4 cm (2 to 2.5 in) the first night. All transverse cuts were deepened to a depth of 10.2 cm (4 in) during the afternoon of the second day.   Seven groups of thermocouples, as well as three vibrating-wire strain gages were inserted into the concrete. Two thermocouples were used to collect the ambient temperatures during the slab investigation, one located at either end of the instrumented slab section. Temperature and stain readings and transverse joint crack developments were recorded continuously for 72 hrs.

## RESULTS AND OBSERVATIONS

### Temperature Histories

A plot comparing the ambient temperature histories recorded for each of the site investigations can be seen in Figure 4. The ambient temperature histories of the I-80 and Corridor H sites are very similar in behavior and value. The I-79 site ambient temperatures experienced less fluctuation than the other sites, but the behavior did follow the same warm day and cool night temperature pattern. This milder temperature fluctuation, however, did have a direct effect on the temperature development in the slab. Figure 5 shows the ambient and slab temperature data collected from thermocouple Group #1 from Corridor H, whose location along the slab length, x, was approximately 16 m (52.5 ft) and was placed (Age=0) at approximately 8:30 AM. It can be seen from this plot that the core (middle) temperature drop during the first night was approximately 26°F (14.4°C), from about 90°F (32.2°C) at an age of 10 hrs to about 64°F (17.8°C) at an age of 25 hrs. As can be seen in Figure 6, thermocouple Group #1 (location x=15.3 m, or 50 ft) from the I-79 slab (Age=0 at approximately 7:30 AM) experienced milder temperature fluctuations; it only underwent a core temperature drop of approximately 16°F the first night, from about 90°F at a time of 11 hrs (age approximately equal to 10.5 hrs) to about 74°F (23.3°F) at a time of 25.5 hrs (approximate age 25 hrs). With a core temperature drop which is 10°F (5.6°C) higher than that of those experienced by the I-79 slab, the Corridor H slab certainly underwent significantly higher stresses due to thermal contraction during the first night after placement.

The time of placement also has a great effect on the temperature history of the slab at a given location. Because these pavements are generally cast throughout the course of a day, there will undoubtedly be variations in temperature profiles along the length of the slab, especially within the first 30 hours after placement. Figure 7 is a plot of the temperature data recorded at thermocouple Group #7 of the Corridor H investigation, located at x=542.1 m (1777.5 ft), near the end of the slab which was cast latest in the day (Age = 0 at about 6:30 PM). It can be seen that the ambient temperature is already starting to fall as the slab is being placed. The core temperature, however, does rise slightly during the first ten hours after placement due to the heat gain from hydration. Shortly thereafter, the ambient temperature begins to climb, so there is little heat lost from the slab at the end of the first day, resulting in a much smaller core temperature drop (only 6°F, from 84°F at 10 hrs to 78°F at 16.5 hrs) when compared to the 26°F drop of Group #1 of the same slab. This is because for Group #1, which was cast in the morning hours, the ambient temperature rise coincided with the heat gain from hydration, raising the temperature very high during the day. At night, however, the ambient temperature declined and the heat of hydration was lost from the slab, resulting in a large core temperature drop.

This varying ambient temperature/heat of hydration relationship is evident throughout the slab at early ages, but time of casting becomes less important when looking at later temperature patterns. Figure 8 plots the middle, or core, temperature readings versus time (Time=0 at 8:30 AM) for all seven Corridor H thermocouple groups (the relative locations of all groups can also be seen in this figure). It can be seen from this plot that the heat of hydration process can be either enhanced or inhibited by the ambient temperature, depending on time of casting. At about 30 hours, though, the heat of hydration is less of a factor, but the ambient temperature still plays an important part. Each investigation showed similar slab temperature behaviors which consistently following the ambient temperature history with early differences along the slab length. These fluctuations in the slab temperature cause expansions and contractions of the slab, which cause longitudinal stresses in the slab.

Ambient temperature changes not only cause longitudinal stresses in the slab, but they also cause differential stresses, also known as curling stresses. This type of stress is caused by a temperature gradient between the top and bottom of the slab. It can be seen from Fig. 5 that when temperatures are falling at night the surface temperature of the concrete slab falls faster than the interior temperatures, creating a temperature difference of up to about 15°F. When this happens, stresses due to thermal contraction are larger at the top of the slab than at the bottom, creating a bending moment in the slab. Ambient temperature sometimes has the opposite effect on the slab; in the case where the slab is heating up, the surface temperature is higher than the inside of the slab so the opposite moment is created.

Crack Development

Crack width data was recorded at various times throughout the site investigations and then plotted for a given time versus location along the slab. At the onset of this study, the joint crack development was thought to be non-uniform and sporadic in nature. The results from this investigation, however, show that a pattern does exist and is somewhat consistent between sites.

Actual joint crack initiations were observed to begin cracking at the base of the saw-cuts and progressed down the slab depth. The crack formations were seen to occur during the nights when the slab surfaces were cooler than the slab bottoms, showing differential slab deformation with respect to depth, or curling behavior. The curling behavior between the saw-cuts can be visualized as a wishbone effect. The top of the saw-cut opens and the bottom can not. Similar behavior was observed at night after cracking had occurred; the top of the joint crack often opened wider than the bottom. Figure 9 shows a plot of the crack data collected at night during the Corridor H slab. These measurements were taken at approximately 22 hours after the onset of construction (Time=0 at 8:30 AM).

Figure 10 was created from the Corridor H joint crack data plotted at various times. The joint cracks were observed to form initially at the most matured end of the concrete slab section and propagate towards the less matured end. It can be seen that joint cracking begins around 16 hours on the most mature end of the slab at fairly even spacings. Not all of the slab experiences joint cracks during the first night, however, as only just over half of the slab experiences cracked joints within the first 33 hrs. Those cracks that form early widen throughout the night. It is not until temperatures drop the second night that the less mature end of the slab cracks. The spacings of these cracks do not appear to be as consistent as the first night's.

Similar trends were observed for the other sites, as well. Figures 11 and 12 show the plots of the I-79 and I-80 slabs' crack developments, respectively.

The I-79 pavement experienced cracked joints along less than one-half of its length during the first night. The crack widths at the end of observation (the widest being only 1.6 mm at about 41.5 hrs) were much smaller than those of the Corridor H slab at a similar time (the widest being greater than 2.7 mm at only 38 hrs). The I-80 pavement experienced joint cracks along more than two-thirds of its length during the first night and its crack widths were similar to those of Corridor H at similar times.

The average joint-crack widths and crack spacing were determined from each investigated site. The I-79 joint-cracks were found to be approximately 1.5 mm wide and averaged every $3^{rd}$ to $5^{th}$ saw-cut by a time of 43 hours. Another previously placed section of I-79 was also observed to have similar crack frequency, every $3^{rd}$ to $5^{th}$ joint, but had increased in average width up to over 4 mm at an age of around 120 hours. The instrumented I-80 slab section joint-cracks averaged about 3 to 5 mm wide and spaced every $4^{th}$ to $6^{th}$ joint by an age of 56 hours. A non-instrumented slab section of I-80 was also observed, and was found to have similar frequency of cracking but had widths ranging from 6 to 9 mm at an age of 147 hours. The Corridor H pavements also produced similar behaviors. The average crack spacing was around every $6^{th}$ to $8^{th}$ saw-cut joint and had widened to 5 to 7 mm at an age of 72 hours. The slab section of Corridor H placed the day after the instrumented section was also checked for joint-crack development. The spacing and widths at ages of 48 hrs were again similar to the adjoining slab sections, being every $6^{th}$ to $8^{th}$ joint and very similar in magnitude at around 3 to 4 mm. Evaluating all of the slab sections' crack developments, many have exceeded the $1/8^{th}$ inch, or 3.175-mm, allowable crack-width specification.

## ANALYSIS AND DISCUSSION

Variation of joint-crack locations is dependent on the slab tensile strength and region of slab end movement. The end of the region of slab end movement is the location of maximum tensile stress, and is a function of concrete density and the coefficient of friction. The spacing of joint-crack formations varied between sites, and was dictated by the frictional resistance from the underlying courses. An increase of the frictional coefficients would decrease the region of slab end movement, thereby increasing the frequency of initial cracking, or the number of cracks per length of pavement. This increase in initial crack frequency would ultimately decrease the average crack width of the cracked joints, but it should be noted that too much frictional resistance might lead to intermediate cracking between saw cuts.

The spacing of the joint cracks was different for each site but the sequenced behavior was similar. In all cases, cracking was only able to occur for about one-half to two-thirds of the slab on the first night. This is because, due to the ambient temperature influence which was discussed before, the first part of the slab attained a very high temperature during the day, so when the ambient temperature dropped at night the slab developed large stresses due to thermal contraction and base-course friction, as well as due to curling. On the other hand, the JPCP paved in the late afternoon was hydrating during the evening. As a result, it did not attain such high initial temperatures; in addition, its heat of hydration counteracted the cold ambient temperatures throughout most of the night, so no large temperature drop was present, and therefore no large stresses were developed. It should be noted, however, that in all cases the temperature drop the second night was sufficient to induce cracking in the remainder of the slab, and in most cases these "second-night cracks" eventually grew to a size comparable with the "first-night cracks."

During all of the site investigations, it was seen that the joint-crack propagated through the concrete slab's depth and also through the base courses, so the frictional resistance to deformation could also be attributed to what lies beneath the base course as well as to the base course itself. Each site had different base/sub-base designs. The base course to sub-base interaction would be different, producing different joint-crack frequencies.

Numerical Analysis

Using the temperature data and material properties established in the Corridor H investigation, a preliminary stress analysis was conducted which estimates the time of joint cracking and the average spacing between initial joint cracks. This analysis utilized a segmented approximation of the variation of temperature and material properties along the slab. The seven thermocouple group locations acted as the dividing points between segments in this analysis. The entire slab length of 553.6 m (1,815ft) is divided into 7 regions, each containing a thermocouple group. Region #1 corresponds to the section containing Group #1, which is the beginning of the pavement. The first region was 61 m (200 ft) long, Regions #2 to #6 were 91.5 m (300 ft) long, and the seventh region was 35.1 m (115 ft). The age-dependent variables, such as concrete material properties, the resistive strength of dowel bar bond, the frictional coefficient of the interface between concrete and the asphalt base, as well as the temperature profiles were considered. These variables are assumed to be the same along the length of each region at any given time [7].

Figure 13 shows the results of the analysis performed using the temperature and material behaviors obtained from the Corridor H investigation. For each region, the time and spacing of initial crack formations were calculated and plotted. It was estimated that the first three regions would crack by about 18 hours after the start of construction with a crack spacing of about 7 to 8 joints. The fourth region would see joint cracking at 24 hrs with a spacing of 12 joints. When comparing these numerical results to the actual data from Figure 10, one can see that the estimations are quite reasonable. The actual observed joint crack pattern was approximately one-third of the slab cracking in the first 18 hrs and about one-half of the slab was cracked by the end of the first night (24 hrs). The approximations for crack spacing were also very close, estimating cracks at every 7 to 12 joints the first night, while the actual joint crack spacing ranged from 8 to 12 joints was observed (Fig.10).

## CONCLUSIONS

The field investigations performed during this study produced useful information for the understanding of newly constructed pavement behavior. The results of this study are summarized as: (1) The ambient temperature has a dramatic effect on the slab temperature profile at early-ages, either amplifying or reducing the heat gain from hydration, depending on the time of casting. After the initial heat of hydration dissipated, the slab temperature behaviors mimicked and approached the ambient temperatures. (2) The transverse saw-cut joint-crack developments were observed to follow propagation sequencing from the earliest placed most mature slab end to the youngest slab end. (3) At all three site investigations, the base courses failed integrally with the concrete slabs, producing somewhat composite-like behaviors. (4) A time and location dependent analysis was developed for early-age JPCP sections. Variations of the thermal behaviors and concrete properties due to age differences along the slab lengths were considered. The analysis results correlate well with the field observations.

## ACKNOWLEDGMENTS

The authors acknowledge the support from West Virginia Department of Transportation (WVDOH RP#149) for this study. The technical assistance and comments provided by Aaron Gillespie, Project Monitor, Barney Stinnett, Director, Materials Control, Soils and Testing Division, WVDOH, and Jack Justice, Federal Highway Administration, are greatly appreciated. The authors would also like to thank Gary Robson of ACPA and the Pennsylvania DOT for assisting us in the field investigation.

## REFERENCES

1. Chen, H. L. and Channell, B.C. "Study of the Joint Movements of 15-ft (4.57m) Highway Concrete Pavements," WVDOH RP#107 Final Report, CFC-97-256, West Virginia University, 1998.

2. Okamoto, P.A., Nussbaum, P.J. et al. "Guidelines for Timing Contraction Joint Sawing and Earliest Loading for Concrete Pavements (FHWA-RD-91-079)," Federal Highways Administration, 1994.

3.  Schoppel, K., Plannerer, M.,and Springenschmid, R. "Determination of Restraint Stresses and Material Properties during Hydration of Concrete with the Temperature Stress Testing Machine." Thermal Cracking in Concrete at Early Ages. Edited by R. Springenschmid, London, E & FN Spon, 1994. 153-160.

4.  Chui, J.J. and W.H. Dilger. "Temperature, Stress, and Cracking Due to Hydration Heat." Creep and Shrinkage of Concrete. Edited by Z. P. Bazant and I. Carol. London, E & FN Spon, 1994. 271-276.

5.  Nagy, A. "Determination of Elastic Modulus of Young Concrete with Nondestructive Method." *Journal of Materials in Civil Engineering*, Vol. 9, No.1, Feb. 1997. 15-20.

6.  Al Rawi, R.S. and G.F. Kheder. "Control of Cracking Due to Volume Change in Base Restrained Concrete Members." *ACI Structural Journal,* July-August 1990, 397-405.

7.  Schell, T. and H. L. Chen, "Field Observation of the Early-Age Behavior of Jointed Plain Concrete Pavements," WVDOH RP#149 Draft Final Report, West Virginia University, 2001.

Table 1 - Mix Designs for (a) I-80, (b) I-79, and (c) Corridor H (per $yd^3$)

|  | (a) | (b) | (c) |
|---|---|---|---|
| Portland Cement | 500 lbs | 525 lbs | 470 lbs |
| Coarse Aggregate | 1779 lbs | 1776 lbs | 1911 lbs |
| Fine Aggregate (sand) | 1255 lbs | 1249 lbs | 1194 lbs |
| Flyash | 88 lbs | 93 lbs | 72 lbs |
| Water | 252 lbs | 248 lbs | 227 lbs |
| Air Entraining | (VR-C) 7.7 oz | (MBVR) 8 oz | (Duravair-1000) 0.5 oz |
| Water Reducer | (1000 N) 12 oz | (220-N) 19 oz | (WDRA-82) 4 oz |
| SLUMP | 3 in | 1.5 in | 1.5 in |

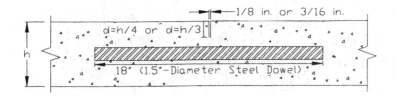

Figure 1 - Schematic Saw-Cut Joint Profile

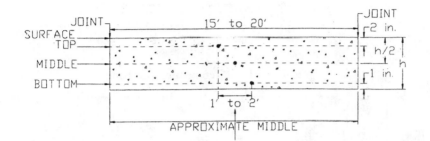

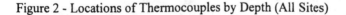

Figure 2 - Locations of Thermocouples by Depth (All Sites)

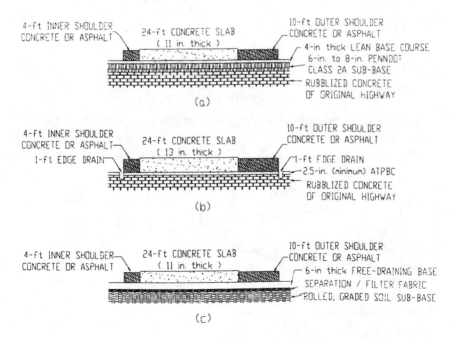

Figure 3 - Pavement Cross Sections (a) I-79, (b) I-80, and (c) Corridor H

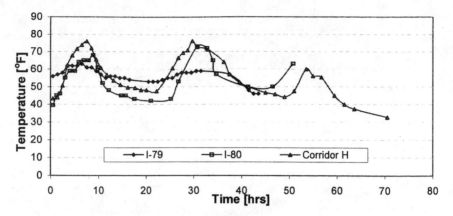

Figure 4 - Ambient Temperature vs. Time

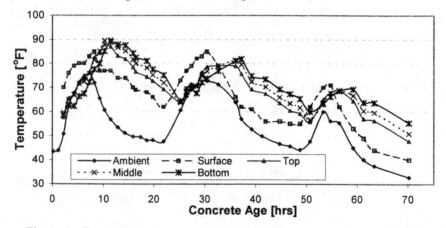

Figure 5 - Temperature vs. Age, Group #1 Corridor H (Age=0 at 8:30 AM)

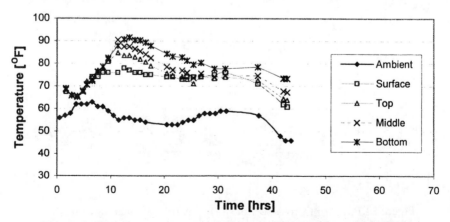

Figure 6 - Temperature vs. Age, Group #1 I-79 (Time=0 at 7:30 AM)

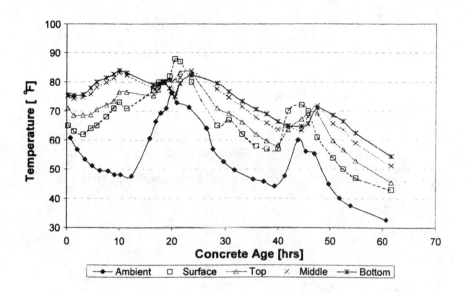

Figure 7 - Temperature vs. Age, Group #7 Corridor H (Age=0 at 6:30 PM)

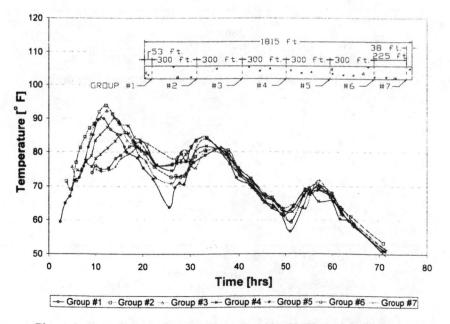

Figure 8 - Core Temperature vs. Time, Corridor H (Time=0 at 8:30 AM)

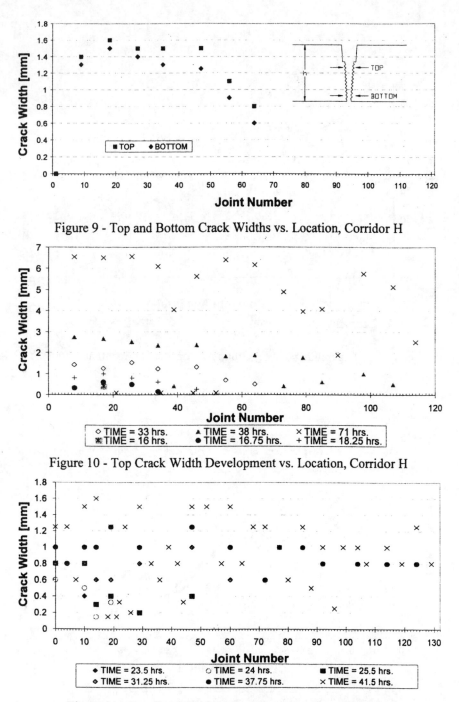

Figure 9 - Top and Bottom Crack Widths vs. Location, Corridor H

Figure 10 - Top Crack Width Development vs. Location, Corridor H

Figure 11 - Top Crack Width Development vs. Location, I-79

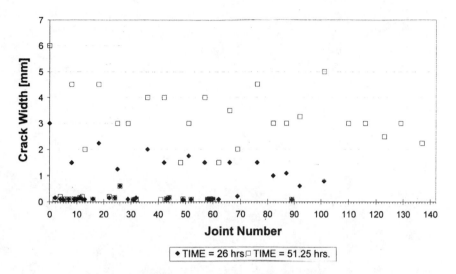

Figure 12 - Crack Width Development vs. Location, I-80 (Time=0 at 8:30 AM)

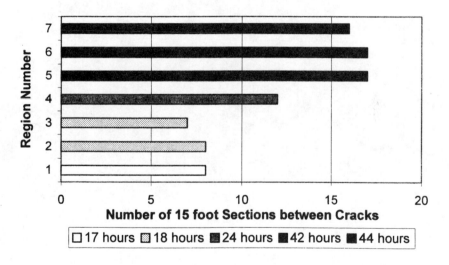

Figure 13 - Estimated Time and Spacing of Initial Cracking for Corridor H

# Premature Transverse Slab Cracking of Jointed Plain Concrete Pavement—Environmental and Traffic Effects

by W. Hansen, D. L. Smiley, Y. Peng, and E. A. Jensen

Synopsis: Top-down premature mid-slab transverse cracking was investigated for a jointed plain concrete pavement project with joint spacing of 4.88 m and located on I-96 in southeastern Michigan.

The environmental (curling/warping) stresses were evaluated using conventional linear temperature gradient analysis (1) and a recent developed method for non-linear gradient analysis (2).

Slab deflection profiles and temperature gradients for different times of day demonstrated that a built-in upward slab curling was present, equivalent to a linear negative temperature gradient of 0.03 $^0$C/mm or greater. This condition increases curling stresses at mid-slab and outer edge during morning hour temperature conditions as the built-in curling condition provides added negative thermal gradients. In addition, increased joint and corner uplift occurs, a condition, which favors loss of slab-base support. For these conditions, finite element analysis for truck tandem axle loading at the edge of transverse joints predicts substantial increased slab deflection and top tensile stresses. Further, loss of contact moves the maximum tensile stress towards the mid-slab region along the outer edge, where also curling stresses are highest. The combined tensile stresses were found to be significant and can initiate top-down transverse cracking. Once surface cracks are initiated they tend to propagate inward and downward from repeated truck loading.

Keywords: built-in curling stress; concrete pavement; loss of slab support; truck loading

**Will Hansen** is Associate Professor of Civil Engineering, University of Michigan. He serves on several ACI committees including ACI 231 Early Age Properties, 224 Cracking, and 209 Creep & Shrinkage. Current research areas include modeling: curling and warping stresses in concrete pavement, fracture and aggregate interlock of highway concrete, reaction kinetics of hydration, early age creep and stress relaxation.

**David L. Smiley** is Pavement Research Engineer, Michigan Department of Transportation, Construction & Technology (C&T). He is managing the materials and pavement research efforts.

**Yanfei Peng** is Graduate Student Research Assistant of Civil Engineering, University of Michigan. Current research area is analytical evaluation of loss of support in PCC pavements from curling and warping.

**Elin A. Jensen** is Graduate Student Research Assistant of Civil Engineering, University of Michigan, and received her M.Sc. degree in 1995 at Aalborg University, Denmark. Current research areas are experimental and analytical evaluation of highway concrete pavements with particular focus on concretes' fracture and aggregate interlock properties.

## INTRODUCTION

Premature cracking of jointed plain concrete pavements (JPCP's) starting as top-down edge cracks have recently been observed in several new construction projects. State Highway Agencies (SHA's) have reported these edge cracks (e.g., Michigan, Illinois, Indiana, and Pennsylvania). The edge cracks, when first observed, vary from partial width and depth to partial width and full depth. Several of these edge cracks propagated within a few service years into fully developed transverse cracks.

Eisenmann and Leykauff, (3) and Yu et al., (4) found that summertime construction during adverse climatic conditions, can leave a pavement not in full contact with the base as long as the differences in temperature and moisture gradients are lower than the built-in zero gradient condition. For these temperature conditions the slab curls upward at the joints and particularly at the joint corners, and in these areas the slab looses base contact. Loss of slab-base contact in a region extending out from the joint greatly increases the slab deflection due to joint loading, and creates a negative bending moment, analogous to a cantilever beam, with large tensile stresses at the slab top at the point of maximum moment.

This phenomenon is influenced by many factors, including the pavement's joint spacing, the concrete mix properties, the stiffness and confinement of the base, the friction between the slab and the base, and the environmental conditions during curing.

In addition, the axle configuration and spacing of multi-axle trucks can match the JPCP's joint spacing of 4.8 m, which will increase the tensile stress at mid-slab.

The purpose of this study was to evaluate curling and truck loading effects in JPCP.

## EVALUATION OF BUILT-IN CURLING

Pavement slabs are not necessarily flat at zero temperature gradient. The factors which can cause built-in upward curling in JPCP slabs include:

- Temperature gradient at the time of set. During the early hydration stages the heat generated by Portland cement combined with solar radiation effects on a warm sunny day, and the higher concrete placement temperatures normally encountered in the field, combine to create a positive temperature gradient (slab surface warmer than the bottom) at time of set. When the actual temperature gradient reaches zero the slab will curl up at edges and joints.

- Drying shrinkage gradient in concrete after time of set. Evaporation of pore-water in concrete during hardening will also cause upward curvature (warping). Drying is slow (diffusion controlled) and highly non-linear with depth below the surface. Due to the complexities involved in modeling warping stresses this property is often treated as an equivalent negative temperature gradient and included in curling stresses.

Due to early age creep and stress relaxation effects (pronounced within the first week after construction) the initial built-in temperature gradient is reduced long-term. However, as long as a built-in curling is present the actual temperature gradients must be added. Surface profile evaluation of older pavements by Byrum (5), and by Yu et al. (6) have shown that built-in upward curling is prevalent and is about 0.02 C/mm (1 F/in.) thickness for slabs in the wet-freeze climate such as in Michigan.

The falling weight deflectometer (FWD) profile results for an intact slab in figure 1 show that the slab is in an upward-curled position during mid-morning, while the temperature profile across the slab thickness is still positive (figure 2). As the temperature gradient increases further during the mid-afternoon condition the slab curvature is reversed with the slab edges and joint curved downward as compared to the mid-slab elevation. Thus, the slab is flat at a positive temperature differential greater than about 8 $^0$C from top to bottom. Conversely, it has a negative built-in temperature gradient of at least 8 $^0$C /250 mm (top to bottom) at an actual temperature difference of zero. This increases the upward curling stress and corner uplift substantially for actual negative temperature gradients and increases the loss of slab support at the joint as well.

A built-in negative temperature gradient, therefore, increases the sensitivity of the slab to joint deflections for morning hour loading conditions when the slab surface is cooler than the bottom. This is illustrated in figure 3. For surface temperatures below about 20 $^0$C a substantial increase in joint deflection was measured due to loss of slab-base support at the joint from upward curling. Using the Westergaard expression for estimating the corner deflection from curling an increase in joint deflection of about 250 microns can be expected for concrete with a normal range CTE (10 $\cdot10^{-6}$ / $^0$C), increasing to 350 microns for a PCC with a high CTE (15$\cdot10^{-6}$ / $^0$C). A modulus of subgrade reaction with a value of 70 kPa/mm was estimated from FWD morning results. The elastic modulus used in calculating radius of relative stiffness was 24 GPa. These values are based on curling deflection alone. Further, it should be kept in mind that corner curling is higher than in the outer wheel-path 0.9 m from the edge. Higher values demonstrate loss of slab support and/or greater built-in thermal gradient. These calculations further support the FWD and temperature profile results that a substantial built-in upward curling is present.

## MID-SLAB CURLING STRESSES AND JOINT CORNER DEFLECTION

### Theory for Curling stress Calculations

Curling stress solution for concrete slab on grade subjected to a linear temperature gradient were derived by Westergaard (7 & 8) and Bradbury (1) for the cases of infinite, semi-infinite, infinite with constant width, and finite slabs.

It is well established from field measurements that the actual distribution of the temperature gradients through the thickness of a slab may be highly nonlinear (8-10). In this paper a recent developed closed-form solution technique is used for calculating the stresses developed in a pavement slab due to nonlinear gradients (2). The analysis is separated into two parts. In the first part, the self-equilibrated stresses within a cross section due to internal restraint are determined (i.e., satisfying equilibrium conditions and continuity of the strain field within the cross section). These stresses are independent of slab dimensions and boundary conditions. The expression for the self-equilibrated stress, $\sigma_{equi}$, is

$$\sigma_{equi} = \frac{E}{(1-\upsilon)}\left[ -\varepsilon(z) + \frac{12M^*}{h^3}(z) + \frac{N^*}{h} \right],$$  (1)

The strain due to a temperature gradient, $\varepsilon(z)$, is
$$\varepsilon(z) = \alpha_t \cdot dT(z)$$  (2)

where $E$ is the PCC elastic modulus, $v$ is the PCC Poisson's ratio, $\alpha_t$ is the PCC coefficient of thermal expansion, $dT$ is the temperature gradient, $z$ is the distance from the slab centerline, $h$ is the slab thickness, and $N^*$ and $M^*$ are the internal normal force and moment required to prevent the thermal deformation due to $\varepsilon(z)$. The expressions for N* and M* are

$$N^* = \int_{-h/2}^{h/2} \varepsilon(z)dz \tag{3}$$

$$M^* = \int_{-h/2}^{h/2} z \cdot \varepsilon(z)dz \tag{4}$$

In step two, the stresses due to external restraint (i.e., self-weight and subgrade reaction) are calculated using an equivalent linear temperature gradient obtained from the first part, and using existing closed-form solutions by Westergaard or Bradbury. The solution to this step includes slab length and boundary conditions. Total internal stresses due to nonlinear gradients are obtained using the superposition principle. The Bradbury equation for interior thermal stresses is

$$\sigma_\alpha = \frac{E\alpha_t \Delta T}{2(1-v^2)}\left(C_\alpha + vC_\beta\right) \tag{5}$$

where $\Delta T_{eq}$ is the equivalent temperature gradient between top and bottom

$$\Delta T_{eq} = -\frac{12M^*}{\alpha h^2} \tag{6}$$

$C_\alpha$ is correction factors for a finite slab, $\alpha$ and $\beta$ directional indices (x,y). If the temperature difference is positive the slab is curled downward at the joints causing tensile stresses at the top of the slab, and vice versa. The Bradbury equation for edge thermal stresses is

$$\sigma_\alpha = \frac{C_\alpha E\alpha \Delta T}{2(1-v^2)} \tag{7}$$

The correction factors, $C_\alpha$, depends on the ratio of $L_\alpha/l$, where $L_\alpha$ is the slab length or width, and $l$ is the radius of relative stiffness defined as

$$l = \left[\frac{Eh^3}{12(1-v^2)k}\right]^{0.25} \tag{8}$$

When linear gradient analysis is used $\Delta T$ is the absolute temperature differential between slab top and bottom.

The magnitude of the upward deflection at the corner, $d_0$, for a linear negative temperature gradient can be estimated based on the Westergaard equation:

$$d_0 = \frac{(1+0.15)}{h}\alpha_t \Delta T \cdot l^2 \tag{9}$$

The predicted maximum curling stresses associated with a built-in negative temperature gradient and for the mid-afternoon temperature conditions are shown in Table 1. Mid-slab edge tensile stress ranges from 1.14 MPa to 1.85 MPa for CTE of $10 \cdot 10^{-6} / \,^0C$ and $15 \cdot 10^{-6} / \,^0C$, respectively. These internal curling stresses would be even higher for surface temperatures lower than the bottom temperature.

The curling stress profiles within the cross section are shown in Figures 4 and 5 for the built-in temperature gradient condition and the afternoon temperature condition, respectively. For afternoon conditions maximum slab tensile is found at the bottom irrespective of whether linear or non-linear gradient analysis is used. This stress-state is expected. Linear gradient theory stresses appear unrealistic.

## EFFECT OF LOSS OF SLAB SUPPORT AND TRUCK LOADING AT JOINTS ON TOTAL DEFECTION AND SLAB STRESSES

ISLAB2000 program (developed by ERES Consultants, Illinois) was used for determining the combined effects of multiple wheel loading and loss of slab support on the maximum slab stresses and for determining their location. The wheel loading investigated here was a rear tandem axle and single front axle configuration matching the joint spacing. Maximum tensile stresses were obtained by placing the tandem axle on either side of a joint (with center to center spacing of 1.07 m ) and front axle up to the other joint. The axle configuration was placed along the outer edge. For this loading configuration, slab length and loss of support, the maximum tensile stresses were found at midslab along the outer edge and at the slab top.

As seen from figure 6 loss of slab support from upward curling and truck loading at the joint substantially increases the joint deflection and the slab bending stress as compared to a flat slab condition. The results also show effects of increasing truck tandem-axle loading from the legal limit in Michigan of 118 kN (26 kip) to 155 kN (34 kip). The very large outer edge deflections of about 1000 microns concur with field observations of trucks traveling across a joint. Joint deflections were visible and slab-base impact could be heard.

The loss of slab support was found to increase the distance from the slab corner to the point where the maximum load stress appears. Thus, the curling stresses combine directly with truck-load stresses when negative temperature gradients are present.

The loss of support was modeled by determining a variable modulus of sub-grade reaction, k, versus slab length, which would predict similar slab deflection responses as observed in the field for mid morning conditions (field testing performed on July 2, 2001). A number of ISLAB2000 runs were made to match the FWD deflection profiles shown in figure 1 for joint deflection corresponding to morning conditions.

The single axle had a maximum load of 62.3 kN. The tandem axle had a maximum load of 118 kN to 155 kN. The pertinent input parameters based on the road test conditions were: elastic modulus of 24 GPa; joint spacing of 4.88 m; slab width of 3.66 m; slab thickness of 254 mm. The k-profiles as obtained from FWD were dynamic values. In the FEM analysis static values were estimated as 50% of the dynamic values. The joint load transfer was set to 75%, which was consistent with the observed field load transfer values.

## TOP-DOWN MID-SLAB CRACKING FROM COMBINED CURLING AND TRUCK LOADING AT JOINTS

The results in figure 6 demonstrate that loss of slab support and high truck axle loading at joints substantially increases mid-slab tensile stresses at the top near mid-slab. For early morning conditions, when the temperature at the slab surface is below the bottom temperature, total curling stresses exceed the built-in curling stresses. It is therefore realistic that combined effects curling and truck loading can in certain cases change the conventional failure mode from bottom up to top down mid-slab cracking.

## SUMMARY

FWD slab deflection profiles and temperature gradients for different times of day demonstrated that a built-in upward slab curling was present, equivalent to a linear negative temperature gradient of 0.03 $^0$C/mm (1.5 $^0$F/in.) or greater.

Built-in curling can be caused by differential drying shrinkage and/or a positive thermal gradient during time of set. This is consistent with paving conditions during hot summer days. Also, the coefficient of thermal expansion (CTE) of concrete has a significant effect on curling stress and loss of slab support.

This condition increases curling stresses at mid-slab outer edge location during morning hour temperature conditions as the built-in curling condition provides added negative thermal gradients. In addition, increased joint and corner uplift occurs, a condition, which favors loss of slab-base support. For these environmental conditions, truck tandem axle loading at a joint and outer edge increased joint deflections and tensile stresses substantially. Further, the loss of joint support caused tensile stresses near midslab at the top.

The combined tensile stresses from curling and truck loading were found to be significant and can initiate top-down transverse cracking. Once surface cracks are initiated they tend to propagate inward and downward from repeated truck loading.

ACKNOWLEDGEMENTS

The authors would like to acknowledge the support from the Michigan Department of Transportation to the joint MSU-UM Pavement Research Center of Excellence under a project entitled, "Qualify Transverse Cracking in Michigan JPCP".

REFERENCES

(1) Bradbury, R. D. (1938) "Reinforced Concrete Pavements," Wire Reinforcement Institute, Washington D.C.

(2) Mohamed, A.R., and Hansen, W. (1996) "Prediction of Stresses in Concrete Pavements Subjected to Non-linear Gradients." Cement and Concrete Composite 18, pp. 381-387.

(3) Eisenmann, J., and G. Leykauf (1990), "Simplified Calculation Method of Slab Curling Caused by Surface Shrinkage", Proceedings, 2nd International Workshop on Theoretical Design of Concrete Pavements. Madrid, Spain.

(4) Yu, H.T., M.I. Darter, K.D. Smith, J. Jiang, and L. Khazanovich. Performance of Concrete Pavements, Volume III – Improving Concrete Pavement Performance, Final Report. Contract No. DTFH61-91-C-00053. Federal Highway Administration, McLean, VA, 1996.

(5) Byrum C.R., "Analysis of LTPP JCP Slab Curvatures Using High Speed Profiles". Paper presented at the 79th Annual Meeting of Transportation Research Board, Washington DC.( 2000).

(6) Yu, H.T., Smith, K.D., Darter, M.I., Jiang, J. and Khazanovics, L., "Performance of Concrete Pavements, Volume III: Improving Concrete Pavement Performance." Report FHWA-RD-95-111. Federal Highway Administration, McLean, VA, 1998.

(7) Westergaard, H. M., (1926) "Computation of Stresses in Concrete Roads,"Proceedings of the Fifth Annual Meeting, Vol. 5, Part I, Highway Research Board, pp. 90-112.

(8) Westergaard, H. M. (1947) "New Formulas for Stresses in Concrete Pavements of Airfields,"American Society of Civil Engineers, Transactions, Vol.113, 1947, pp. 425- 444.

TABLE 1 Curling stresses for linear and nonlinear gradient analysis.

| | | | CTE = 10·10⁻⁶ (mm/mm/°C) | | | | CTE = 15·10⁻⁶ (mm/mm/°C) | | | |
|---|---|---|---|---|---|---|---|---|---|---|
| | | | Midslab Center | | Midslab Edge | | Midslab Center | | Midslab Edge | |
| | | | $\sigma_x$ (MPa) | $\sigma_y$ (MPa) | $\sigma_x$ (MPa) | $\sigma_y$ (MPa) | $\sigma_x$ (MPa) | $\sigma_y$ (MPa) | $\sigma_x$ (MPa) | $\sigma_y$ (MPa) |
| Zero Temperature Gradient (Upward Curvature) | Nonlinear -7.7C | top | 1.24 | 0.96 | 1.14 | 0.82 | 1.85 | 1.44 | 1.71 | 1.23 |
| | | bottom | -0.63 | -0.36 | -0.53 | -0.22 | -0.94 | -0.53 | -0.80 | -0.32 |
| | Linear -8.0C | top | 0.97 | 0.69 | 0.87 | 0.54 | 1.46 | 1.03 | 1.30 | 0.81 |
| | | bottom | -0.97 | -0.69 | -0.87 | -0.54 | -1.46 | -1.03 | -1.30 | -0.81 |
| Afternoon (Downward curvature) | Nonlinear 10.3C | top | -0.52 | -0.20 | -0.46 | -0.08 | -0.78 | -0.29 | -0.68 | -0.12 |
| | | bottom | 1.00 | 0.68 | 0.93 | 0.56 | 1.50 | 1.01 | 1.40 | 0.84 |
| | Linear 20.1C | top | -1.48 | -0.85 | -1.35 | -0.63 | -2.22 | -1.27 | -2.03 | -0.94 |
| | | bottom | 1.48 | 0.85 | 1.35 | 0.63 | 2.22 | 1.27 | 2.03 | 0.94 |

# 268 Hansen et al.

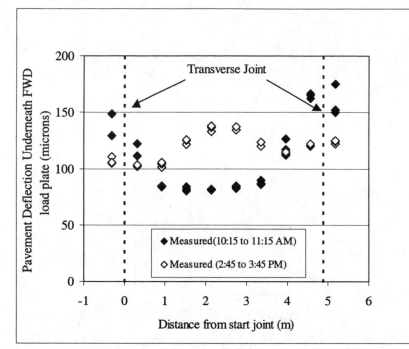

Figure 1. Slab deflection profile in outer wheel path for JPC, I-96, EB, Howell, Michigan. Tested July 2, 2001. 4.88m (16 ft) joint spacing.

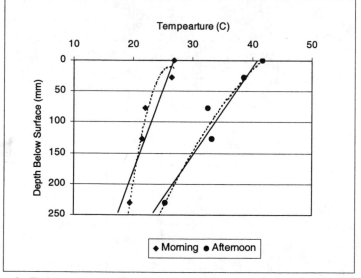

Figure 2. Temperature profiles through a highway concrete slab during summertime conditions for JPC, I-96, EB, Howell, Michigan. Tested July 2, 2001. 4.88m (16 ft) joint spacing.

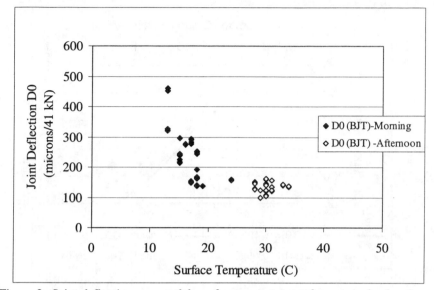

Figure 3. Joint deflection versus slab surface temperatures for outer wheel path on JPCP, EB I-96, Howell, Michigan. Tested September 27, 2000. 4.88m joint spacing.

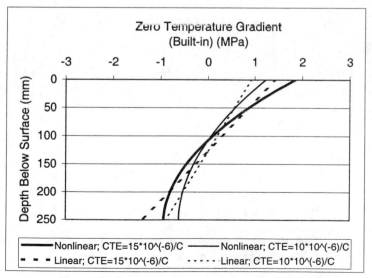

Figure 4.   Zero temperature gradient (Built-in).

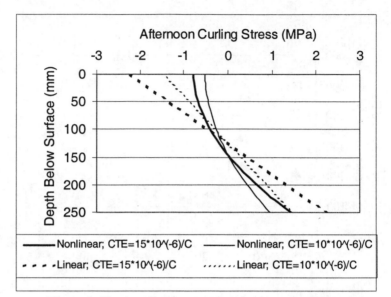

Figure 5.  Total curling stresses for afternoon conditions.

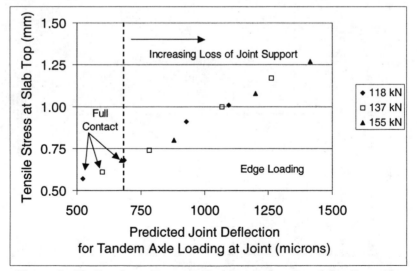

Figure 6.   Predicted maximum tensile stresses due to truck loading with
tandem axle at transverse joint.

# Evaluation of Bond-Slip Behavior of Twisted Wire Strand Steel Fibers Embedded in Cement Matrix

by C. Sujivorakul and A. E. Naaman

**Synopsis**: The bond stress versus slip behavior of steel fibers cut from twisted wire strands (which are made from at least two round steel wires wrapped helically around each other) is investigated and compared to the behavior of a single triangular steel fiber (Torex) twisted along its axis. Single fiber pullout tests simulating fiber pullout in a cracked tensile specimen were used. Parameters investigated are: 1) the embedded length of the fiber in the matrix; 2) the number of individual wires making a strand fiber; 3) the pitch distance of the fiber; 4) the compressive strength of the mortar matrix (44 and 84 MPa); and 5) the tensile strength of the fiber. It is observed that, depending on the combination of parameters (embedded length, matrix, strand type and pitch distance) the bond stress-slip response can be elasto-plastic in shape or slip softening after the peak. However, the twisted wire strand fibers are less efficient than the single Torex fiber, in term of peak pullout load and pullout energy.

Keywords: bond; bond-slip; fiber reinforced concrete; frictional; mechanical; pullout test; shear stress

**Chuchai Sujivorakul** is currently a doctoral student in the Department of Civil and Environmental Engineering at the University of Michigan, Ann Arbor. He received his ME (1995) from Asian Institute of Technology and is a lecturer at King Mongkut's University of Technology Thonburi, Thailand. His research focuses on the development of high performance fiber reinforced cement composites.

**Antoine E. Naaman** is a Professor of Civil Engineering at the University of Michigan, Ann Arbor. He is a member of ACI Committees 440, Fiber-Reinforced Polymer Reinforcement; 544, Fiber Reinforced Concrete; 549, Thin Reinforced Cement Products; and Joint ACI-ASCE Committee 423, Prestressed Concrete. His research interests include high performance fiber reinforced cement composites and prestressed concrete.

## INTRODUCTION AND BACKGROUND

The last three decades of the 20th century have seen an overwhelming number of research studies on fiber reinforced concrete with the largest proportion devoted to the use of steel fibers. In contrast, few studies related to the developments in the fiber themselves can be found [ref. (1), (2), (3)]. Indeed the steel fibers mostly available on the market today (hooked, crimped, straight steel fibers) have been conceived and introduced more than twenty years ago. More recently, Naaman (1998), (4), Naaman (1999), (5), and Naaman (2000), (6), introduced a new steel fiber of optimized geometry called the Torex fiber. It is essentially polygonal in cross section (primarily triangular or square) and twisted along its longitudinal axis. The key feature of this fiber is that it shows a pullout load versus slip response that is nearly elastic-plastic in shape up to about 70%-90% of embedded length; when the pullout load is transformed into an equivalent bond stress, the bond stress-slip response has a unique slip-hardening shape; that is the bond stress seems to increase with the slip. Several studies have described this behavior in details [Guerrero (1999), (7), Guerrero and Naaman (2000), (8)].

Pullout tests, whereby a single fiber is pulled out from a cement matrix, are generally used as an indirect method to evaluate bond. They lead to a pullout load versus end slip relationship from which various bond parameters can be extracted. The main objective of this study is to compare the pullout behavior of the single wire triangular twisted (Torex) fiber with twisted wire strand fibers made from at least two round steel wires wrapped (or coiled) helically around each other. Some background is given next.

### Twisted Polygonal Steel Fibers (Torex Fibers)

US patent No. 5989713, (9), and subsequent studies by Naaman (1999), (5), and Guerrero (1999), (7), at the University of Michigan have led to the development of steel fibers, identified as Torex (or Tor-x) fibers (Figure 1a), with optimized geometry (low order polygonal or substantially polygonal in section) that offer a ratio of lateral surface area to cross sectional area larger than that of round fibers. This ratio has been defined as the fiber intrinsic efficiency ratio (*FIER*). An increase in the *FIER* of a fiber leads to a direct increase in the contribution of the adhesive and frictional components of bond. In order to increase the *FIER* of a fiber a non-round polygonal shape is needed, with surfaces that can be indented. Moreover, unlike a fiber with a round section, a fiber of polygonal section can be twisted along its longitudinal axis, developing ribs, thus creating a very effective mechanical bond component. One of the key features of the new fibers is that they show, under direct pullout, a unique bond stress versus slip response not observed to date with any other fiber on the market. Figure 2 shows a typical pullout load versus end slip response of a twisted polygonal steel fiber, which has been identified and described by Naaman (1999), (5). The initial ascending part is almost linear up to point A (Figure 2b) mainly due to combined adhesive-frictional bond. The adhesive bond vanishes after point A. As a result, a sudden drop is observed from point A to point B. Then the pullout load recovers and keeps increasing up to a certain level (Point C) at which it tends to stabilize. Point D represents the load prior to failure. The most surprising behavior of Torex fibers is that a high level of pullout load is maintained up to very large slips, about 70 to 90% of the embedded length. This behavior is due to successive untwisting (Figure 1b) and locking of the embedded portion of the fiber during slip, and can be described as a "pseudo-plastic" response. A recent investigation by Naaman and Sujivorakul (2001), (10), describes a set of parameters influencing the particular bond characteristics of the twisted polygonal steel fibers. These parameters are: (a) the cross-sectional shape of the fiber, that is, a triangular or a square cross section; (b) the number of ribs per unit length of the fiber; (c) the compressive strength of the matrix and the tensile strength of the fiber; and (d) the embedded length of the fiber. They also reported that the twisted fibers show a higher peak pullout load and significantly higher pullout energy than hooked and smooth fibers.

## Twisted Wire Strand Steel Fibers

Twisted wire strands have been widely used as tension elements such as cord, cable, and prestressing strand. The basic elements of a twisted wire strand are individual wires (at least two wires) which are wrapped (coiled) helically together along their length (Figure 3). One of the main advantages of this system is that small individual wires can be merged in parallel in order to form a single big strand with strength and ductility properties far superior to those of a single bar of equivalent diameter (Costello (1997), (11)). However, to the best of the author's knowledge, no study is available in the technical literature on the pullout response of short discontinuous fibers cut from small diameter strands. Figure 3 shows the configuration of twisted wire strand fibers used in this study and the definition of pitch distance. These were made from wires having 0.3

mm and 0.5 mm diameter, respectively. The pitch distance is defined as the distance at which an individual wire rotates helically 360 degree. Figure 3a shows a strand made with two individual wires wrapped helically with different pitch distances. The smaller the pitch distances the higher the observed number of ribs per unit length. Figure 3b shows strand fibers made with 2, 3 and 4 individual wires having the same pitch distance.

## OBJECTIVE

The main objective of this research is to investigate the bond behavior of twisted wire strand fibers embedded in cement based matrices, and compare their response to that of a single wire polygonal (triangular) twisted fiber (Torex). Parameters investigated are the pitch distance, the fiber embedded length, the compressive strength of the matrix (44 and 84 MPa), the tensile strength of the fiber, and the number (2, 3, and 4) of individual wires forming a strand fiber.

## RESEARCH SIGNIFICANCE

Fibers with improved bond characteristics, particularly slip-hardening bond, are key to the development of high performance fiber reinforced cement composites (HPFRCCs) which exhibit multiple cracking and strain-hardening behavior. These advanced fibers are expected to increase the limits of structural and non-structural applications of HPFRCCs while approaching least cost for equal performance.

## TEST PROGRAM

The experimental program carried out for this study is summarized in Figure 4. It comprises sixteen series of pullout tests covering the type of twisted steel wire strand fiber, the matrix compressive strength, the pitch distance, and the fiber embedded length. At least 3 specimens were tested for each series. Five different types of twisted wire strand steel fibers were used; the fibers are identified by their tensile strength, number of individual wires and diameter of individual wire. The first ID number denotes the number of individual wires of the fiber (2 = two individual wires, 3 = three individual wires, 4 = four individual wires); the second set of letter denotes the diameter of the individual wire (d3 = 0.3 mm. diameter wire, d5 = 0.5 mm. diameter wire); and the last letter denotes the tensile strength of the fibers (A = 2618 MPa, B = 2412 MPa, C = 760 MPa). Two different mortar matrices (Mix-1 and Mix-2) with compressive strength equal to 44 and 84 MPa were used; the mix proportions are given in Table 1. The component materials were Portland cement Type III, and silica sand ASTM 50-70 (Ottawa silica sand). The average compressive strength was obtained from 50 mm. cube specimens tested at 14 days.

## SPECIMEN PREPARATION

The test specimens were prepared using plexiglas molds.  They had a half dog-bone shape, $25 \times 23$ mm. in cross section and 70 mm. in length; the fiber is embedded with a specific embedded length as shown in Figure 5a, and maintained in that position using a sandwich of styrofoam pads (Figure 5b). At least three specimens were prepared for each test series. The mortar matrices were prepared in a food type (Hobart) mixer and poured into the mold mounted on a vibration table; very low vibration was applied.  After finishing, the molds were placed in a 100 percent relative humidity environment at room temperature for 24 hours; then the specimens were removed and cured in a water tank for 14 days at room temperature. Before testing, the specimens were removed from the water tank and kept in laboratory air environment for at least 48 hours.

## TEST SETUP

The test setup is shown in Figure 6.  The top end of the fiber is held by a specially designed grip attached to the load cell of a testing machine while the body of the half dog-bone specimen is restrained by a grip used to test standard ASTM tensile briquettes of mortar. It is noted that the free length of the fiber (Figure 6) is minimized during gripping in order to simulate a fiber pulled in a typical cracked tensile specimen and to minimize the effect of free elongation of the fiber. A linear variable differential transducer (LVDT) is placed as shown in the figure in order to measure the slip at the free end of the fiber, defined here as the differential movement between the fiber end and the fixed matrix specimen.  The values of the pullout force and the corresponding end slip were recorded by a data acquisition system, and stored in a computer file. The pullout load versus end slip relationship was then plotted using this data file. All tests were performed under displacement control. The load was applied at a rate of 0.2 mm/min. at the beginning of the test up to an end slip of 2.5 mm. After that the rate was changed to 1.25 mm/minute until the fiber was totally pulled out.

## ANALYSIS OF RESULTS

### Typical Pull-Out Behavior of Twisted Wire Strand Fibers

Figure 7a shows a typical pullout load versus end slip response of a twisted wire strand fiber; it is very similar to the response observed with the triangular Torex fibers shown in Figure 2a. Most of the results obtained in this study resemble this type of response. Interpretation of the response given in Figure 2b for Torex fibers can also be used for twisted wire strand steel fibers. The initial ascending part up to point A is mainly due to combined adhesive-frictional bond. The sudden drop observed from point A to point B is due to the vanishing adhesive bond, if any. Then the pullout load recovers and keeps increasing up to a certain level (Point C) at which it tends to stabilize. Point D represents the load prior to failure. A pseudo-plastic pullout load versus slip response is also observed for this fiber. This behavior is believed to be due to successive unwrapping and locking of the wires in the embedded portion of the strand fiber during slip. Experimental results discussed next illustrate the influence of several parameters on the pullout load versus slip response of the strand fibers. For comparison purposes, a representative curve for each series of tests was selected; for instance in Figure 7a, curve-A was selected as a representative curve for this series.

### Pitch Distance of the Fiber

Three different pitch distances (4.2, 6.3 and 12.7 mm.) of 2-d5 Strand-B (made of two 0.5 mm.-diameter wires) are compared for the matrix Mix-1 (f'c = 44 MPa) and the matrix Mix-2 (f'c = 84 MPa) in Figures 8a and 8b, respectively. It is first observed that a decrease in pitch distance (up to the values used) increases the peak pullout load. However, in Mix-1 (Figure 8a), the pseudo-plastic behavior is not observed for all pitch distances. Typically the pseudo-plastic response becomes less evident with an increase in the peak pullout load. This suggests, as anticipated, that a very small pitch will eventually negate the benefits of twisting. For instance, the pullout load drops suddenly after the peak for the 4.2 mm. pitch, because the pullout load, is higher than the shearing resistance of the tunnel of matrix around the fiber. The tunnel of matrix fails before the fiber unwraps (or unravels), a process that seems necessary to lead to pseudo-plastic behavior. The shearing resistance of the matrix can be enhanced if the matrix strength is increased as in the case of Mix-2 (Figure 8b). As a result, the pseudo-plastic behavior is observed up to 80 % of the embedded length for all the pitch distances used.

## Embedded Length of the Fiber

The influence of embedded length of fiber on the pullout response of 2-d3 Strand-A and 2-d5 Strand-B is shown in Figures 9a and 9b, respectively. As expected, an increase in embedded length leads to an increase in the peak pullout load and the initial slope of the ascending branch. A pseudo-plastic response is observed for the cases of 2-d5 Strand-B. However, it is not observed for the 2-d3 Strand-A, especially for Le = 12.7 mm. (Figure 9a) for which the pullout load softens after the peak with increasing slip.   This behavior is discussed next in relation to Fig. 10a.

## Matrix Compressive Strength

The pullout versus slip curves of 2-d3 Strand-A and 2-d5 Strand-B embedded in matrices of two different strengths (44 and 84 MPa) are shown in Figures 10a and 10b, respectively.  It is observed that an increase in matrix strength lead to an increase in the first peak load (see point A in Figure 2b) and the initial slope of ascending branch. However, if the first peak load is higher than the pullout load that needs to unwrap the wires of the stand during pullout, a pseudo plastic response will not be observed. This is the case of fiber 2-d3 Strand-A embedded in Matrix Mix-2 (Fig. 10a). On the other hand, Figure 10b shows a pseudo-plastic behavior of 2-d5 Strand-B fibers for the two compressive strengths. The peak value of pullout load is almost the same for the two matrices because the unwrapping (unraveling) resistance of the strand is the same. As long as the tunnel of matrix does not fail, the pullout load will reach at least that value.

## Tensile Strength of the Fiber

Figure 11 shows a comparison of the pullout load versus slip response of 2-d5 Strand fibers having two different tensile strengths (B = 2412 MPa and C = 965 MPa). The fibers have the same size and pitch distance.  It is observed that an increase in the tensile strength of the fiber increases the peak pullout load. This is due to an increase in the resistance necessary to unwrap the wires of a strand fiber during pullout. Thus, the contribution of the mechanical component of bond increases with fiber tensile strength. Note that since the fiber must remain highly stressed and not fail during unwrapping, it is safer to say that the mechanical contribution increases with the yield strength of the fiber. A non-ductile fiber (say made from a glass strand) will not show a similar behavior.

## Number of Individual Wires per Strand

The pullout response of twisted wire strand fibers having different numbers of individual wires (2, 3 and 4 wires) is compared in Figure 12 for a matrix compressive strength of 84 MPa. It is noted that these fibers have the same pitch distance and the same diameter of individual wires. However, they have different cross-sectional area. Thus in order to provide a basis of comparison, the tensile stress developed in the fiber during pullout is used. It is observed that the peak pullout stress is almost the same for all cases, at about 840 MPa. Although the number of tests is small, this result may indicate that there is no significant benefit in using a multiple-wires strand versus a 2-wires strand.

## Comparison of Torex and Twisted Wire Strand Fibers

Since both types of the fibers have different cross-sectional area, the tensile stress developed during pullout is used to compare their pullout response as shown in Figure 13 for a matrix compressive strength of 84 MPa. It is seen that both single wire triangular Torex fibers and twisted wire strand fibers develop a pseudo-plastic behavior up to large slip. However, the peak pullout stress and initial slope of the ascending branch of the triangular Torex fibers are higher than those of the twisted wire strand fibers. The peak pullout stress for the 1-d3 and 1-d5 triangular Torex fibers are about 45 % and 76% higher than that of the 2-d3 and 2-d5 strand fibers, respectively. Assuming that the production costs of each fiber system (that is single triangular twisted fiber versus multiple-wires twisted strands) are similar, the above result favors the use of a single wire triangular twisted fiber.

## CONCLUDING REMARKS

The main objective of this study was to evaluate the pull-out load versus slip response of steel strand fibers and compare it to that of a triangular twisted single wire steel fibers (Torex). Depending on the combination of parameters, three types of pullout response are observed, differentiated by the portion after the first peak load:

(a) Pseudo-plastic behavior: a maximum or plateau pullout load is maintained up to about 70% - 90% of the fiber embedded length. This behavior is believed due to successive unwrapping (unraveling) and locking of the strand wires in the embedded portion of the fiber during slip. It is valid provided the tunnel of matrix surrounding the embedded length of fiber does not fail in shear.

(b) Sudden drop of pullout load after the peak: this behavior is observed when the tunnel of matrix surrounding the fiber embedded length fails in shear, and before the load necessary to unwrap (unravel) the wires of the strand fiber is attained. It occurs mostly for low matrix strength and small pitch distance of the fiber.

(c) <u>Softening of pullout load after the first peak</u>: this behavior is observed when the first peak load due to the combined adhesive-frictional bond is significantly higher than the pullout load needed to unwrap the stand fiber during pullout. Some pseudo-plastic behavior may be observed if the difference between the two loads is small.

A pullout load versus slip response exhibiting a pseudo-plastic behavior leads to a slip-hardening bond; that is, the average bond stress increases with the slip. Fibers with slip-hardening bond are key to the development of high performance fiber reinforced cement composites (HPFRCCs) which exhibit multiple cracking and strain-hardening behavior. Similarly to single wire triangular twisted steel fibers, strand steel fibers can exhibit slip-hardening bond. However, based on this study, everything else being equal, strand steel fibers are less efficient (45% to 75%) than single wire triangular twisted steel fibers in terms of both peak stress and pullout energy. Note that slip-hardening bond was also observed by the authors for inclined fibers being pulled-out from a cement matrix; angles of inclination tested for the triangular twisted fiber were 0, 15, 30, 45, and 60 degrees. It was observed in these tests that the peak pullout load of the inclined fibers is not less than that of the fibers pulled out in parallel to the fiber axis, except for the case of the 60 degrees inclination; in this last case the surface of matrix spalls off prematurely causing a significant reduction in the embedded length of the fibers and thus in pull-out load.

## ACKNOWLEDGMENTS

This research was supported in part by a grant from Center for Advanced Cement-Based Materials and by the University of Michigan. The first author acknowledges the Ministry of University Affairs, Thailand, for giving a fellowship to pursue his doctoral studies at the University of Michigan. Any opinions, findings, and conclusions expressed in this study are those of the authors, and do not necessarily reflect the views of NSF or the ACBM Center.

## REFERENCES

1. Banthia, N. and Trottier, J.F., "Concrete reinforced with deformed steel fibers, Part I: bond-slip mechanisms," *ACI Materials Journal*, Vol. 91, Sept.-Oct. 1994, pp. 435-446.

2. Chanvillard, G., and Aitcin, P. C., "Pull-Out Behavior of Corrugated Steel Fibers", *Advanced Cement Based Materials* (ACBM), No. 4, pp. 28-41, 1996.

3. Rossi, P., and Chanvillard, G., "A New Geometry of Steel Fibers for Fiber Reinforced Concretes," in *"High Performance Fiber Reinforced Cement Composites,"* H.W. Reinhardt and A.E. Naaman, Editors, E and FN SPON, an Imprint of Chapman and Hall, London, U.K, 1992, pp. 129-139.

4. Naaman, A. E., "New Fiber Technology: Cement, Ceramic, and Polymeric Composites," *Concrete International* V. 19, No. 7, 1998, pp. 57-62.

5. Naaman, A. E., "Fibers with Slip Hardening Bond," *High Performance Fiber Reinforced Cement Composites - HPFRCC 3*. H. W. Reinhardt and A. E. Naaman, Editors, RILEM Pro 6, RILEM Publications S.A.R.L., Cachan, France, May 1999, pp. 371-385.

6. Naaman, A.E., "Fiber Reinforcement for Concrete: Looking Back, Looking Ahead," in Proceedings of Fifth RILEM Symposium on Fiber Reinforced Concretes (FRC), BEFIB' 2000, Edited by P. Rossi and G. Chanvillard, September 2000, Rilem Publications, S.A.R.L., Cachan, France, pp. 65-86

7. Guerrero, P., "Bond Stress-Slip Mechanisms in High Performance Fiber Reinforced Cement Composites," *Ph.D. Thesis*, 1999, University of Michigan, Ann Arbor, 249 pages.

8. Guerrero, P., and Naaman, A.E., "Effect of mortar fineness and adhesive agents on the pull-out response of steel fibers," *ACI Materials Journal*, Vol. 97, No. 1, January-February, 2000, pp. 12-20.

9. Naaman, A.E., U.S. Patent No. 5,989,713, "Optimized Geometries of Fiber Reinforcements of Cement, Ceramic and Polymeric Based Composites," Nov. 23, 1999. Divisional: US Patent No. 6,060,163; May 9, 2000.

10. Naaman, A. E., and Sujivorakul, C., "Pull-out Mechanisms of Twisted Steel Fibers Embedded in Concrete," *Proceeding of the International Conference on Engineering Developments in Shotcrete*, April 2-4, 2001, Hobart, Tasmania Australia, pp. 197-203.

11. Costello, A. G., "*Theory of Wire Rope*. 2nd Edn.," Springer-Verlag, New York, Inc., 1997, 122 pages.

Table 1—Composition of Matrix Mixtures by Weight Ratio and Their
Compressive Strength

| Matrix ID | Cement | Fly Ash | Sand[A] | Super-plasticizer[B] | Water | $f'c$ (MPa)[C] |
|-----------|--------|---------|---------|----------------------|-------|----------------|
| Mix-1 | 0.8 | 0.2 | 1.0 | None | 0.45 | 44 |
| Mix-2 | 0.8 | 0.2 | 1.0 | 0.04 | 0.26 | 84 |

[A] Ottawa Silica Sand ASTM 50-70
[B] Melamine based superplasticizer
[C] Compressive strength of the mortar obtained from $50 \times 50 \times 50$ mm. cube
specimens

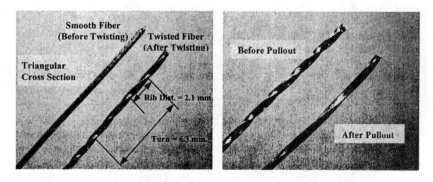

(a) Comparison of Smooth and Torex
Fibers Having Triangular Cross Section

(b) Successive Untwisting of the Fiber
Observed after Pullout

Figure 1—Configuration of Triangular Torex Fibers, de = 0.5 mm., Turn = 6.3
mm.

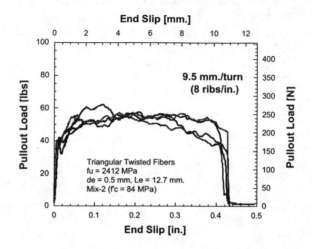

Figure 2(a)—An Example of Pullout Response Obtained from the Tests of Single Wire Triangular Torex Fibers

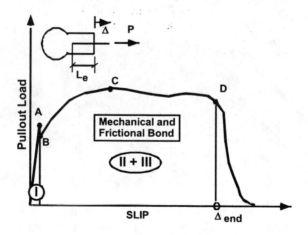

Figure 2(b)—Typical Pullout Response of Torex Fibers Defined by Naaman (1999), (5)

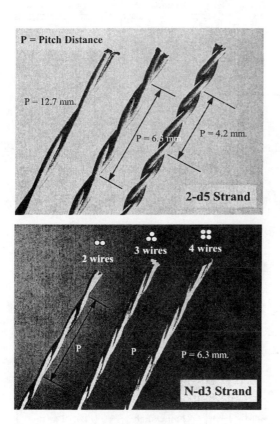

(a) 2-d5 Strand

(b) N-d5 Strand
where N = 2, 3 and 4

Figure 3—Configuration of Twisted Wire Strand Fibers and Definition of Pitch Distance

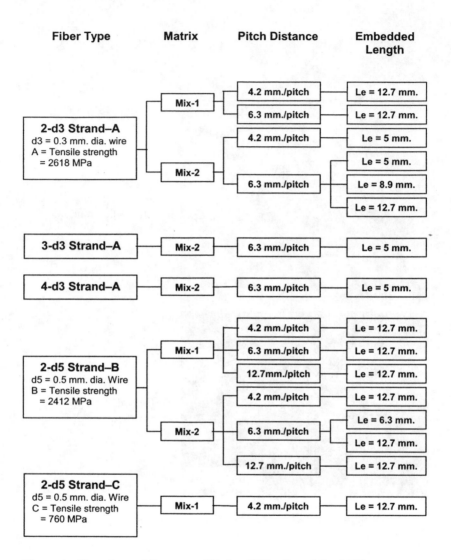

Figure 4—Experimental Program of Twisted Wire Strand Steel Fibers

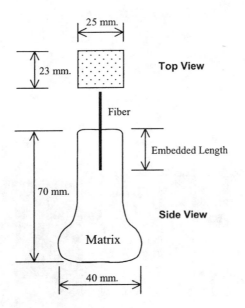

Figure 5a—Layout of Typical Pullout Specimens

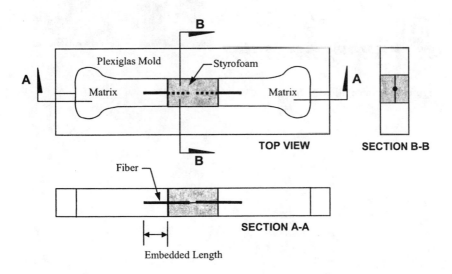

Figure 5b—Layout of Specimen Preparation Using Plexiglas Mold

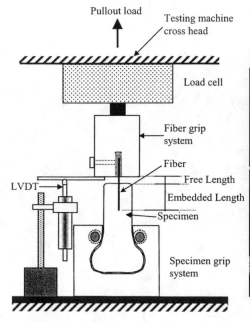

Figure 6—Fiber Pullout Setup

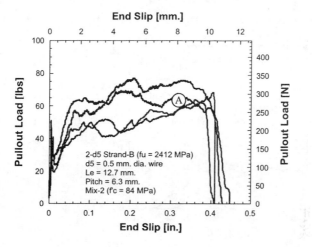

Figure 7(a)—Typical Pullout Load versus End Slip Response of 2-d5 Strand-B Fibers

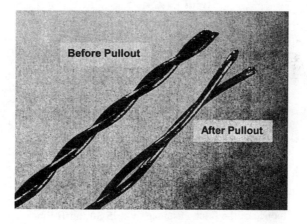

Figure 7(b)—Unwrapping of the 2-d5 Strand-B Fibers Observed after Pullout

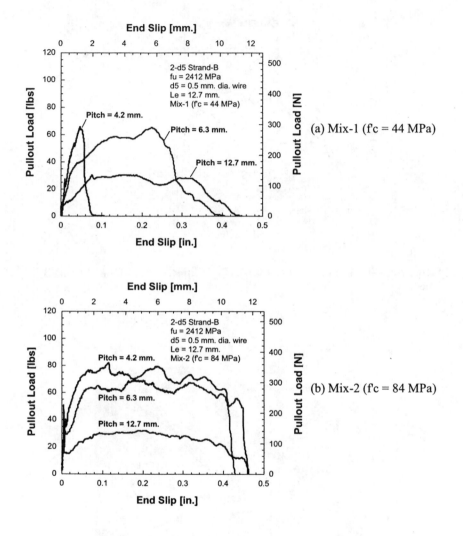

(a) Mix-1 (f'c = 44 MPa)

(b) Mix-2 (f'c = 84 MPa)

Figure 8—Influence of the Pitch Distance in 2-d5 Strand-B Fibers

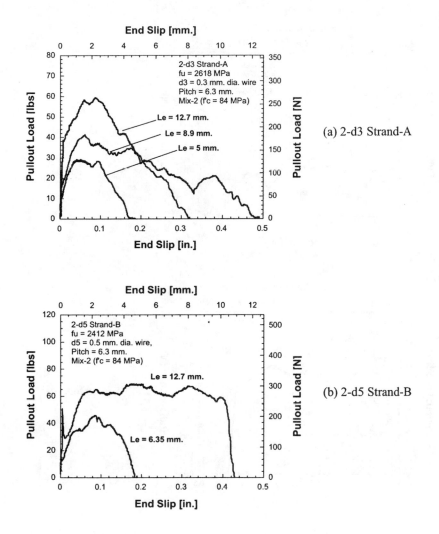

(a) 2-d3 Strand-A

(b) 2-d5 Strand-B

Figure 9—Influence of the Fiber Embedded Length in the Matrix Mix- 2 (f'c = 84 MPa)

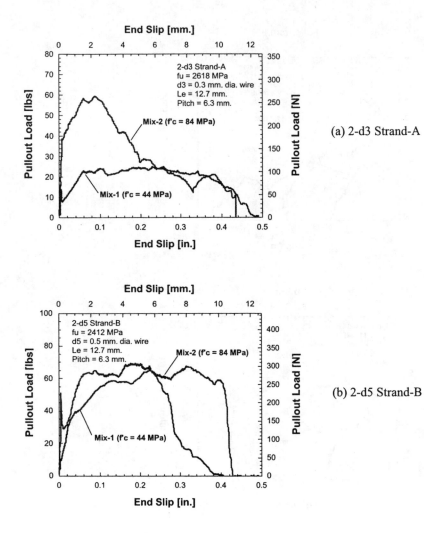

(a) 2-d3 Strand-A

(b) 2-d5 Strand-B

Figure 10—Influence of the Compressive Strength of the Matrix

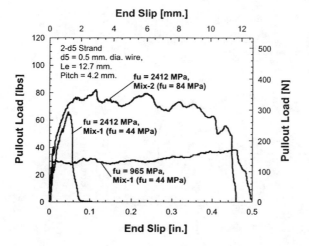

Figure 11—Influence of the Tensile Strength of the Wire in 2-d5 Strand Fibers

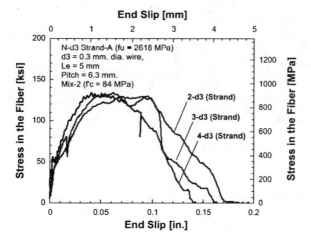

Figure 12—Comparison of the Number of Individual Wires in Twisted Wire
Strand Fibers

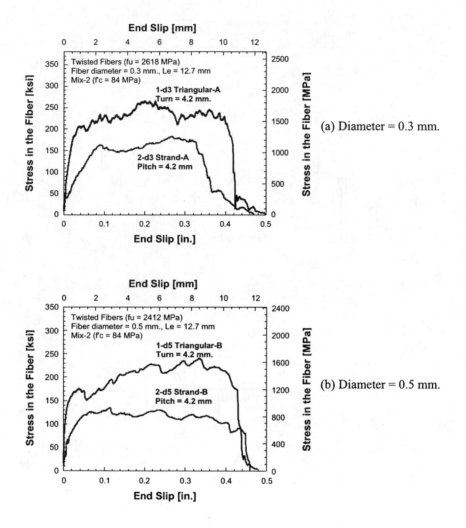

(a) Diameter = 0.3 mm.

(b) Diameter = 0.5 mm.

Figure 13—Comparison of Triangular Torex Fibers and Twisted Wire Strand Fibers

# Early Age Stress—Hydration Kinetics and Thermomechanics

## by I. Pane and W. Hansen

Synopsis: A current research project on hydration kinetics, mechanical properties and early age stress behavior of blended cement conducted at the University of Michigan is reviewed in this paper. A number of experiments including calorimetry and differential thermal analysis were performed to investigate hydration kinetics. The mechanical properties investigated included the compressive strength, splitting tensile strength, Young's modulus, creep compliance, relaxation modulus, and coefficient of thermal dilation. The early age stress behavior was studied by measuring the stress developed in a uniaxially restrained concrete member. In addition, the deformation due to autogeneous shrinkage was also measured experimentally. The experimental data could be used to quantify degree of hydration, and temperature effects on hydration, and could be used as inputs for predicting the early age stress development in concrete.

Keywords: blended cements; heat; hydration; non-isothermal; strength

**Ivindra Pane** is a graduate research assistant at University of Michigan. His research interests include hydration kinetics, viscoelastic behavior, and fracture mechanics.
**Will Hansen** is an Associate Professor at Department of Civil Engineering, University of Michigan. He is a member of several technical committees on concrete materials research (i.e. ACI, TRB, RILEM). His current research include hydration kinetics, early age stress prediction, curling and warping stress prediction in PCC pavement, effects of mineral additives on PCC durability.

## INTRODUCTION

The issue on thermal cracking at early ages has received relatively little attention compared to the cracking caused by the applied load, most likely because there are so many complex phenomena involved. The early age stress develops with time and involves time dependent properties like creep while other mechanical properties are strongly influenced by hydration. Several investigations have been dedicated to studying creep of concrete (1),(2),(3). Yet, only a few investigations have been conducted to obtain early age creep properties (4),(5). In addition, most of creep compliance data were obtained under compression. Despite the argument that reports that creep of concrete is the same in tension and compression, there have been several studies reporting the difference between them. Analyzing the risk of cracking at early ages favors the use of tensile creep since concrete is under tension during most of the early period.

The prediction of the temperature effects on the property development is often done using the maturity concept together with the activation energy. Unfortunately, the activation energy is not constant. It varies with the degree of hydration (6). A more fundamental quantity that relates unambiguously to mechanical properties of concrete is degree of hydration, because all age dependent properties must strongly tie in with the microstructure development, which in turn depends on the degree of hydration (7),(8),(9).

In the early age cracking problem, restraints to hydration-induced deformations cause the stress to develop. Sources of hydration-induced deformation including temperature change and self-dessication are known to be affected by the cement type (10),(11). Blended cements are often thought to be beneficial in reducing the risk of early age cracking. Because some blended systems can reduce the autogeneous shrinkage (12),(13). In order to effectively use blended cement and to optimize its composition, behavior such as like hydration, properties development, and early age stress development, must be analyzed and the relations among them must be understood.

The objectives of this paper is to summarize effects of hydration, temperature and viscoelastic properties on early age stresses. The effects of mineral additives are discussed.

## RESEARCH ON CONCRETE BEHAVIOR AT EARLY AGES

### Experiments

To investigate concrete behavior at early ages, several important properties need to be determined. In the research conducted at University of Michigan, the mechanical properties, hydration kinetics, autogeneous deformations, and stress development were obtained experimentally. Concrete mixes tested are listed in Table 1.

The mechanical properties obtained experimentally included the compressive strength, Young's modulus, splitting tensile strength, and coefficient of thermal dilation (CTD). The strengths and Young's modulus were obtained using a servo-hydraulic machine. CTD was measured using a system consisting of a frame made of invar steel, a water container, and a temperature-controlled water bath. Hydration kinetics were studied using calorimetry and differential thermal analysis (DTA). To investigate the effect of temperature the hydration experiments were conducted on specimens cured at different temperatures isothermally. The autogeneous shrinkage was measured using a steel mold with an LVDT. To maintain the isothermal condition during the shrinkage measurement, a temperature-controlled water bath was also used. The early age stress experiment was conducted using a horizontal servo-hydraulic machine. The stress was measured in a uniaxially restrained concrete specimen under a controlled temperature history.

### Analytical

The mechanical properties such as the compressive strength, splitting tensile strength, and Young's modulus were determined since they are basic concrete properties that are useful for monitoring concrete quality and are also used in designing concrete structures. The other properties obtained in this research are creep compliance and coefficient of thermal dilation (CTD). These two properties together with the autogenous deformation are very important since they can be used to characterize the stress relaxation behavior of concrete at early ages. Hydration kinetics is also important since it describes the aging behavior of concrete properties. The relations between the degree or heat of hydration and concrete properties can be considered unique. In addition, the effect of temperature on concrete properties can be incorporated through degree of hydration.

Since temperature strongly affects degree of hydration and therefore, also affects the mechanical properties, it needs to be incorporated in calculations or predictions of properties development and early age stress development. The stress development without the temperature effect can be calculated using the following form:

$$\sigma(t) = \int_0^t R(t,t')d\varepsilon_{tot}(t')$$

$$\varepsilon_{tot}(t) = \varepsilon_{thermal}(t) + \varepsilon_{autog.}(t)$$

$$= \alpha_{CTD}\Delta T(t) + \varepsilon_{autog.}(t)$$

Meanwhile, stress development including the effect of temperature on hardening behavior is calculated from:

$$\sigma(t) = \int_0^t R(t,Q(t'))d\left[\alpha_{CTD}\Delta T(t') + \varepsilon_{autog}(Q(t'))\right]$$

$$R(t,Q(t')) = E(Q)\left[1 - \frac{\left[(t-t')/m_2(Q)\right]^p}{1+\left[(t-t')/m_2(Q)\right]^p}\right]$$

$$E(Q) = k_1 Q^{k_2}$$

$$m_2(Q) = \lambda\left(\ln\frac{\beta}{Q}\right)^\gamma$$

In the above equations, $R$ is the relaxation modulus, $Q$ is the heat of hydration, and $T$ is temperature. The aging parameter $m_2$ as a function of $Q$ is obtained experimentally, $k_1$ and $k_2$ are obtained experimentally as well. Two deformation components are considered here, thermal and autogeneous deformations. The detail derivation for the procedure of the early age stress calculation can be found in (14).

RESULTS

As mentioned above, relations between mechanical properties can be uniquely related to a measure of hydration. As shown in Fig.1 and 2, the splitting tensile strength ($f_{sp}$) and Young's modulus (E) appear to follow some unique trends when plotted against normalized heat of hydration. Data points shown in Fig.1 and 2 correspond to concrete mixes made of ordinary portland cement (OPC), OPC and 25% fly ash (FA), OPC and 25% blast furnace slag (GGB), and OPC and 10% silica fume (SF). These blended cements seem to have only minor effects on $f_{sp}$ and E

In this research, an analytical method was developed to calculate the degree or heat of hydration in concrete undergoing non-isothermal curing. The method has been applied to semi-adiabatic heat data shown in Fig.3. As indicated, the prediction results seem to agree well with the measured data. The developed method can also be extended for predicting the time development of mechanical properties under non-isothermal curing conditions.

As mentioned previously, the creep property of concrete is a very important factor in determining the stress relaxation behavior. To determine the relaxation behavior, the creep data obtained were converted to relaxation modulus shown in Fig.4. The conversion was done numerically using the trapezoidal integration scheme (1).

Using the relaxation modulus, the measured CTD, and the measured autogeneous deformation, the stress development at early ages can be calculated using the equations given previously. To incorporate the temperature effect on the hardening behavior of concrete on the early age stress development, a computational procedure was developed, which was a numerical calculation of the integrals in the above equations. The results of the stress calculation are shown in Fig.5. The temperature history and the measured stress history are also shown. Two predictions were made: pred.1 was calculated considering the effect of temperature on hardening of concrete, while pred.2 was calculated without such an effect. It is clear that the stress predicted incorporating the effect of temperature on hardening gives a more accurate result.

## SUMMARY

Major findings obtained here suggest that the concrete properties like compressive strength, Young's modulus, splitting tensile strength, creep compliance, and relaxation modulus are strongly tied to degree of hydration. Thus, the degree or heat of hydration can be used as the fundamental parameter describing aging or hardening of concrete. Furthermore, accurate early age stress prediction requires incorporating the temperature effects on hardening of concrete.

## ACKNOWLEDGEMENTS

The authors would like to acknowledge the support from the Center for Advanced Cement-Based Materials (ACBM) and its director Prof. Surendra Shah.

## REFERENCES

1. Bazant, Z.P. (Editor), ``Material Models for Structural Creep Analysis,'' RILEM Proc., Mahematical Modeling of Creep and Shrinkage of Concrete, John Wiley \& Son, 1988.
2. Neville, A. M., Dilger, W. H., and Brooks, J. J.,``Creep of Plain and Structural Concrete,'' Construction Press, New York, 1983.
3. Illston, J. M., ``The Creep of Concrete Under Uniaxial Tension,'' Mag. of Concrete Res. 17 (100), 1965.
4. Emborg, M., ``Thermal Stresses in Concrete Structures at Early Ages,''Doctoral Thesis, Lulea Univ. of Tech., 1989.
5. Khan, A.A., Cook, W.D., and Mitchell, D., ``Creep, Shrinkage, and Thermal Strains in Normal, Medium, and High-strength Concretes During Hydration,'' ACI Mat. J. 94 (2), pp.156, 1997.
6. Abdel-Jawad, Y.A., ``The Relationships of Cement Hydration and Concrete Compressive Strength to Maturity,'' Ph.D Thesis, Univ. of Michigan, 1988.
7. Mindess, S., and Young, J.F., ``Concrete,'' Prentice-Hall, Englewood Cliffs, N. J., 1981.
8. Mehta, P.K., and Monteiro, P.J., ``Concrete,'' 2nd ed., Prentice-Hall, 1993.
9. Granju, J.L., and Grandet, J., ``Relation between the hydration state and the compressive strength of hardened portland cement pastes,'' Cem. Conc. Res. 19, pp.579, 1989.
10. Koenders, E.A., ``Simulation of Volume Changes in Hardening Cement-Based Materials,'' Ph.D Thesis, Delft Univ. Tech., 1997.
11. Jensen, O.M., and Hansen, P.F., ``Autogenous Deformation and Change of the Relative Humidity in Silica Fume-Modified Cement Paste,'' ACI Mat. J. 93 (6), pp.539, 1996.
12. Tangtermsirikul, S., ``Effect of Chemical Composition and Particle Size of Fly Ash on Autogeneous Shrinkage of Paste,'' Proc. JCI Workshop Autog. Shrink. of Conc., Japan, E \& FN Spon, pp.175, 1999.
13. Hori, A., Morioka, M., Sakai, E., and Daimon, M., ``Influence of Expansive Additives on Autogeneous Shrinkage,'' Proc. JCI Workshop Autog. Shrink. of Conc., Japan, E \& FN Spon, pp.187, 1999.
14. Pane, I., "Kinetics, Microstructure, and Thermomechanics of Hydrating Blended Cements," Ph.D thesis, University of Michigan, 2001.

Table 1-Concrete mixes and compositions

| Mix | %OPC | %Additives | w/b |
|------|------|-----------|------|
| 45-1 | 100 | 0 | 0.45 |
| 45-2 | 75 | 25 (FA) | 0.45 |
| 45-3 | 75 | 25 (GGB) | 0.45 |
| 45-4 | 90 | 10 (SF) | 0.45 |
| 35-1 | 100 | 0 | 0.35 |
| 35-2 | 75 | 25 (FA) | 0.35 |
| 35-3 | 75 | 25 (GGB) | 0.35 |
| 35-4 | 90 | 10 (SF) | 0.35 |

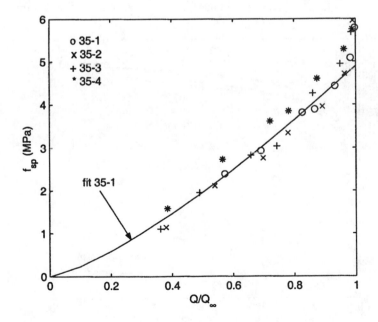

Figure 1-Variations of splitting tensile strength with normalized heat of hydration.

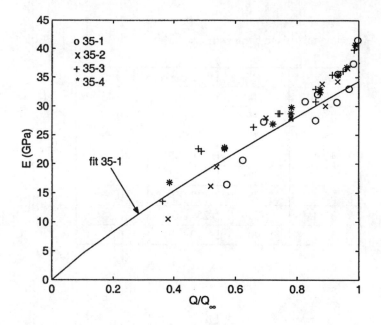

Figure 2-Variations of Young's modulus with normalized heat of hydration.

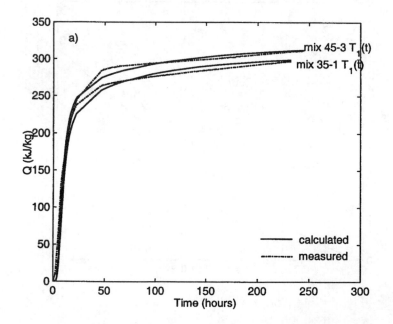

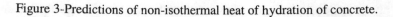

Figure 3-Predictions of non-isothermal heat of hydration of concrete.

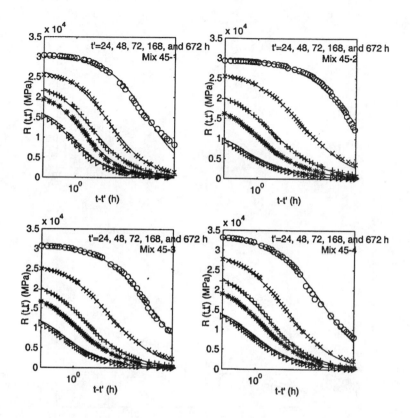

Figure 4-Relaxation modulus of concrete containing blended cements.

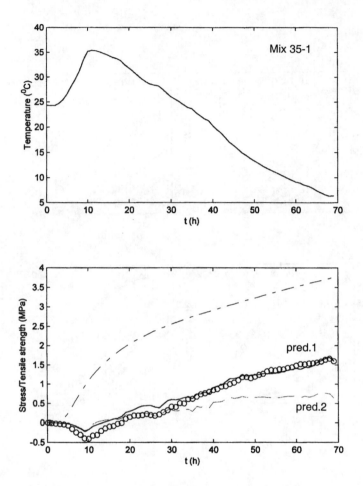

Figure 5-Prediction and verification of early age stress development.

# Thermal and Chemical Activation of High Flyash Content Cement Based Materials

by S. Mane and B. Mobasher

**Synopsis:** An experimental study was conducted to evaluate the mechanical properties of mortar containing flyash subjected to both thermal and chemical activation. Mortar specimens containing class F flyash and various activators were prepared. Up to 50% by weight of cement was replaced with flyash and the results were compared with the control mixture. In order to activate the hydration reactions, additives such as sodium hydroxide (NaOH) and potassium hydroxide (KOH) were used at a rate of 2.5% of total binder weight. Thermal activation was achieved using an autoclave curing process. Both the strengthening and toughening mechanisms were studied in compression, flexure. The fracture results are analyzed using the fracture energy method, the two-parameter facture model of Jenq and Shah, and also the R-curve approach. Experimental data indicate that increasing the flyash content from 20% to 50% results in a favorable influence on compressive strength whereas the flexural and fracture properties remain virtually at the same level. R-Curves provide a more descriptive measure of fracture response. Autoclave curing of high flyash mortar samples results in a marked increase in the strength but a marginal reduction in ductility.

**Keywords:** activators; flyash; fracture mechanism; hydration; physical and chemical activation; toughness

**Sandeep Mane** is a Graduate Student Assistant in the Department of Civil and Environmental Engineering at Arizona State University in Tempe, Arizona.

ACI member **Barzin Mobasher**, Ph.D., is an associate professor of Civil and Environmental Engineering at Arizona State University.  He is a member of ACI Committee 544, Fiber Reinforced Concrete, and ACI 446, Fracture Mechanics. His research activities include fiber reinforced concrete, toughening mechanisms, and modeling of durability.

## INTRODUCTION

During the past several years, use of flyash in concrete has become inevitable. This is partly due to the fact that a large body of research work conducted in the area of blended cements has pointed out the beneficial aspects of the use of flyash in improving the performance of concrete materials.  The use of mineral admixtures such as flyash, silica fumes, and superplasticizers with careful selection of constituent materials has made the production of HPC easier and more economical. A major area of research in the study of influence of flyash on the properties of concrete has developed. Significant number of papers has been published in this area and the effect of flyash on the strength, durability, chemical and pozzolanic reaction, porosity, hydration, etc. has been well documented [1].

Several investigations have been conducted to evaluate the possibility of accelerating the strength development of concrete with blended cements usingactivators other than portland cement or gypsum. Strongly alkaline compounds such as sodium hydroxide, sodium carbonate, potassium hydroxide and calcium hydroxide are frequently used. Palomo [2] studied the mechanism of activation of flyash with highly alkaline solutions NaOH, KOH, etc. and showed a notable influence in the development of the mechanical strength of the final product.  Dongxu [3] describes the influence of alkalinity on the activation and microstructure of flyash with NaOH, $Ca(OH)_2$ and $Na_2SO_4$. Mesto and Kajaus [4] and others [5] [6] [7] reported the effect of various alkali activators on the strength development of slag concrete. However, limited data is available on the correlation of fracture properties with strength.  It is expected that the fracture parameters do not scale proportional to the strength of these formulations.

## RESEARCH SIGNIFICANCE

An experimental study on the influence of fly ash level and various alkali activators on the toughening properties of cement mortar was conducted. Various methods to activate the hydration reaction need to be comparatively evaluated.  From a chemical activation perspective, influence of the type and amount of alkali activators on the mechanical properties of flyash mortar mixtures needs to be studied.   From a thermal activation perspective, the performance of the flyash cement mortar cured by high steam curing was also studied, and both methodologies compared.   Finally, use of R-Curves to characterize the toughening in cement-based materials was verified.

## RESEARCH METHODOLOGY

It is well known that flyash results in low early age strength. Therefore chemical activators were used to activate the hydration reaction of concrete. Alternatively, it is possible to increase the rate of reaction of flyash by means of autoclave curing. A comparative analysis of both these methods is needed. The strengthening and toughening mechanisms of fly ash cement mortar under compression, normal flexure, and cyclic loading-unloading tests were studied. In addition to compressive strength tests, a three-point bending test geometry was used. The beam is subjected to loading-unloading cycles under closed loop crack mouth opening control and the test results were used to construct the R-Curve response of the specimen.  The results of the R-curves were used in the comparative analysis.

## MINERALOGICAL EXAMINATION

Samples of flyash were obtained from the Salt River Project (SRP) Coronado power Station.   The ash was categorized as Class F and it was obtained from a blended source of coals.   The chemical and mineralogical composition of flyash depends upon the characteristics and composition of the coal burned in the power plant.  Owing to the rapid cooling of material, flyash is composed mainly (50-90%) of mineral matter in the form of glassy particles. Qualitative X-ray Diffraction (XRD) was used to determine the nature of the mineralogical species present in flyash samples.     Several samples were analyzed and a typical pattern obtained [8] is reported in Figure 1.  The most notable peaks are identified as Quartz ($SiO_2$), Mullite ($Al_6Si_2O_{13}$) and hematite ($Fe_2O_3$).

## MIXTURE PROPERTIES

The mortar specimens were prepared according to the ASTM C-109 standard.  This procedure specifies a water binder ratio of 0.40 for Portland cement mortars, and an amount of mixing water to produce a flow of $110 \pm 5$ for mortars of binders other than 100% portland cement.  Six mortar samples were prepared with control and various combinations of flyash (20-30%).  In addition, mixtures with activators were also used at a dosage rate of 2.5% by weight of total binder. Sodium and potassium hydroxide (NaOH & KOH) were used as activators.  The dosage of activators was 2.5% by weight of total cementitious materials. The 1N solution of all the activators was prepared and used as a part of the total water content to maintain a constant w/c ratio in the mix. The activator solution was thoroughly mixed with water before it was added to the dry mix. Six mortar samples with 40% and 50% fly ash and activators were also prepared and cured with high pressure steam curing.

Table 2 presents the mix design of mortar mixtures. For each of the mortar blends two batches (A and B) were prepared containing various levels of flyash. Three replicate samples from each batch were tested at 7 days and 28 days. Results of this study were used to evaluate the effect of age on the strength development of mortars.

## SPECIMEN PREPARATION

All material preparation, proportioning, mixing, and testing in laboratory were in accordance with the ASTM C39 and C109 Standards. A vibration table was used to help with the consolidation of the fresh mixture in the molds. The specimens were filled in two layers with proper compaction in between the layers. From each mix, the following specimens were made:

- 4 replicate samples of $342.9 \times 76.2 \times 25.4$ mm ($13.5 \times 3 \times 1$ in.) fracture beams.

- 6 replicate samples of $50 \times 50 \times 50$ mm ($2 \times 2 \times 2$ in.) mortar cubes.

After 24 hrs, specimens were placed in a curing room at 90% (RH) and $23\,^{\circ}$C ($70\,^{\circ}$F) for 7 and 28 days of curing. After the initial curing, the autoclave specimens were placed in an autoclave and procedure followed as specified in [9]. A pressure of 1MPa (150 psi) and a temperature of $182\text{-}185\,^{\circ}$C ($360\text{-}365\,^{\circ}$F) were used in the autoclave curing process. According to the standard guidelines of autoclave curing, the product obtained by this method has compressive strengths, which are normally at least as high as those obtained after 28 days curing at room temperature [10]. A water-cooled diamond blade circular saw was used to cut a 12.7 mm (0.5) in notch at the mid-span of the specimens for three point bending flexure or fracture tests. The specimens were allowed to dry in the laboratory for 12 hrs prior to the tests.

## COMPRESSION TEST

The compression test is perhaps the most common tool used for characterizing concrete mixtures. The average compressive strength of the cement mortar cubes for all the mixtures is given in Table 3.

It is well known that the partial replacement of cement by flyash may result in lower compressive strength at early ages, followed by the development of greater strength at later stages. From Figure 2 we can notice that addition of up to 30% flyash reduces the compressive strength by 50% at the age of 7 days. At 28 days, although the average strength is more than 34 MPa (5000 psi), compared to the control mixture, there is a strength reduction of the order of 25%. With the normal curing, the NaOH activated samples have a significant influence on the compressive strength at 28 days. The activators have an adverse effect at 7 days but finally at the 28 days, the strength gain is almost double, and we obtain a higher compressive strength than that with only flyash. The autoclave-cured cubes have a higher strength as compared to normal cured specimens. At the 40% and 50% flyash levels, the samples yield compressive strengths of about 40 MPa (5800 psi). The activators (KOH and NaOH) do not influence significantly for 40% and 50% flyash mortar. In fact the strength drops by 15-25% with the addition of activators.

## THREE-POINT BEND FLEXURE TEST

Three-point bend flexural tests were performed on $342.9 \times 76.2 \times 25.4$ mm ($13.5 \times 3 \times 1$ in.) beam specimens with an initial notch of 12.7 mm (0.5 in.) A span of 304.8mm (12 in.) was used. The behavior of the specimen in flexure is dominated by the cracking that initiates at the notch and grows along the depth of the specimen. As the test progresses, the deformation localizes at the notch and is dominated by crack propagation. Since the critical deformations are the opening of the crack tip, measured at the base of the notch, the best-controlled variable in flexure tests is the crack opening or a similar displacement. The test is performed under closed loop control with crack mouth opening deformation (CMOD) as the controlled variable.

Figure 3 represents the flexural test set up. In this test, the crack mouth opening displacement (CMOD) was measured across the face of notch using an extensometer. The deflection of the beam was also measured using a spring-loaded linear variable differential transformer (LVDT) with a 2.54 mm (0.1 in) range. A displacement measuring yoke was developed to measure the centerline deflection of the beam with respect to the supports. This device eliminates extraneous deformations such as support settlements and specimen rotations.

## R-CURVE ANALYSIS USING THE COMPLIANCE APPROACH

It is generally accepted that due to existence of a relatively large fracture process zone, which results in the stable crack propagation, LEFM cannot be directly applied to cement-based composites. One alternative is to conduct a stable three-point bend test on a notched beam to obtain a continuous load-deflection curve. This curve is further used to obtain the fracture energy as the area under the curve. A primary characteristic in fracture is the existence of stable crack growth prior to the crack reaching its critical length. The length of process zone depends on microstructure (size of aggregate) as well as on the geometry of the specimens. R-Curves present a methodology to characterize the fracture and take into account the effect of geometry, material properties, and the size of the process zone. R-Curve models integrate the energy dissipation in the process zone as a toughening component of the matrix material. Approaches that are based on the energy principle and the unloading-reloading methods have been quite convenient for evaluating nonlinear fracture toughness parameters as a function of crack length [11]. These ideas relate the energy dissipation in the process zone to an effective elastic crack length.   Ouyang et al. studied the influence of geometry on the R-Curves and on other characteristics of the fracture response [12]. An experimental procedure was developed for the measurement of R-Curves based on loading-unloading curves by Li, Arino, and Mobasher [13].

The strain energy release rate, G, represents the energy available for incremental crack extension. Once it reaches a critical value $G_{IC}$, an instability condition is reached and crack propagation occurs. This is shown as the horizontal line in Figure 5. To characterize fracture toughness using a single parameter $G_{IC}$, only the peak load of a notched specimen tested under mode I condition is required. Quasi-brittle materials dissipate energy due to frictional sliding, aggregate interlock, and crack surface tortuosity. After an initiated crack begins to propagate, the dissipating mechanisms evolve. The increase in the apparent toughness can be related to the stable crack growth by means of an R-Curve. This is shown in Figure 5 for quasi brittle and FRC materials.   The condition for stable crack growth is:

$$G(a) = R(a) \qquad \frac{\partial G(a)}{\partial a} < \frac{\partial R(a)}{\partial a} \tag{1}$$

The condition for crack instability can be defined as:

$$G(a_c) = R(a_c) \qquad \frac{\partial G}{\partial a} = \frac{\partial R}{\partial a} \quad @ \, a = a_c \tag{2}$$

The procedure for calculation of R-Curves using the compliance approach for loading unloading cycles is as follows: A closed loop control test is conducted on a three point specimen with geometrical dimensions b, t, $a_0$, and S representing the depth, thickness, initial notch length, and span. Several cycles of loading-unloading are recorded and the load vs. deformation data is collected. Three measurements were taken for each loop of loading-unloading representing the load (P) where the unloading of a cycle starts, the compliance ($C_u$) for each unlading curve, and displacement (u) at a point where each loop ends. Sets of three readings were used as an input to calculate the R-Curve and its parameters.   First the experimentally measured compliance is used to calculate the crack extension $\Delta a$ for that cycle according to equation 3. From the initial first loading curve ($\Delta a = 0$) the Modulus of Elasticity (E) is calculated and used for future calculations.

$$C_u = \frac{6\,S\,(a_0 + \Delta a)\,V(\alpha)}{E\,b^2\,t}\ ,\ \alpha = \frac{a_0 + \Delta a}{b} \tag{3}$$

Next, the rate of change of compliance can be used to obtain the strain energy release rate as shown in equation 4. The compliance is plotted as a function of the crack extension and a curve fit algorithm is applied to fit the response. Numerical differentiation of the compliance-crack length results in an average value of the rate since several compliance measures are used to calculate the rate. In the presence of residual displacements, additional terms are needed to account for the rate of change of inelastic displacement with respect to crack growth as well. Once all the parameters are obtained, the results are compiled according to equation 4.

$$G^*(a) = \frac{1}{2t}\frac{\partial C}{\partial a}\,P^2 + \frac{1}{2t}\frac{\partial \delta_r}{\partial a}\,P \tag{4}$$

The R curve is then obtained using the well-known relationship of equation (5).

$$K^R(a) = \sqrt{E_c'\,G^*(a)} \tag{5}$$

Where, $E'_c = E_c\,/\,(1 - \upsilon_c{}^2)$ for plane strain and $E_c$ for plane stress. $E_c$ and $\upsilon_c$ represent the elastic modulus and the Poisson's ratio. The R-Curve is plotted as a function of the incremental crack growth.

In addition, two other measures of fracture response were measured from these experiments. The area under the load-deflection curve for the specimen may be related to amount of energy dissipated in the process of fracture, and is referred to as the fracture energy or Toughness ($G_f$) [14]. Furthermore, Critical Stress Intensity Factor ($K_{Ic}{}^s$), and the Critical Crack Tip Opening Displacement (CTOD$_c$) were also measured. Parameter $K_{Ic}{}^s$ is measured for an effective crack length at which unstable crack growth takes place (peak load). [15], while The CTOD$_c$ is a measure of brittleness of the material and corresponds to the crack opening at which the unstable crack growth takes place. A limit value of zero represents the LEFM conditions that crack propagation is possible with infinitesimal crack growth. [18].

## RESULTS OF CONVENTIONAL FRACTURE TEST

Analysis of the flexural test results suggests that flyash has the significant influence on the flexural properties. Figure 6 represents the load vs. crack mouth opening response for a series of specimens made with several levels of flyash. In the pre-peak region all curves behave in a similar manner. Concrete with 30% flyash yields higher strength as compared to control, 15%, 20%, and

25% flyash concrete. There is a 25% increase in flexural strength with 30% flyash as compared to control mixtures. The control concrete responds in a ductile manner in the post-peak region as compared to the others. A comprehensive set of test results pertaining to the effect of flyash content on the flexural response are reported in [8].

It is also noted that concrete with 15% and 20% flyash has a reduction in flexural capacity, whereas with 25% and 30% flyash, the flexural strength is improved by 15% and 30% respectively.   One of the functions that flyash serves is to strengthen the aggregate and paste region commonly referred to as the interface transition zone (ITZ).  By reducing the porosity at the ITZ region, a stronger, denser concrete with fewer voids is produced.  Cracks would then have a tendency to go through the aggregates as opposed to the around the aggregate.  The effect of flyash on the brittleness is clearly demonstrated by lower values of deflection in the 30% flyash concrete.  It is noticed from Figure 6 that the crack mouth opening displacement (CMOD) in the post-peak regime with 30% flyash is significantly less than the comparable mixtures.

## EFFECT OF FLYASH AND ACTIVATORS ON R-CURVE

Figure 7 shows a typical loading-unloading curve used for the analysis of fracture parameters. Various fracture parameters such as toughness, E, CTOD, etc. were given in table 4 and indicates that the toughness values are adversely affected by the flyash addition as compared to control mix. As more flyash is added, the specimen becomes more brittle and exhibits a reduction in toughness. Use of activators has a minor influence on the toughness properties of mortar specimen.

The $K_{Ic}^s$ is also influenced by the addition of flyash. The controlled specimen has a higher $K_{Ic}^s$ than the flyash mortar specimen. Based on the data presented in Table 4, it can be conclude that the flyash mortar specimens are more brittle and susceptible to the crack propagation. The KOH activators improve the $K_{Ic}^s$, whereas the NaOH activator does not have any effect on mortar mix.

Figure 8, 9 and 10 shows the influence of flyash and activators on the R-curve.  The control concrete has a higher plateau level compared to the blended cement mixtures irrespective of the size of the specimen tested. It is interesting to note that with the addition of flyash, more energy is required for the initiation of the crack. However, once the crack is formed less energy is required to propagate the same crack. Therefore one can conclude that the addition of flyash decreases the toughness, making the concrete more brittle in nature. Figure 8 shows that with the addition of flyash (15% to 30%), the R-curve is reduced.  The KOH activator has a significant influence on the fracture parameters. Figure 9 shows that 30% flyash with KOH activator has the better resistance than any other activators. It can be noticed from all the figures that the initiation of cracks needs more energy with the addition flyash and the activators.

Figure 10 shows the effect of flyash and activators on the autoclave cured mortar specimen. In the initial period the flyash and activated mix has less brittleness than the control mix. However, the control has more resistance to crack propagation when the load reach the peak. For the 40% flyash mortar specimen the brittleness is the same when KOH activators are added. Whereas, the 40% flyash mortar with the NaOH activators is less brittle as compared to KOH activators. The 50% flyash mortar mix has more resistance to crack propagation as compared to the 40% flyash mix. The NaOH activator has less brittleness as compared to KOH activators.

## CONCLUSION

A study is presented to show the effect of flyash with both thermal and chemical activators. Results indicate that flyash cement based mixtures with up to 50% cement replacement can be manufactured with high strength and ductility. Compressive strength levels as high as 40 MPa were obtained using the autoclave curing procedure. The addition of flyash decreases the toughness, making the concrete more brittle in nature. The ductility of these mixtures may be characterized using an R-Curve approach, which clearly shows the reduction in stable crack growth.

## ACKNOWLEDGEMENTS

The Authors are grateful to the financial and technical support of Salt River Project (SRP). The technical guidance of Mr. Mark Bailey of SRP is greatly appreciated.

## REFERENCES

1    Tixier, R. and Mobasher, B.," *Development and Application of High Performance Concrete Materials with Blended Coal Flyash*"- Literature Review, TECHNICAL REPORT No 00-1, Submitted to SRP, Arizona State University, July 2000.

2    Paloma, A., Grutzeck, M. W., Blanco, M. T. *"Alkali-activated fly ashes A cement for the future"*, Cem. & Conc. Res., 1998; pp. 1323-1329.

3    D. Li., Y. Chen, J. Shen, J. Su, X. Wu, *"The influence of alkalinity on activation and microstructure of fly ash"* Cem. & Conc. Res, 2000; pp. 881-886.

Table 1- Chemical composition of flyash and cement, in % reported in [9]

| Reference | $SiO_2$ | $Al_2O_3$ | $Fe_2O_3$ | $SiO_2 + Al_2O_3 + Fe_2O_3$ | CaO | MgO | $K_2O$ | $Na_2O$ | $SO_3$ | LOI |
|---|---|---|---|---|---|---|---|---|---|---|
| Cement | 21.1 | 4.53 | 3.29 | 28.92 | | 2.21 | | | 3.09 | 1.40 |
| Flyash | 56 | 23 | 5 | 84 | 8.7 | 2.4 | 1.2 | 1.3 | 0.5 | 0.55 |

Table 2- Portland cement mortar mix design, (* N=NaOH, K=KOH)

| Mix ID # | % FA | W/B ratio | S/C ratio | Dry weight of materials, g | | | | Plasti-cizer | Activator, |
|---|---|---|---|---|---|---|---|---|---|
| | | | | Cement | Flyash | Sand | Water | ml | g |
| 25FN | 25 | 0.4 | 2 | 525 | 175 | 1400 | 262.5 | 10 | 17.5 N* |
| 30FN | 30 | 0.4 | 2 | 560 | 240 | 1600 | 300 | 15 | 20N |
| Con | - | 0.4 | 2 | 900 | 0 | 1800 | 360 | 18 | - |
| 20FA | 20 | 0.4 | 2 | 720 | 180 | 1800 | 360 | 18 | - |
| 25FA | 25 | 0.4 | 2 | 825 | 275 | 2200 | 440 | 18 | - |
| 30FA | 30 | 0.4 | 2 | 770 | 330 | 2200 | 440 | 18 | - |
| 40FA | 40 | 0.4 | 2 | 270 | 180 | 900 | 180 | 10 | - |
| 50FA | 50 | 0.4 | 2 | 225 | 225 | 900 | 180 | 10 | |
| 40FK | 40 | 0.4 | 2 | 270 | 180 | 900 | 168.8 | 10 | 11.2K |
| 50FK | 50 | 0.4 | 2 | 225 | 225 | 900 | 168.8 | 10 | 11.2K |
| 40FN | 50 | 0.4 | 2 | 270 | 180 | 900 | 168.8 | 10 | 11.2N |
| 50FN | 50 | 0.4 | 2 | 225 | 225 | 900 | 168.8 | 10 | 11.2N |

Table 3- Compression test results of the autoclave cured cement mortars.

| Identification | Chemical Activator | Thermal Curing | Average Compressive Strength, MPa (psi) | |
|---|---|---|---|---|
| | | | 7 Days | 28 Days |
| Control | - | Normal | 43 (6257) | 48 (6960) |
| 20FA | - | Normal | 31.5 (4578) | 34 (4927) |
| 25FA | - | Normal | 31 (4477) | 35 (5073) |
| 30FA | - | Normal | 26.75 (3882) | 35.25 (5113) |
| 25FN | NaOH | Normal | 18.5 (2675) | 26 (3788) |
| 30FN | NaOH | Normal | 17.5 (2544) | 35.75 (5202) |
| 40FA | - | Autoclave | 39.75 (5768) | |
| 50FA | - | Autoclave | 40.25 (5832) | |
| 40FK | KOH | Autoclave | 37 (5394) | |
| 50FK | KOH | Autoclave | 35.75 (5171) | |
| 40FN | NaOH | Autoclave | 37 (5459) | |
| 50FN | NaOH | Autoclave | 31.75 (4602) | |

Table 4- Measured fracture properties from three-point bending test

| Mix Id # | %Flyash | Activators | Load | At the peak | | | E | $G_f$ | $a_c$ | $K_{Is}^c$ | $CTOD_c$ |
|---|---|---|---|---|---|---|---|---|---|---|---|
| | | | | Effective stress | CMOD | Deflection | | | | | |
| | | | (lbs.) | (psi) | (in.), $10^{-3}$ | (in.), $10^{-3}$ | (psi), $10^6$ | lbs./in | in | psi in $^{-1/2}$ | in |
| 3 x 1 x 13.5 in specimen | | | | | | | | | | | |
| 15FA1 | 15 | - | 269 | 129.12 | 1.22 | 1.66 | 2.72 | 2.14 | 0.657 | 733.30 | 4.81E-04 |
| 20FA1 | 20 | - | 200 | 96.00 | 0.85 | 1.42 | 2.59 | 1.07 | 0.582 | 517.39 | 2.51E-04 |
| 25FA1 | 25 | - | 240 | 115.20 | 0.50 | 1.02 | 3.99 | 1.48 | 0.530 | 593.43 | 1.11E-04 |
| 30FAL | 30 | 2.5%L | 218 | 104.64 | 0.81 | 0.88 | 2.82 | 1.19 | 0.655 | 614.68 | 3.84E-04 |
| 30FAK | 30 | 2.5%K | 300 | 144.00 | 1.36 | 2.01 | 2.85 | 1.07 | 0.681 | 793.36 | 5.34E-04 |
| 30FAN | 30 | 2.5%N | 197 | 94.56 | 0.62 | 0.10 | 2.75 | 0.77 | 0.705 | 579.40 | 4.33E-04 |
| 30FA | 30 | - | 143 | 68.64 | 1.00 | 1.12 | 2.98 | 0.85 | 0.84 | 472.80 | 4.37E-04 |
| Control | - | - | 199 | 95.52 | 1.34 | 2.21 | 3.16 | 1.63 | 0.906 | 701.2 | 6.81E-04 |
| Autoclave cured mortar specimen | | | | | | | | | | | |
| 40FA | 40 | - | 186 | 89.28 | 1.04 | 1.64 | 2.77 | 1.10 | 0.713 | 549.77 | 4.16E-04 |
| 40FK | 40 | 2.5%K | 173 | 83.04 | 1.30 | 1.44 | 2.40 | 0.71 | 0.740 | 421.06 | 4.90E-04 |
| 40FN | 40 | 2.5%N | 213 | 102.24 | 1.38 | 1.72 | 2.4 | 0.85 | 0.839 | 704.29 | 7.94E-04 |
| 50FA | 50 | - | 192 | 92.16 | 0.95 | 1.47 | 2.5 | 0.80 | 0.725 | 574.83 | 4.89E-04 |
| 50FK | 50 | 2.5%K | 178 | 85.44 | 1.06 | 1.53 | 2.39 | 0.73 | 0.77 | 541.30 | 5.45E-04 |
| 50FN | 50 | 2.5%N | 196 | 94.08 | 0.90 | 1.37 | 2.26 | 0.91 | 0.635 | 538.06 | 3.89E-04 |

**Note:**

Potassium Hydroxide, KOH = K

Sodium Hydroxide, NaOH = N

Lime, $Ca(OH)_2$, = L

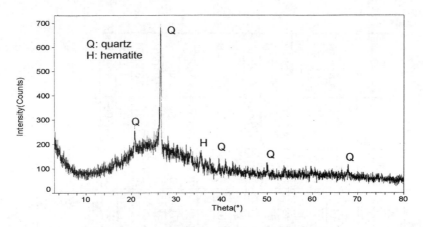

Figure 1- XRD pattern for class F flyash used

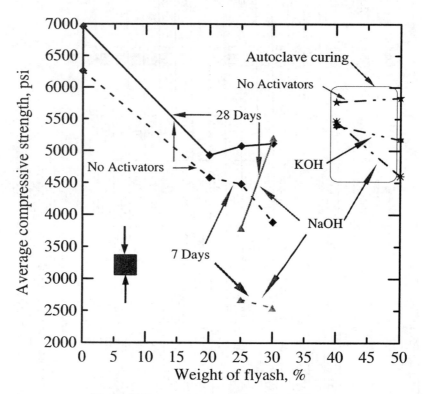

Figure 2- Compression test on mortar cubes

Figure 3- Three-point bend flexural test setup

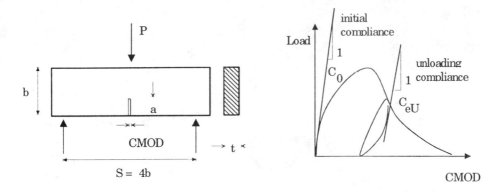

Figure 4- Testing Configuration and Geometry of Specimen for the Fracture Test

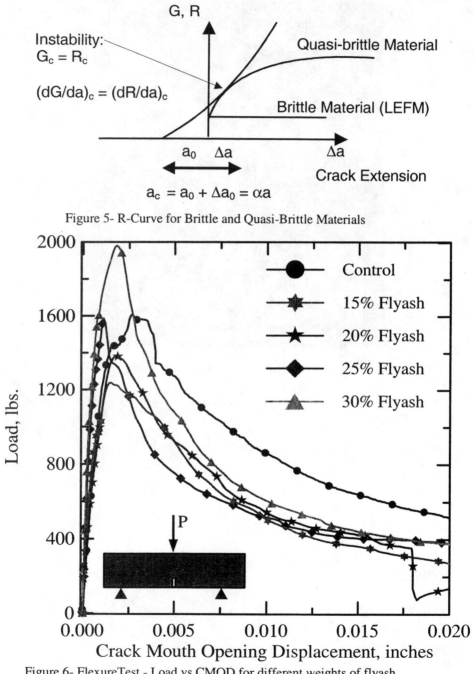

Figure 5- R-Curve for Brittle and Quasi-Brittle Materials

Figure 6- FlexureTest - Load vs CMOD for different weights of flyash

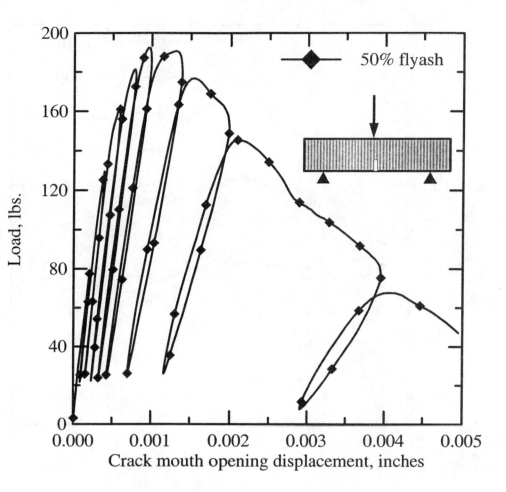

Figure 7- Typical Load vs CMOD curve.  Note that the compliance increases
with the growth of the crack.

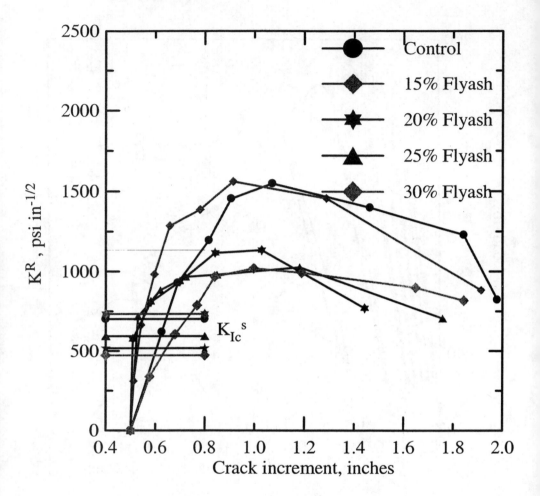

Figure 8- Effect of flyash on R-Curve of concrete with the location of the $K^s_{Ic}$

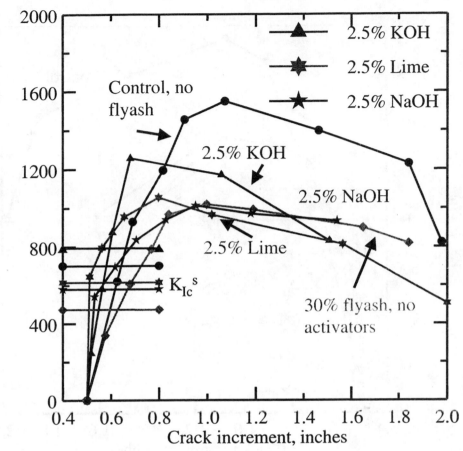

Figure 9- Effect of activators on R-Curve of 30% flyash concrete

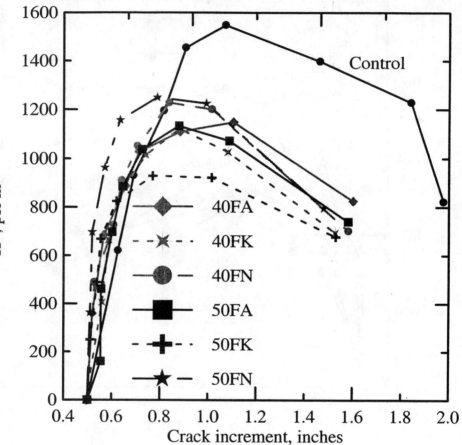

Figure 10- Effect of flyash and activators on R-Curve of autoclave cured
specimen (FA = Flyash only, FN= Flyash activated with NaOH, and
FK=Flyash activated with KOH)

# LAMINATED AND FIBER REINFORCED CEMENT COMPOSITES

# Ferrocement: International Revival

## by A. E. Naaman

**Synopsis:** Ferrocement and more generally laminated cement based composites are in a state of revival worldwide. Looking back and looking ahead, this paper focuses on three aspects of development that are at the basis of the current revival of ferrocement: 1) the application side, where some new limits and new daring ideas were demonstrated; 2) the professional side, particularly the activities of the International Ferrocement Society and the publication of the first Ferrocement Model Code, and 3) the technical side, where high performnace fiber reinforced polymeric (FRP) meshes are introduced either alone or in combination with fibers or micro-fibres leading to hybrid composites with improved performance and reduced cost. Prospects for the near future are also discussed.

**Keywords:** ferrocement; fiber reinforced concrete; fiber reinforced polymeric meshes; high performance fibers; hybrid composites, modulus of rupture; processing; steel meshes; textile concrete; 3 D meshes

**Antoine E. Naaman**, FACI, is a professor of Civil Engineering in the Department of Civil and Environmental Engineering at the University of Michigan, Ann Arbor. He obtained his Ph.D. in Civil Engineering from the Massachusetts Institute of Technology in 1972. He is an active member of ACI Committees 544, Fiber Reinforced Concrete; 549, Ferrocement; 440, FRP Reinforcements; and joint ACI-ASCE Committee 423, Prestressed Concrete. His research interests include prestressed and partially prestressed concrete, high performance fiber reinforced cement composites, and the integration-tailoring of new construction materials in structural applications.

## INTRODUCTION

Ferrocement was truly the first invention of reinforced concrete; the first application of ferrocement dates from 1849 and created the real impetus for reinforced concrete initial development around 1860 [1. 2, 9, 14, 19]. Since this initial period, ferrocement and reinforced concrete have followed their own and separate path of development; the rapid development of reinforced concrete stifled that of ferrocement which underwent a relatively dormant period until about the early1960's, time at which a more scientific inquiry of its behavior, properties and particular characteristics was initiated.

For all practical purposes, the definition of ferrocement has changed little since its invention [14]. The definition was clarified and adopted by ACI Committee 549 in its first State-of-the-Art Report on Ferrocement [1]. It stated that *"Ferrocement is a type of thin wall reinforced concrete commonly constructed of hydraulic cement mortar reinforced with closely spaced layers of continuous and relatively small size wire mesh. The mesh may be made of metallic or other suitable materials."*

However, at the beginning of the twenty first century, the basic definition of ferrocement is expanding in scope. Ferrocement is an extension of reinforced concrete but also is now considered a member of the family of laminated composites; it can be reinforced with steel, or non-metallic meshes such as fiber reinforced polymeric (FRP) meshes, leading to the term "textile concrete". The addition of fibers or micro-fibers as complementary reinforcement in the cement matrix, to improve performance, makes ferrocement a hybrid composite. An extended definition accommodating the above variations is suggested in [14].

Thin reinforced concrete products or cement-based composites designed for structural applications are possible either as ferrocement or fiber reinforced mortar. The main difference between them is in the reinforcement. Ferrocement uses continuous reinforcement or meshes while discontinuous reinforcement in the form of fibers is used in FRC composites. At one boundary, we have ferrocement and at the other boundary we have fiber-reinforced mortar. The whole spectrum of reinforcements between the two boundaries is covered by hybrid combinations, where both continuous reinforcements or meshes and discontinuous fibers are used.

The early 1970's brought increased interest in ferrocement. Today it can be said that ferrocement is in a state of revival and renaissance because it has achieved the status of structural material on its own merit and because it has solidly established its particular niche as the main extension of reinforced concrete in thin products applications.

## FERROCEMENT REVIVAL: THE PROFESSIONAL SIDE

### Historical Milestones

The key feature of ferrocement when it was invented is that it could be used as a water proof or impermeable container that will hold liquids and be durable (i.e., will not rot) in comparison to wood; so the initial applications were Lambot's boat or canoe, large flower pots, and orange containers. In the mid twentieth century, with input from Nervi [19] the use of ferrocement extended into relatively larger boats or yachts, docks, dome structures and shells. In such applications strength and strength to weight ratio, ductility, crack resistance (i.e. crack width limits) are as important as impermeability and durability.

In 1972 the US National Academy of Science formed a panel to report on the applications of ferrocement in developing countries. Subsequently two major events occurred: in 1975 the American Concrete Institute formed ACI Committee 549, Ferrocement, and in 1976 the International Ferrocement Information Center (IFIC) was established at the Asian Institute of Technology in Bangkok. The IFIC acts as a repository of all information related to ferrocement including research, applications, guidelines, educational material and codes. ACI Committee 549 produced two important documents: the State-of-the-Art Report on Ferrocement (first published in 1982) and the Guide for Ferrocement Construction, Design and Repair (first published in 1988) [1, 2]. These documents were disseminated and received acceptance worldwide. However in 1997, ACI Committee 549 changed its name to reflect development in non-metallic reinforcements of thin reinforced concrete products and the term "Ferrocement" was dropped from the title of the Committee. While this seemed unfortunate for the benefit of ferrocement, it may have created another opportunity. Indeed, in 1991, the International Ferrocement Society was formed with headquarters at the Asian Institute of Technology in Bangkok. The International Ferrocement Society (IFS) was founded in 1991 to coordinate and to cater to the needs of practitioners, architects, engineers and researchers on application, development, utilization and research on ferrocement. The IFS has essentially taken the lead in anything "ferrocement"; it organizes every three to four years an International Symposium on Ferrocement, disseminates knowledge through educational short courses and seminars, and has established an number of technical committees to help respond to the most urgent needs of the profession. They are described next with their mission.

### Committees of the International Ferrocement Society

IFS Committee on Ferrocement Model Code; IFS-10. Ever since ferrocement was recognized as a structural material, again and again, architects, engineers, and practitioners all over the world have demanded a code for ferrocement. Construction and inspection permits could not be obtained without a code of reference. When applying conventional reinforced concrete codes to ferrocement, satisfying certain design criteria, such as minimum concrete cover or minimum fire rating, was not possible. Yet stucco has been used as cladding in housing for decades, and houses entirely made from wood and plywood do pass the fire rating requirements and are being built on a large scale in the US. This contradiction can only be explained by the fact that each material is sponsored by a user's group, a professional society, whose main initial objective is, and has always been, a code of practice. Thus the IFS Committee on the Ferrocement Model Code was formed in 1995, and has completed the first code on ferrocement in existence worldwide, *Ferrocement Model Code: Building Code Recommendations for Ferrocement, IFS 10-01* [6]. The main mission of this Committee is to pursue the development, improvement, revision, expansion, etc. of the Ferrocement Model Code and any related matter.

IFS Committee on Education: IFS-20. The main mission of this Committee is to provide educational support for the dissemination of ferrocement knowledge worldwide, that is, 1) organize seminars, short courses, and technology transfer workshops, 2) identify local or regional needs and help prepare targeted educational material, and 3) maintain information on resource faculty and act as a liaison with local organizers. The Committee is to work with country representatives of the Ferrocement International Network to identify the needs, and planning and local organization. First on the agenda is the immediate dissemination and proper use of the Ferrocement Model Code.

IFS Committee on Corrosion and Durability: IFS-30. The main mission of this Committee is to develop guidelines for the design of ferrocement to prevent reinforcement corrosion and to improve durability under various environmental conditions. This includes the use of stainless steel meshes and FRP reinforcements, as well as adding anti-corrosion agents to the matrix.

IFS Committee on Housing and Terrestrial Structure: IFS-40. Housing is potentially the largest application of ferrocement. The main mission of this committee is to identify housing alternatives where ferrocement, and laminated cementitious composites in general, can be competitive as the primary structural material, and to provide guidelines and recommendations on such applications. The Committee also evaluates applications of ferrocement as a support material in housing, such as for water tanks, roofing elements, and sunscreens, etc.

IFS Committee on Tools and Equipment: IFS-50. The main objectives of this committee are: 1) to assemble a document describing the tools and light equipment needed for the successful and cost efficient construction of ferrocement structures; and 2) to recommend the development of new and improved tools that would help make construction more efficient and cost effective in the future. Examples include a pneumatic tool to tie together several layers of mesh.

IFS Committee on Advanced Materials: IFS-60. The main mission of this committee is to develop and report information, and develop and maintain guidelines for the proper use of advanced materials in ferrocement. For instance, it is clear today that the addition of fly ash to the mortar matrix offers a number of short- and long-term benefits. The committee could recommend the use of fly ash and make sure that fly ash is available either separately or as a cement additive, or recommend that cement manufacturers make available cement bags containing blended cement and fly ash in the proper proportions. Also, ferrocement can benefit from the use of steel meshes with higher yield strength. The committee is examining such a variable in order to make recommendations to the mesh manufacturers to help them develop an optimum mesh system for ferrocement.

IFS Committee on Seismic Applications; IFS-70. This committee has two complementary missions: 1) provide guidelines and standards for planning, designing, building and evaluating ferrocement structures subjected to seismic loads; and 2) provide guidelines and standards for the use of ferrocement as a retrofit and strengthening material to structures damaged by earthquakes. Examples include jacketing of RC columns to improve their ductility and shear resistance, and cladding of masonry walls to minimize loss of human life should failure of the wall occur.

## FERROCEMENT REVIVAL: THE APPLICATION SIDE

The use of ferrocement in industrialized countries, particularly the North America, has been minimal. This is perhaps due to the ready availability and relatively low cost of other competing construction materials such as wood and plywood, which can be easily worked with unskilled labor.

There is, however, increasing evidence that even in industrialized countries, ferrocement can be cost competitive through mechanized fabrication and the proper choice of mesh reinforcement. For instance, a factory produced ferrocement element using expanded metal mesh instead of woven wire mesh, may cost two to three times less than conventional manually produced ferrocement elements of equivalent performance. The increasing availability of fiber reinforced polymer (FRP) or plastic meshes and the use of combinations of meshes and fibers, will likely lead to further cost reductions.

In order to be competitive in the North America, ferrocement must find applications where other materials are not cost competitive, or where its properties are uniquely needed, or where labor cost is not a constraint. Examples include the cement sheet business, applications where fire resistance is required, marine applications, and increasingly, repair, and retrofitting work.

## Notable Applications

It is informative to review the size and type of some structures or structural elements built with ferrocement in order to provide an idea of their scope and reach. Following are some examples documented in Refs. 14 and 20:

1. Boats of up to 33 meters in length (China, New Zealand); boats using combined ferrocement and reinforced concrete of up to 90 meters in China.
2. Double cantilever V shaped roofing beams spanning 33 m and having a thickness of only 50 mm (reported by de Hanai in Brazil).
3. Structural shell elements spanning 16 m, such as for the group of "coupolas" built by V. Barberio, to house a fish farm in Abruzzo, Italy.
4. A 55-meter long by 15.9-m wide oil tanker to carry up to 1100 tons of fuel (Pertaminal Fuel Oil Barge) built in Indonesia by Douglas Alexander of New Zealand.
5. A series of domes built for the mausoleum of Queen Alia in Amman, Jordan, the largest having a diameter of 16 m with a 10-m height as reported by Jennings.
6. Ribbed precast ferrocement coffers (3.6 x 3.6 m and 50 mm thick) used as permanent formwork for the reinforced concrete roof of the Schlumberger laboratories in Cambridge, England.
7. Lining for an olympic size swimming pool in Cuba, reported by Wainshtok Rivas.
8. Thinnest ferrocement shell built at the University of Sidney for their ferrocement canoe (Aurora Australis); it had a thickness of about 2 mm and used the concept of an origami folded structure.
9. A six meters span hollow cored box section bridge built for one way traffic and light trucks weighting up to 8 tons in Mexico, reported by Fernandez.
10. A 150 m$^3$ elevated water tank in Brazil, reported by de Hanai.
11. Prefabricated water tanks, 3.6 m in diameter and 16 m$^3$ in capacity, developed in Singapore by Paramasivam.
12. Sunscreen L shaped elements 5 m long and 40 mm thick also developed in Singapore by Paramasivam and Mansur.

Some of the most daring structures using ferrocement were built by D. Alexander in New Zealand [3]. He combined the concepts of ferrocement, fiber reinforced concrete, and prestressing with high strength wires to build relatively large scale oil barges and wharves at competitive cost. In particular, he combined the use of high tensile wires and fibers with wire mesh reinforcement to improve crack distribution, crack width, and overall mechanical properties. In a similar manner, another engineer from Hong Kong and Australia, Peter Allen, built large scale roofing elements (hangars) of up to 30 m in span, using ferrocement, high strength steel wires, and prestressing. He also built a 22-m long ocean racing yacht, the *Helsal,* using ferrocement and post-tensioning. *Helsal* received several line honors and held several records in Australia. Curved wing like external wall panels, 9 m long and 3 m wide, were built as part of the Technical Building for the Yambu Cement Company in Saudi Arabia.

A new generation of engineers and architects is discovering the benefits, versatility, and unique characteristics of ferrocement in special applications. A case in point is the main gate to the Yambu Cement Company, which resembles a piece of fabric blowing in the wind, to form the roofing structure, which finishes at one end by a ribbon-like ferrocement that undergoes a steep rotation and becomes the tower of the structure as if to provide a veil for it (Fig. 1). Another recent example reported by R. Alexander is the use of ferrocement for building a 7.3 m diameter prototype diffuser augmented wind turbine for generating electricity from wind (Fig. 2). A warehouse semicylindrical structure of about sixty meters in length was built by Milinkovic in Yougoslavia (Fig. 3). And recently a pedestrian ferrocement bridge using a hollow-cored box deck was built in the United Kingdom and reported by Nedwell. These accomplishments should be considered at the boundaries of today's ferrocement technology. It is hoped that they will become widely utilized in the future, while new daring frontiers will be attained. The reader is referred to Ref. 14 where extensive information, detailed references and photos of most of the above-described applications can be found.

**Function Evolution of Ferrocement**

The function of ferrocement in construction applications has also evolved with time leading to increased applications and versatility. Figure 4 illustrates the evolution of the function of ferrocement in various structural applications. Ferrocement can be seen as a stand-alone structural material for applications such as boats, shells, and housing systems. Ferrocement can also be viewed as a support material in special applications with new structures of reinforced and prestressed concrete such as for instance in confinement or for permanent molds. Finally ferrocement can be seen as a repair and strengthening material particularly in light structural applications, such as strengthening masonry walls, providing a lining for swimming pools, and lining for sewage tunnels.

## FERROCEMENT REVIVAL: THE TECHNICAL SIDE

For all practical purposes, the definition of ferrocement has changed little since its invention (see Introduction). An extended definition accommodating hybrid combinations of reinforcements and the fineness of the mortar matrix is suggested in [14].

The use of FRP mesh and hybrid reinforcements of meshes and fibers are two relatively recent development that can be key for the revival of ferrocement at the technical level. The evolution of the reinforcement system of ferrocement is described in Fig. 5. It can be observed that other ideas are also being pursued, such as the use of three-dimensional meshes, and the combination of at least two extreme meshes taking in sandwich a fiber mat. Fortunately the various reinforcement systems described in Fig. 5 are not only hypothetical but also they have been tried in various research studies and proven successful.

### FRP Meshes and Hybrids

The use of FRP meshes in ferrocement (leading to what is also described as textile concrete), seems logical in applications where corrosion and weight of the structure are of concern, where non-magnetic properties are desired, and where the production process can be simplified.

From a mechanical performance viewpoint, FRP meshes (or fabrics or textiles) can be successfully used in ferrocement and other thin reinforced concrete (mortar) applications, as observed by several researchers [4, 5, 7, 8, 10, 11, 12, 13, 15, 16, 21, 24]. An extensive analytical investigation by Parra and Naaman [21] showed that high performance (particularly high modulus) FRP meshes can lead to bending strengths far superior to those achieved using conventional steel meshes; moreover, a good cracking distribution and a good ductility or energy absorption before failure can be achieved. Similar conclusions can be drawn from experimental observations. Typically the modulus of rupture of ferrocement plates using conventional low yield strength steel meshes does not exceed 50-60 MPa at volume fractions of mesh of up to 8% [14].

Figures 6a and 6b illustrate the bending response of ferrocement plates reinforced reinforced with two layers of FRP meshes without and with fibers. Figure 6a shows that equivalent elastic bending strength close to 26 MPa was obtained using only two layers of either carbon mesh, Kevlar mesh, or Spectra mesh. The total volume fraction of reinforcement was between 1.15% and 1.7%. Increasing the number of reinforcing mesh layers in the composite does not lead to a proportional increase in bending resistance. The use of hybrid composites where intermediate meshes are replaced by discontinuous fibers may offer, in some cases, a better and cost effective solution.

Figure 7 shows that the modulus of rupture increased from 26 MPa to more than 38 MPa when in addition to two layers of Kevlar mesh ($V_r = 1.15$ percent), short discontinuous PVA fibers were added to the matrix in amounts equal to 1 percent by volume. Similarly, the modulus of rupture increased to about 43 MPa when 1.5% Spectra fibers by volume were added (Spectra fibers are made of a high molecular weight polyethylene). Such strength values are sufficient for a large proportion of applications. The combined use of only two layers of mesh reinforcement, placed near the extreme surfaces, with discontinuous fibers offered an optimum combination for bending behavior and should be further explored in future studies. The use of fibers improved significantly the interlaminar shear resistance of the matrix and thus allowed the extreme layer of mesh reinforcement to contribute its full capacity to the resistance of the member.

Other benefits of using fibers in a hybrid configuration, include increased cracking strength, finer microcracks or shrinkage cracks, smaller crack widths under load, and improved of the composite to be drilled through (like wood or steel) without significant fracture and fragmentation. Figure 8 shows the effect of fiber addition on the first cracking strength and the initial multiple cracking region for specimens with Kevlar meshes. It can be observed that, with the addition of fibers, the first cracking strength is higher, and the curve is much less rugged; that is, the crack width and spacing are significantly smaller.

The results of Figs. 6, 7, and 8 are surprisingly encouraging. Moduli of rupture close to 40 MPa are attained with less than 3% total volume fraction of reinforcement (meshes and fibers).   Moreover, the addition of fibers leads to improved ductility, higher first cracking strength, extensive multiple cracking, smaller crack widths, smaller crack spacing, and improved shear resistance. Figure 8 shows the effect of fiber addition on the first cracking strength and the initial multiple cracking region. The curve is less rugged when fibers are added leading to finer and well distributed cracks.

**Steel Meshes and Hybrids**

The benefits of hybrid compositions whereas fibers are added to the matrix to replace the intermediate layers of mesh apply also to steel meshes.   Figure 9 illustrates this statement.   Generally it is not desirable to use a relatively heavy mesh in ferrocement. The requirement for a high specific surface forces the use of relatively finer wire mesh. The main reason is to improve multiple cracking, minimize spalling of the concrete cover, and not exceed interlaminar shear resistance.  However, in bending, intermediate layers of mesh are significantly less effective than the outer layers, particularly with systems that do not yield.  Thus using a heavier mesh at the extreme surfaces may be more cost effective.  This can be achieved provided other properties are not impaired.

Figure 9 illustrates the effect of using fibers with two types of expanded metal mesh.  The first mesh provides a volume fraction of 3.73% with only two layers. The bending resistance of the ferrocement composite increases from about 23 MPa to about 39 MPa simply by adding 1.5% Spectra fibers to the matrix.   Note that the same figure shows that with a heavier mesh, leading to 6.7% for two layers, the addition of fibers does not influence as much the bending resistance; this is because the section becomes essentially over reinforced.  However, other benefits of the fibers include much finer cracks and reduced spalling of the cover as described in [25].

The use of hybrid compositions with continuous (meshes) and discontinuous (fibers) reinforcements offer numerous advantages beside cost efficiency.   The addition of micro-fibers improves crack width and spacing, but most importantly keeps the mortar cover from spalling under large deformation and thus preserves the structural integrity of the composite under overload, that is, beyond service loads up to ultimate.

## Looking Ahead

We are witnessing a considerable evolution in materials and materials technology. At the matrix level, while the basic cement and sand components remain the same, an increasing number of additives are becoming available. These additives allow for many properties to be easily obtainable, such as improved workability, flowability, setting time, strength, bonding, high durability, and high impermeability. Mortar compressive strengths exceeding 100 MPa can be readily achieved using superplasticizers and microsilica (silica fumes). Fibers or micro fibers can be added to the matrix to improve its intrinsic properties.

At the reinforcement level, while steel meshes and steel fibers remain the primary reinforcing material, other materials such as carbon fiber meshes and organic synthetic meshes (Kevlar, Spectra) offer many advantages. Synthetic meshes can be produced to specification and delivered to site, some at a cost competitive with organic natural meshes (sisal, jute, and bamboo). More recently, three-dimensional meshes of either steel or polymeric materials (fabrics, textiles) have become available. Since a single three-dimensional mesh may replace several layers of plain mesh, substantial savings in labor cost are expected. Further savings can be achieved when the mortar matrix is mechanically applied as in shotcreting, extrusion, or pultrusion. Also, given modern trends in manufacturing, it is likely that robotics techniques will be applied to ferrocement and FRC production eventually leading to substantial reduction in labor cost. In the case of ferrocement, it is also likely that small efficient pneumatic or electrical tools will be developed for tying the armature system in an effective way.

Given what we have learned so far and from the above described review, the ideal ferrocement composite of the future is likely to be one that is described in Fig. 10. The trade-off between cost and performance favors hybrid composites with a fiber reinforced matrix and only two extreme layers of mesh. The reinforcement could be prefabricated all together in the form of two meshes taking in sandwich a fiber mat. Improved interlaminar shear resistance, and improved construction time can be achieved with three-dimensional meshes. Even in this case, the addition of fibers to the matrix will improve performance such as finer cracking, reduced spalling and the like.

## CONCLUDING REMARKS

Because corrosion studies have not been undertaken so far, and although good performance can be inferred, there is need to investigate the durability and long term stability of ferrocement and other thin reinforced concrete products using FRP meshes. It is very likely that hybrid ferrocement composites with high strength, high toughness and high durability will be developed. There is also a need to develop FRP meshes with parameters especially optimized for such applications. These include mechanical parameters such as strength, modulus, and ultimate elongation; geometric parameters such as yarn diameter, yarn spacing, and specific surface; and practical parameters such as rigidity of the mesh for handling and placing; and thickness and width of the mesh. In the case of steel meshes, there is need to develop meshes with significantly higher tensile or yield strength; steel wires used in tire cord and to make strands are readily available and can be used. In the case of natural meshes, such as made from sisal or jute, there is need to ascertain the compatibility (mechanical and durability) of the mesh material with the matrix to produce a acceptable composites. Current thinking points toward the use of lightweight and very ligthtweight mortar matrices (with or without fibers) to develop successful products.

New materials and new concepts will make ferrocement and laminated cementitious composites increasingly competitive for particular applications such as in cement sheets, cement boards, claddings, thin reinforced concrete products, and housing.

However, it should be observed that the optimization of composite performance should involve the manipulation of not only the fundamental composite parameters (matrix and mesh parameters), but also variables related to the production process, the rheology of the fresh mix, the properties of the hardening composite and the final application of the material. Recent advances in self-compacting and self-leveling matrices with and without fibers will greatly impact the production-fabrication process of ferrocement composites. A great deal of research and development is needed and is justified by current market potential worldwide.

The history of ferrocement as a modern construction material is longer than that of reinforced concrete, prestressed concrete and steel. Its path for the future as a laminated cementitious composite combining advanced cement based matrices, high performance reinforcing meshes and fibers, and new construction techniques, promises to be as bright. It is hoped that engineers, architects and other professionals will increasingly discover the superb characteristics of ferrocement and laminated and hybrid cementitious composites, and will expand their use worldwide for the benefit of the public.

## DEDICATION

This paper is dedicated to S.P. Shah for his continuous effort in exploring the science of cement based materials including ferrocement; for initially introducing this writer to ferrocement in 1969; and for the fruitful and continuous collaboration between them since then.

## ACKNOWLEDGEMENTS

This research was supported in part by a grant from the NSF Center for Advanced Cement Based Materials, the University of Michigan, and NSF Grant No. CMS-9908308 with S. Gopu as program director. The author is very grateful for their support. Opinions and conclusions expressed in this paper are solely those of the author and do not necessarily reflect the opinion of the sponsors.

## REFERENCES

1. ACI Committee 549, "State-of-the-Art Report on Ferrocement," ACI 549-R97, in *Manual of Concrete Practice*, American Concrete Institute, Farmington Hills, Michigan, 1997, 26 pages.

2. ACI Committee 549-1R-88 and 1R-93, "Guide for the Design Construction, and Repair of Ferrocement," ACI 549-1R-88 and 1R-93, in *Manual of Concrete Practice*, American Concrete Institute, Farmington Hills, Michigan, 1988 and 1993, 27 pages.

3. Alexander, D. J., and Atcheson, M.G.A., "Fibrous Ferrocement for Commercial Vessels," *Journal of Ferrocement* , Vol. 5, No. 2-3, Mar-May, 1976, pp. 25-48.

4. Balaguru, P.N., Hammel, J. and Lyon, R., "Applications of Ferrocement Principles for the Analysis of Advanced Fiber Composites," in *"Ferrocement 6 - Lambot Symposium,"* Proceedings of Sixth International Symposium on Ferrocement, A.E. Naaman, Editor, University of Michigan, CEE Department, June, 1998.

5. El Debs, M.K. and Naaman, A.E., "Bending Behavior of Ferrocement Reinforced with Steel Meshes and Polymeric Fibers," *Journal of Cement and Concrete Composites*, Vol. 17, No. 4, December 1995, pp. 327-328.

6. *Ferrocement Model Code: Building Code Recommendations for Ferrocement*, International Ferrocement Society, Asian Institute of Technology, Bangkok, 2001, 91 pages.

7. Haupt, G. J. and Mobahser,B., "Tensile and Shear Response of Angle Ply Cement Based Composites,"in *"Ferrocement 6 - Lambot Symposium,"* Proceedings of Sixth International Symposium on Ferrocement, A.E. Naaman, Editor, University of Michigan, CEE Department, June, 1998.

8. Kurbatov, O. A., Mironkov, B. A., and Sterin, V. S. (1994). "Ferrocement Structures with Reinforced Fabrics Made of Polymer Fibers," in *Ferrocement: Proceedings of the Fifth International Symposium*, Edited by P.J. Nedwell and R.N. Swamy, E. and F.N. Spon, London, pp. 485-497.

9. *Lambot Symposium,* Proceedings of Sixth International Symposium on Ferrocement, A.E. Naaman, Editor, University of Michigan, Ann Arbor, June, 1998, pp. 139-151.

10. Mobasher, B., Pivacck, A., and Haupt, G.J., "Cement Based Cross-Ply Laminates," *Journal of Advanced Cement Based Materials*, No. 6, 1997, pp. 144-152.

11. Naaman, A.E. and Al-Shannag, J., "Ferrocement with Fiber Reinforced Plastic Meshes: Preliminary Investigation," in *Ferrocement: Proceedings of the Fifth International Symposium*, P.J. Nedwell and R.N. Swamy, Editors, Manchester, E & FN Spon, London, 1994, pp. 435-445.

12. Naaman, A.E., and Guerrero, P., "Bending Behavior of Thin Cement Composites Reinforced with FRP Meshes," *Proceedings of First International Conference on Fiber Composites in Infrastructures, ICCI 96*, H. Saadatmanesh and M. Ehsani, Editors, University of Arizona, Tucson, Jan., 1996, pp. 178-189.

13. Naaman, A.E., and Chandrangsu, K., "Bending Behavior of Laminated Cementitious Composites Reinforced with FRP Meshes," ACI Symposium on *High Performance Fiber-Reinforced Concrete Thin Sheet Products*, Edited by A. Peled, S.P. Shah and N. Banthia, American Concrete Institute, Farmington Hills, ACI SP 190, 2000, pp. 97-116.

14. Naaman, A.E., *Ferrocement and Laminated Cementitious Composites*, Techno Press 3000 (www.technopress3000.com), Ann Arbor, Michigan, ISBN 09674939-0-0, 2000, 372 pages.

15. Lopez, M., and Naaman, A.E., "Study of Shear Joints in Fiber Reinforced Plastic (FRP) Ferrocement Bolted Connections," in Ferrocement 6 - Lambot Symposium, proceedings of the*Sixth International Symposium On Ferrocement*, A.E. Naaman, Editor, Ann Arbor, Michigan, June, 1998.

16. Naaman, A.E., S.P. Shah and J.L. Throne, "Some Developments in Polypropylene Fibers for Concrete," in Proceedings of the ACI International Symposium on Fiber Reinforced Concrete, Special Publication SP-81, American Concrete Institute, Detroit, 1984, pp. 375-396.

17. Naaman, A.E., "Fiber Reinforcement for Concrete: Looking Back, Looking Ahead," in Proceedings of Fifth RILEM Symposium on Fiber Reinforced Concretes (FRC), BEFIB' 2000, Edited by P. Rossi and G. Chanvillard, September 2000, Rilem Publications, S.A.R.L., Cachan, France, pp. 65-86.

18. Naaman, A.E. "Ferrocement and Thin Fiber Reinforced Cement Composites: Looking Back, Looking Ahead," Proceedings 7th International Symposium on Ferrocement and Thin Reinforced Cement Composites, Singapore, June 2001.

19. Nervi, P.L.., "Ferrocement: Its Characteristics and Potentialities," Library Translation No. 60, Cement and Concrete Association, London, July, 1956, 17 pages.

20. Paramasivam, Mansur, et al., Editors, Proceedings Proceedings 7th International Symposium on Ferrocement and Thin Reinforced Cement Composites, Singapore, June 2001.

21. Parra-Montesinos, G., and Naaman, A.E., "Parametric Evaluation of the Bending Response of Ferrocement and Hybrid Composites with FRP Reinforcements" Proceedings 7th International Symposium on Ferrocement and Thin Reinforced Cement Composites, Singapore, June 2001.

22. Peled, A., Bentur, A., Yankelevsky, D., "Effect of Woven Fabric Geometry on the Bonding Performance of Cementitious Composites," *Advanced Cement Based Materials Journal*, Vol. 7, 1998, pp. 20-27.

23. Shah, S.P. and Peled, A., "Advances in Science and Technology of Ferrocement," in Ferrocement 6 - Lambot Symposium, proceedings of the*Sixth International Symposium On Ferrocement,* A.E. Naaman, Editor, Ann Arbor, Michigan, June, 1998, pp. 35-51.

24. Swamy, R. N., Hussin, M. W., "Continuous Woven Polypropylene Mat Reinforced Cement Composites for Applications in Building Construction," in *Textile Composites in Building Construction*, P. Hamelin and G. Verchery (Eds.), Part 1, pp. 57-67, 1990.

25. Wang, S., Naaman, A.E., and Li, V.C., "Bending Response of Hybrid Ferrocement Plates with Meshes and Fibers" Proceedings 7th International Symposium on Ferrocement and Thin Reinforced Cement Composites, Singapore, June 2001.

Fig. 1 -- Main gate of the Yambu Cement Company in Saudi Arabia showing the ferrocement ribbon like roof that ends wrapping around the tower like a vale in the wind. *(courtesy D. Angelotti, Studio 65, Italy.)*

Fig. 2 -- Ferrocement prototype diffuser wind turbine DAWT. *(Courtesy R. Alexander, New Zealand.)*

Fig. 3 -- Sixty-meter long semi-cylindrical ferrocement storage facility in Yougoslavia. *(Courtesy M. Milinkovic.)*

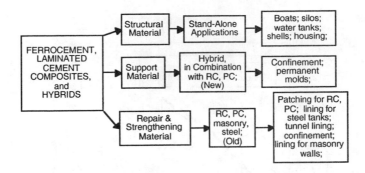

Fig. 4 -- Evolution of the function of ferrocement as a construction material.

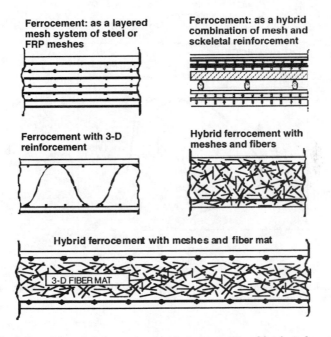

Fig. 5 -- Evolution of the reinforcement system in ferrocement and laminated cementitious composites.

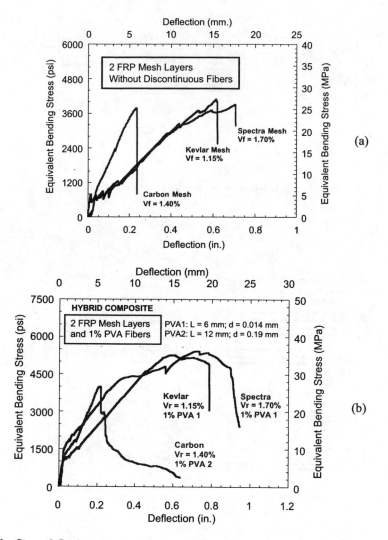

Fig. 6 — Stress-deflection response of ferrocement plates reinforced with FRP meshes a) without fibers; and b) with fibers.

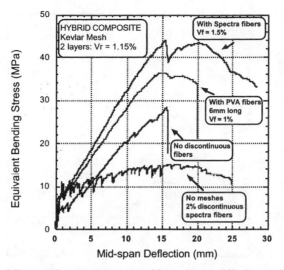

Fig. 7 -- Effect of fibers on ferrocement plates with two layers of Kevlar mesh.

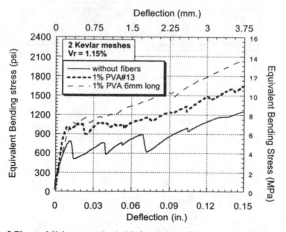

Fig. 8 -- Effect of fiber additions on the initial portion of the stress-deflection response of ferrocement plates reinforced with FRP meshes.

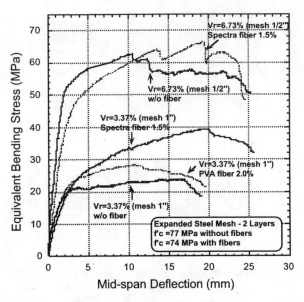

Fig. 9 -- Effect of fiber addition on ferrocement plates with two layers of expanded metal mesh.

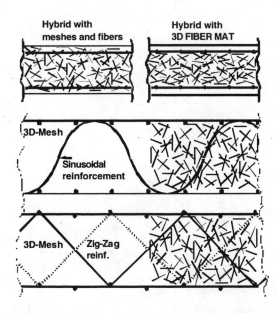

Fig. 10 -- Anticipated logical evolution of the reinforcement for cost and performance competitive ferrocement.

# Textile Fabrics for Cement Composites

## by A. Peled and A. Bentur

**Synopsis:** One of the most efficient ways to obtain a high performance cementitious composite is by reinforcement with continuous fibers. Production of such composites can be accomplished by the use of textile fabrics, which are impregnated with cement paste or mortar. The present paper provides an overview of the major characteristics in predicting the performance of cement composites reinforced with fabrics. The geometry of the fabrics themselves (weft insertion warp knit, short weft warp knit and woven fabrics) as well as the geometrical structure of the yarns within the fabrics are discussed. It was found that the geometry of a given fabric could enhance the bonding and enable one to obtain strain hardening behavior with low modulus yarn fabrics. On the other hand, variations of the geometry in a fabric could drastically reduce the efficiency, resulting in a reduced strengthening effect of the yarns in the fabric relative to single yarns not in a fabric form. The improved bonding in low modulus yarn was found to be mainly the result of the special shape of the yarn induced by the fabric. Therefore, in cement composites, the fabrics cannot be viewed simply as a means for holding together continuous yarns so that they can be readily placed in the matrix.

**Keywords:**  bond; cement; composite; fabric; fiber; flexure; pull-out; yarn

Alva Peled, Faculty of Engineering Science, Ben Gurion University, Beer Sheva Israel. She was a post-doctoral fellow at the Center for Advanced Cement-Based Materials at Northwestern University for three years. Her D.Sc. degree is from the Technion, Israel Technology Institute in Civil Engineering, her M.Sc. is in Chemistry of Polymers and Textile from The Hebrew University, Israel and her B.Tech. is in Textile Technology. Her research interests include fibers and fabrics as reinforcement for composites, processing and fiber-matrix interfaces.

Arnon Bentur, FACI, is Professor of Civil Engineering and holder of the Edwards Chair of Engineering at the Technion, Israel Institute of Technology. He is currently the Director of the Neaman Institute for Advanced Studies of Science and Technology at the Technion. His areas of expertise are cementitious composites, high performance concrete and durability of construction materials. He is on the Board of Editors of several international journals and the co-author of several books on construction materials. At present he the Chair of RILEM Technical Activities Committee.

## INTRODUCTION

Fabrics are extensively used as reinforcements of polymeric matrices (1). They enable the use of high content of fibers in the composite and control their orientation. Fabric structures are more readily handled and placed in precast products. Modern textile technology offers a wide variety of fabrics, with great flexibility in the design of the geometry and the fibers, which make up the fabric. Fabrics are produced from yarns (that is, a longitudinal reinforcing unit made of fibers) by different methods, such as weaving, knitting, breading and non-woven. They differ in the way that the yarns are connected to each other at the junction points. The interlacing of the yarns to form a fabric affects not only the geometry of the fabric itself but also the geometry of the individual yarns, which make up the fabric.

When polymer matrices are reinforced with fabrics, in which the yarns within the fabric do not maintain a straight geometry, reduction in the reinforcing effectiveness was reported (1). Therefore, in polymer composites, the reinforcing fabrics should contain as many straight yarns as possible. Prediction of the reinforcement efficiency of the fabric usually takes into account only the longitudinal yarns in the fabric, which are in the loading direction. The perpendicular yarns are treated as "non structural", whose object is to provide a mechanism to hold the longitudinal yarns in place during the production of the composite.

In cement composites this concept may not be adequate since the micro-mechanics of the interaction between the cement matrix and the fabric and its individual yarns might be more complex. Indeed, previous work has indicated that fabric geometry and the special geometry of its yarns have a significant effect on bonding (2-5).

The object of the present paper is to present an overview of the major characteristics required to predict and control the performance of cement composites reinforced with fabrics. The geometry of the fabrics themselves as well as geometrical structure of the yarns within the fabrics are discussed. This overview is based in part on a series of reports in which specific topics within this framework were studied. (2-5)

## EXPERIMENTAL

Three types of fabrics were examined in this work: woven fabric (plain weave, (Fig. 1a)), weft insertion warp knitted fabric (Fig.1b) and short weft warp knitted fabric (Fig.1c); (weft and warp are the terms used for the yarns in the different orientations of the fabrics). The fabric structures differ by the way the yarns are combined together. In the woven fabric the warp and the fill (weft) yarns pass over and under each other resulting in a crimped shape for the yarns in the fabric (Fig. 1a). In the weft insertion knitted fabric the yarns in the warp direction are knitted into stitches to assemble together the straight yarns in the weft (Fig. 1b). In the short weft knitted fabric the warp yarns are also knitted into stitches but in this case they bind together a set of yarns, which are laid-in intermittently in both the weft and the warp directions (in a zigzag form in Fig.1c). Polyethylene (PE) monofilament (a continuous fiber) was used to produce the woven and short weft knit fabrics. High density polyethylene (HDPE) in the form of bundled yarns was used to produce the weft insertion knit fabrics. The properties of the fibers are presented in Table 1.

An additional parameter evaluated in this series of studies was the density of the woven and the weft insertion knit fabrics on mechanical performance of the composite. The density of the fabrics was varied by changing the density of the yarns perpendicular to the applied load. In the woven fabrics the fills density was 5, 7 or 10 fills per cm and the warps' density was kept constant at 22 warps per cm. It should be noted that the fabric density affects several characteristics: the crimped structure of the warp yarns in the woven fabric, the number of joints and the penetrability of the cementitious matrix into the fabric. These characteristics can influence the bond capacity with the cement matrix. In the weft insertion knitted fabrics the warp density (the stitches) was varied: 0.8, 1.6, 2.4 or 3.2 warps per cm and the wefts' density (straight yarns) was kept constant at 16 weft per cm. In the short weft knit the density of the warps (the stitches) was 3 warps per cm.

In order to better understand the influences of the yarn itself, individual straight yarns and crimped yarns untied from the woven fabrics were used to prepare reinforced cement composites.

## TESTING

The efficiency of the yarns and the fabrics was determined by evaluating their performance in the composite, as determined by characterization of the bond behavior as well as the overall properties of the composite in flexure. In all cases the matrix was a cement paste with a 0.3 water/cement ratio.

The evaluation of the composite bond characteristics was based on pull-out tests of the original straight yarn, yarn untied from the fabric and the fabric itself. The pull-out tests were carried out in an Instron testing machine at a crosshead rate of 15 mm/min as described in detail in Ref. 4. Load-slip curves were recorded. In the present paper, only reference to the maximum pull-out load will be made; typical complete pull-out curves for some of the systems are reported in Ref. 4. The average bond per single filament was used to quantify the interfacial bond taking into account in the calculation the surface area of all the filaments. This can serve as an estimate for the "effective apparent bond", although strictly speaking it is not the actual bond.

Flexural properties of the composite were studied by four point bending test carried out in an MTS testing machine at a crosshead rate of 1.5 mm/minute, with a span of 90 mm. Load-deflection curves were recorded. The specimens for this test were prepared by lay-up of 8 layers of fabrics or yarns in the cement paste. The specimen dimensions were 15 mm thick with length and width of 110 and 20 mm, respectively. The volume content of the reinforcing yarns (i.e. the yarns in the reinforcing direction) are 5.7% for the woven and PE straight yarns, 3.5% for the weft insertion knitted fabrics and HDPE straight yarns, and 2% for the sort weft knit fabrics. Note that in the short weft knit only the part of the yarn that was laid in the reinforcing direction was taken into consideration. Details of specimen preparation and testing are provided in Ref. 5. Two parameters were calculated from the flexural tests: the post cracking flexural strength and the yarn reinforcing efficiency. The latter is defined as the ratio between the post cracking flexural strength and the product of the yarn volume and its tensile strength (composite flexural strength/(fiber volume x fiber strength)). The efficiency factor can serve as a direct estimate of the fabric efficiency since this value is already normalized for the yarn content in the composite, which was not always the same for all the composites prepared here, due to production constraints.

## RESULTS

### Flexural Performance

Comparison of flexural behavior of woven and short weft knit fabric composites and straight yarn composite (all from low modulus PE yarns) is presented in Figure 2a. The positive effect induced by the fabric structure is clear in this figure. Both fabric structures improve strength and toughness of the composite leading to a strain hardening type of response. The improved performance is considerably high at large deflections. The performance of the short weft knit fabric composite is relatively high compares to its low yarn content, only 2%, by volume, whereas in the woven and the straight yarn composites the yarn content is as high as 5.7%, by volume. Figure 2b presents the flexural behavior of the weft insertion knit fabric composite and compares it with a straight yarn composite (both are made from high modulus HDPE yarns). The trend here is opposite to that observed in Fig. 2a. The fabric structure reduces significantly the flexural behavior of the composite compared with that of the straight yarn. Note that the flexural behavior of the weft insertion knit fabric is not much better than that of the low density PE fabrics, especially at high deflections, inspite of the fact that this fabric is made of high modulus yarns and its yarn content is higher (3.5%, by volume) than that of the short weft knit fabric (only 2%, by volume).

The above trends are also observed when comparing the efficiency factor values in Figure 3. The figure clearly shows that the short weft knitted fabric exhibits extremely high efficiency factor, 3.08, which is much greater than the efficiency factor of 1.21 of the woven fabric, and the weft insertion knitted fabrics, in which the value is lower than 1.0, although it is made from high modulus yarn (HDPE). Both, the short weft knit and woven fabrics, exhibit efficiency factor values bigger than that of the straight yarns, This improvement in the efficiency factor is much greater for the short weft knit fabric. On the other hand, opposite trend is observed for the weft insertion knitted fabric: in this case the value of the efficiency factor is lower than the value for the straight yarn.

These characteristics are also demonstrated in Figure 4, in which the effect of the density of the yarns perpendicular to load direction on the flexural strength of woven and weft insertion knitted fabrics is plotted. The positive effect of increasing density is clearly seen for the woven fabric, especially in the low density range. Here again, the situation is quite different for the weft insertion knitted fabric, where an increase in density is associated with an immediate marked reduction in the flexural properties.

It should be kept in mind that the shape and nature of the individual yarn in these fabrics are rather different: in the weft insertion knit fabric, the yarns are multifilament in a straight form, whereas in woven and short weft knit fabrics they are monofilament and do not maintain straight form (they are in crimp geometry in woven fabric and in relatively complicated geometry, "zigzag" in the sort weft knit fabric). Such differences in yarn geometry may affect the above differences in the performance of the composites. The improved flexural strength induced by the crimp geometry of the individual yarn is seen in Figure 5, which compares the flexural strength of woven fabric, individual crimped yarns untied from this fabric and individual straight yarns from the same material. Both, woven fabric and untied crimped yarn composites, are significantly stronger than the straight yarn composite; the crimped yarn composite is the strongest. Moreover, the efficiency values are also higher for the crimped yarn (untied from the fabric) compared to the woven fabric (Figure 3), suggesting that the improved performance of the woven fabric can be accounted for the crimped geometry of its individual yarns.

## Bond

Based on the above results, it can be suggested that the geometry of the individual yarns within the fabric affects the performance of the cement composites. "Non straight" forms improve the composite properties, as observed for the short weft knit and woven fabrics.

Improved flexural performance of fabrics, having individual yarns with relatively complicated shape, might be explained on the basis of enhanced bonding, as these individual yarns apparently induces anchoring effects. Pullout test was performed to resolve these effects, for woven and weft insertion knitted fabrics, as well as for the individual crimped yarns untied of the woven fabrics and straight yarns made up these fabrics. Pullout results are presented in Figure 6 in terms of bond strengths per single filament. It is clear that the bonding in the actual fabric is different from that of the individual yarns. This trend correlates with the flexural results. It should be noted that it was impossible to pullout the short weft knitted fabric due to the complicated shape of its reinforcing yarn.

For the weft insertion knitted fabrics, the bond in the fabric is lower than that of the individual yarn. The decrease in bond is about 75%. Such reduction in bonding may account for the marked reduction in flexural performance of weft insertion fabric compared to straight yarn composite (Figures 2b and 4).

In contrast, the bonding in the woven fabric is enhanced relative to the individual straight yarn (Figure 7), which correlates with the improved flexural performance of this fabric. The bonding of the crimped yarns is greater than that of the straight yarn, as increase in crimp density improves the bond strength. This indicates that the crimped geometry of the individual yarn affects the bond of the woven fabric to the cement matrix; denser fabric means higher crimping, leading to better bonding and improved flexural performance (up to an optimum value, Figure 4).

## DISCUSSION

The geometry of the fabric showed a marked effect on the properties of cement composites, demonstrating opposing influences: improved performance of woven and short weft knit fabrics, and reducing drastically the performance of weft insertion knit fabric. These contradictory influences can be explained on the basis of bond effects. The improvement by the fabric structure can be accounted by the shape of the individual yarn which make up the fabric, which induces anchoring effects. Complex shape (i.e. not a straight form) results in strong anchoring effects, leading to better composite performance for woven and short weft knit fabrics (for low modulus yarns).

In the woven fabric the crimped shape of the reinforcing yarns enhances the anchoring effect, and increase in the density of the fabric enhances the positive influence of the crimping. However, some poor compaction of the cement matrix is seen at the interlacing points of the fabric (Figure 8), mainly for high density fabrics. Therefore, at high densities reversed trend is seen (Figure 4), because of this poor compaction.

The situation is quite different for the weft insertion knitted fabric, where the fabric structure is associated with a marked reduction in flexural properties for all densities (Figure 4). This may be due to the fact that the reinforcing yarns in the weft insertion knitted fabrics are straight, made from bundles and connected at the joint points by stitches. The efficiency of bundles as reinforcement for the cement composite was found to be quite low, due to poor penetration of the matrix between the filaments (Refs. (5,6). When these bundles are kept in a straight form, but as part of a knit fabric structure the penetrability of the matrix is even lower, due to the presence of the bulky stitches themselves, as well as the tightening effect of the stitches, which strongly hold the filaments in the bundle and prevent spaces from being opened between them. Therefore, the weft insertion HDPE knit fabric shows a poor performance relative to the superior properties of its yarn (and it is no better than the low modulus PE fabrics).

## CONCLUSIONS

Fabrics for cement reinforcement cannot be viewed simply as a means for holding together continuous yarns to enable them to be readily placed in the cement matrix. The geometry of the fabric may have a marked effect on the properties of the composites; it may enhance bonding and result in a strain hardening behavior of low modulus yarn composites, or reduce drastically the efficiency of a high performance bundled yarn that is very effective for reinforcement when it is not part of a fabric.

The enhanced bonding of the fabric structure was found to be dependent mainly on the special geometry induced to the individual yarns by the fabric structure. This special geometry may provide mechanical anchoring of the reinforcing yarns in the cement matrix. Thus, fabrics having relatively complex yarn shapes, such as short weft knit, enhance the bonding and improve the composite performance. On the other hand, in weft insertion knitted fabric the reinforcing yarns are in a straight form and any interference with compaction at the junction points of the fabric is not compensated for by any other counteracting mechanism (such as the improved bonding due to crimping or other complicated shape as in the woven and sort weft knit fabrics), resulting in a marked reduction in flexural properties of such fabrics (for high modulus yarns).

## ACKNOLEDGMENTS

The authors would like to thank Karl Mayer Textile Machinery Ltd., Germany and Polysack Ltd., Israel, for their cooperation and for the effort they made to provide the warp knitted fabrics used in this study.

## REFERENCES

1.   C. Zeweben, in C. Zeweben, H.T. Hahn, T. Chou (Eds.), Mechanical Behavior and Properties of Composite Materials, Delaware Composites Design Encyclopedia, 1, Technonic: Lancaster, 1989, pp. 3-45.

2.   A. Peled and A. Bentur, "Geometrical Characteristics and Efficiency of Textile Fabrics for Reinforcing Composites", Cement and Concrete Research, Vol. 30, 2000, pp. 781-790.

3.   A. Peled, A. Bentur, Reinforcement of cementitious matrices by warp knitted fabrics, Mat & Struc 31 (1998) 543-550.

4.   A. Peled, A. Bentur, D. Yankelevsky, Effects of woven fabrics geometry on the bonding performance of cementitious composites: mechanical performance, Advan Cem Bas Comp 7 (1998) 20-27.

5.   A. Peled, A. Bentur, D. Yankelevsky, Flexural performance of cementitious composites reinforced by woven fabrics, ASCE J. Mat. Civ. Eng. Eng. (Nov. 1999) 325-330.

6.   V.C. Li, W. Wang, S. Backer, Effect of inclining angle, bundling and surface treatment on synthetic fiber pull-out from cement matrix, Composites 21 (1990) 132-140.

Table 1: Properties and structure of yarns

| Yarn Type | Yarn Nature | Strength (MPa) | Modulus of Elasticity (MPa) | Yarn Diameter (mm) |
|---|---|---|---|---|
| PE | Monofilament | 260 | 1760 | 0.25 |
| HDPE | Bundle | 1960 | 55000 | 0.25 |

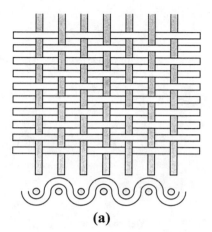

(a)

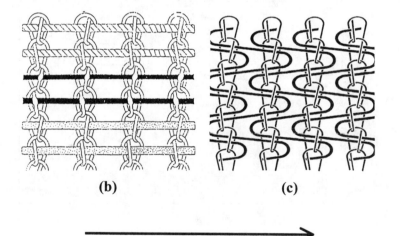

(b)                              (c)

**Reinforcing Direction**

Figure 1 - Structure of fabrics: (a) woven, (b) weft insertion knitted fabric, and (c) short weft knitted fabric

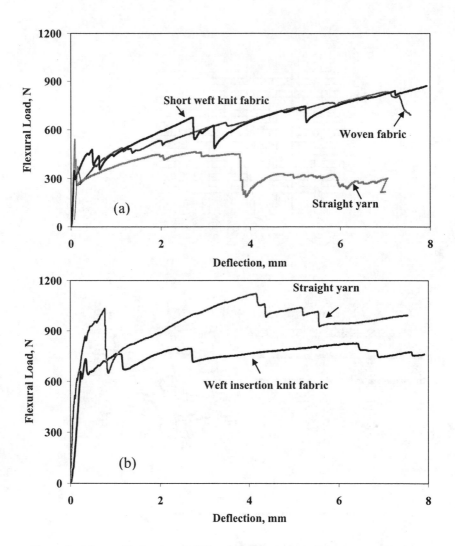

Figure 2 - Flexural behavior of different fabrics and straight yarn composites: (a) woven (with 7 perpendicular yarns per cm) and short weft knit fabrics (PE), and (b) weft insertion knit fabric (with 3 perpendicular stitches per cm, HDPE).

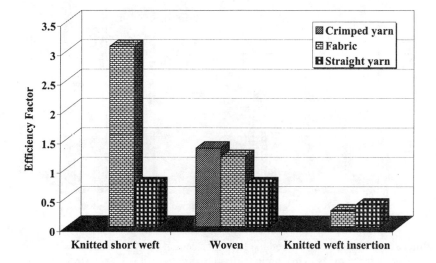

Fig. 3 - Comparison between the efficiency factor of individual yarns and fabrics for woven (with 7 perpendicular yarns per cm) and short weft knit PE, and knitted weft insertion HDPE (with 3 perpendicular stitches per cm)

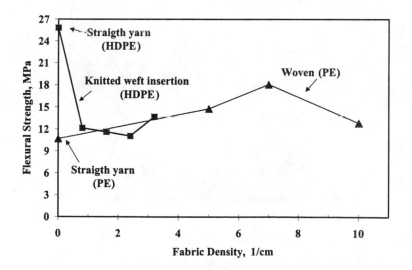

Figure 4 - The influence of the density of yarns perpendicular to the main reinforcement on the flexural strength

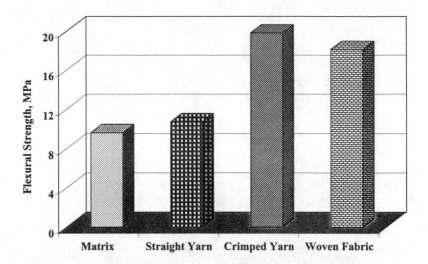

Figure 5 - Comparison between the flexural strength of woven fabric (with 7 perpendicular yarns per cm), untied crimped yarn from this fabric, straight yarn which made up the fabric, and plain cement matrix

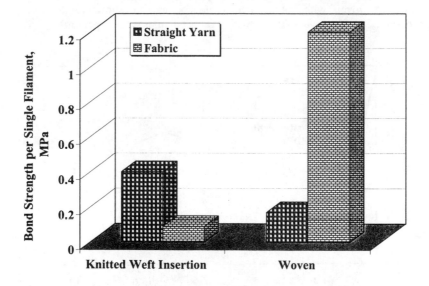

Figure 6 - The effective bond of single yarns and yarns in a fabric for the different fabric structures and yarns (woven with 7 perpendicular yarns per cm, weft insertion knit fabric with 3 perpendicular stitches per cm)

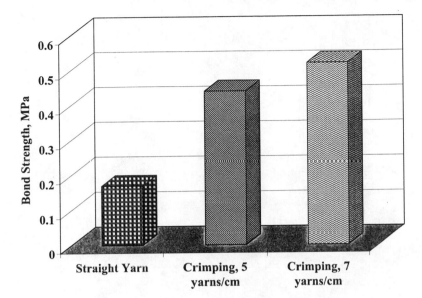

Figure 7 - The effective bond of untied crimped yarns and straight yarn from PE

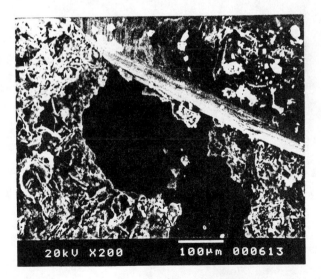

Figure 8 - SEM micrographs of woven fabric with 10 perpendicular yarns per cm in the cement matrix

# Thin Plates Prestressed with Textile Reinforcement

## by H. W. Reinhardt, M. Krüger, and C. U. Grosse

<u>Synopsis:</u>
Tests were carried out on thin plates either reinforced or prestressed with textiles. The textiles had a warp knitted structure with 10 mm mesh size. The two materials AR glass and carbon behaved completely different in the four-point bending experiments. Pull-out tests with acoustic emission measurements showed clearly that the different bond behaviour was the cause for such differences.

<u>Keywords:</u>  AR-glass; bending; carbon; concrete; prestressing; textile reinforcement

ACI Member H.W. Reinhardt holds the chair of Construction materials at the University of Stuttgart, Germany, and is Managing director of the Otto-Graf-Institute (Research and Materials Testing Establishment of Baden-Württemberg). His current research fields include fracture and transport of liquids in concrete, fibre reinforced concrete, non-destructive testing and demountable concrete structures. He is a member of ACI Committee 446 Fracture Mechanics.

Dipl.-Ing. Markus Krüger was born in Essen, Germany, in 1971. Graduated from the University of Dortmund in civil engineering in March 1998. Since April 1998 PhD student at the Institute of Construction Materials, University of Stuttgart. Currently teaching and educating students in concrete technology. Now working on a PhD project "Prestressed Textile Reinforced Concrete".

Dr.-Ing. Christian U. Grosse was born in 1961 and studied Geophysics at the University of Karlsruhe. He finished his thesis in July 1996 ("Quantitative NDT in civil engineering") and is now a research scientist at the Institute of Construction Materials, University of Stuttgart, Germany. He is responsible for the collaborative research project "NDT echo methods" and is secretary of the Rilem TC "Advanced testing of cement based materials during setting and hardening".

## INTRODUCTION

To use textile reinforcement instead of steel has a long tradition nowadays (1, 2, 3). The advantages are obvious. Most textiles do not corrode in carbonated or chloride containing concrete permits reducing the concrete cover just large enough to transfer bond stresses from the reinforcement to the concrete. This allows production of thin reinforced concrete elements. Second, textile reinforcement is lightweight and flexible which makes handling rather easy. Third, textile reinforcement has a very high strength and allows a low reinforcement ratio.

However, one knows from steel reinforced concrete that concrete cracks and that, only after cracking, reinforcement is efficient. Now, the advantage of a low reinforcement ratio and high strength together with a lower modulus of elasticity changes to a disadvantage. The first crack becomes rather wide and subsequent cracks if they develop at all become also wide. This makes a structure less stiff. This may be explained with the following sketch (Fig. 1).

Case 1 refers to an element with low reinforcement ratio and/or low modulus of elasticity of the reinforcement. Case 2 refers to a high reinforcement ratio and/or stiff reinforcing material. One can see that the initial stiffness does not vary much and that the cracking forces are similar. The differences show up only after cracking when the reinforcement ratio and/or stiffness of the reinforcement influence the stiffness and the bearing capacity of the composite elements. The consequence of this behaviour is that either the serviceability state is rather low, i.e. within the range of the uncracked state or, when cracks are accepted, the deformation or deflection may be rather large when no uneconomically high reinforcement ratio is provided. In order to improve the situation a new approach is chosen.

The new approach is to prestress the concrete in order to enlarge the tensile strength of concrete fictitiously such that the cracking force is also increased. The situation is illustrated by the following sketch (Fig. 2). The element is prestressed with a force equal to $N_p$ which means that the uncracked (or elastic) state is enlarged by this value.

The states of decompression and cracking are separated by the tensile strength of concrete. The value of $N_p$ depends on the prestressing stress of the reinforcing material and on the reinforcement ratio. The serviceability state in the uncracked state is enlarged.

There is an additional aspect which refers to a textile fabric i.e. a two-dimensional reinforcement element. In many fabrics are two directions known as warp and weft in a woven fabric. The weft direction is always very soft at first stressing (4). When prestressed this initial elongation has vanished. Further, all textile elements either one-dimensional or two-dimensional show a certain initial elongation and the stiffness increases with stress (or elongation). Prestressing results in a smaller first crack than in the case of reinforced elements.

The following reports on first series of results of prestressed plates made of fine grain concrete and textile reinforcement.

TESTING VARIABLES

The variables of the research were the type of textiles and the prestress. The textile was an alkali resistant (AR) glass and carbon. The prestress was either zero or 1.5 MPa for AR glass and 2.5 MPa for carbon. Two types of tests were carried out, a four-point bending test and a pull-out test.

## TESTING MATERIAL

A fine grain concrete mix was developed which should be self-compacting for various reasons which will be explained later. The maximum grain size was limited to 1.2 mm. The early age strength should be rather high because the prestress was applied after one day. The concrete mix is given in Table 1. A CEM I 42.5 R which is a rapid hardening portland cement was used.

Fly ash from hard coal (low calcium content) was used as an addition for better workability. Silica fume should increase strength, mainly early age strength and bond. To make the mix workable a superplasticizer on the basis of polycarboxylate was added. The total water consisted of the water of the silica slurry, the superplasticizer and the added water which made together 220 dm³. The water-cement ratio was 0.46, the water-binder ratio 0.33. The workability was measured by slump-flow of mortar and amounted to 26 cm.

After one day, the compressive strength was 25 MPa as can be seen from Table 2. It increased to 62 MPa after 7 days and 75 MPa after 28 days. The compressive strength was measured on the half of 40 mm x 40 mm x 160 mm prisms.

Flexural strength from the bending test resulted in 5, 9 and 11.5 MPa after 1, 7 and 28 days. Shrinkage was measured on 40 mm x 40 mm x 160 mm prisms. The final value was nearly reached after 90 days and amounted to 0.65 mm/m.

The textile reinforcement consisted of AR glass and carbon with a warp knitted structure. Table 3 shows the main characteristics. The glass fabric has a unit weight which is almost double of the carbon one but the maximum load per roving is only one half.

The tensile strength which could be reached in the uniaxial tensile test is about 1300 MPa for carbon and 330 MPa for glass (Fig. 3). This rather low strength is due to low static fatigue limit, imperfections and stiffness of the clamping devices and also the damage caused by the textile production. It is not the true strength of the material itself. The structure of the fabric is the most straight forward, i.e. a warp knitted structure with two perpendicular rovings. The rovings are connected to each other by a binder thread. This makes that the initial strain is limited. The pronounced weft effect does not occur.

## SPECIMEN PREPARATION

Plates of 1000 mm x 1000 mm x 10 mm were made where the textile fabric was orientated in the middle to get a uniform prestress. The reinforced plates manufactured by placing a 5 mm thick layer of concrete on the mould and putting the textile layer on it. Further on another layer of 5 mm was cast on it and vibrated by a special vibrator which was developed for the purpose. The specimens were demoulded after 24 h and then wet cured for 26 days. Depending on the experiment, smaller strips were cut out and then stored another day at 20°C and 65% RH until testing at 28 days.

The specimen size of the prestressed plate was the same. The fabric had to be stressed by special clamping devices. For that purpose two saw-tooth shaped steel plates were covered with a thin plastic foil, the fabric was put in between, an epoxy resin was applied and the plates were clamped together. By this procedure it was avoided that the resin stuck to the metal. The idea of this method was adapted from (5). Fig. 4 shows the device.

After prestressing the plate was cast in once and slightly vibrated, which was necessary because of the low thickness of the concrete layer although self-compacting concrete was used. However, to make sure that the whole fabric was embedded in concrete it was decided to vibrate a little.

A square frame was designed for the prestressing procedure (Fig. 5). It allows to prestress an element of 1 m² with 10 hydraulic pistons at each side. The pistons are connected to an oil pump. The oil pressure can be increased to 12 MPa which generated an individual force in the piston of about 4000 N or a total of 40 kN/m. This maximum possible prestress was not used in the experiments. The maximum achieved was only 15 kN/m for the glass reinforcement which is equivalent to 1.5 MPa for a 10 mm thick plate. For the carbon reinforcement it was 25 kN/m.

The specimens for the pull-out test were 80 mm x 80 mm x 40 mm prisms. The prisms were demoulded after one day and kept in water another 5 days. Thereafter they were stored at 20°C and 55% RH. The roving for the pull-out test was held horizontally in a center position.

TESTING PROCEDURE

Bending tests

A electro-mechanical four-point bending machine was used with a span of 250 mm. There was a constant displacement rate of 2 mm/min. The load was applied at the third points. The displacement of the loading point and the center was measured. Crack opening and crack spacing were monitored.

Pull-out tests

The pull-out tests were carried out with an embedment length of 40 mm. The equipment is shown schematically in Fig. 6. The prisms with the roving is positioned on a steel plate which is fixed to four steel bars which are held by the upper cross-head of the testing machine. The roving is pulled out at the bottom end. The relative displacement of the roving with respect to concrete is measured by an LVDT, the strain in the roving by a displacement gauge. The pull-out displacement rate was constant at 1 mm/min.

Acoustic emission measurements

During the pull-out tests 8 sensors were fixed to the specimen such that an acoustic signal could be localized. The sensors have a relatively flat frequency response between 50 and 250 kHz. The software WinPecker (6) was used which allowed a localization accuracy of about ± 5 mm for a single event in practice.

RESULTS

Bending test

The limit of proportionality (LOP) was determined and the modulus of rupture (MOR) which was calculated if the plates behaved elastically, i.e. the crack was not taken into account in the MOR state. The results are summarized in Table 4 in terms of stress and center displacement.

There was a large difference in crack width as can be seen in Table 5.

The following graphs show a few examples of local deflection diagrams. Fig. 7 shows a loading/unloading sequence of a 100 mm wide and 10 mm thick strip with central reinforcement. A similar picture is given in Fig. 8 for carbon reinforced elements. The differences between the two graphs will be discussed later.

Pull-out tests

Pull-out tests were always performed together with acoustic emission measurements. The following diagrams in Fig. 9 show in the upper left the events and the energy of the events, in the middle left the depth from which the acoustic signals come from.

The upper right picture is a 3D localization of the events. The lower left gives two lines for the development of the pull-out force and slip. The lower right finally is a pull-out force displacement diagram. Fig. 9 is a sequence of pictures for a AR glass roving which has been pulled out at a concrete age of 7 days.

Similar pictures have been received for carbon reinforcement at a concrete age of 7 days. The sequence is given in Fig. 10.

After the presentation of the results a discussion follows in the next chapter.

## DISCUSSION OF RESULTS

LOP and MOR

The limit of proportionality (LOP) is directly related to the flexural strength of the concrete. Looking to Table 4 it is obvious that the prestressed specimens had a lower LOP than the reinforced ones. The reason for that can be a large relaxation of the AR glass in the concrete such that the 1.5 MPa prestressing disappeared completely. One should have expected that the flexural strength as measured on prisms (Table 2) would have been reached. However, the curing conditions were much less and the dimensions of the specimen differ which might explain the difference. This is especially true for the specimen either reinforced or prestressed with carbon. There, the prestressed one is superior to the reinforced one but both are inferior to the standard prism (Table 2). Also here, bad curing may be the origin of the low strength.

When we consider the modulus of rupture (MOR) there is an increase of strength to the AR glass due to prestressing. This will be explained later with the aid of force-displacement diagrams. The carbon reinforced element behaves completely different: there is a strong reduction of MOR of the prestressed compared to the reinforced one. As will be shown later this effect is due to the inferior bond and due to the slip of the carbon in the concrete in the anchorage zone. Obviously, the bond is destroyed over a large length so that the roving could slip in the concrete under load.

<u>Load-deflection behaviour</u>

The deflection however was in both prestressed cases much smaller than in the reinforced ones. As mentioned earlier, every textile has an initial strain which is compensated by prestressing. The initial strain is due to imperfection but mainly by the fact that not all filaments are loaded at the same time. This explains also why the actual strength of the roving was considerably lower than the theoretical strength of the roving.

The cyclic response to a load is much dependant on prestress. Comparing the upper and lower figure of Fig. 7 one can see the difference immediately. Whereas the reinforced plate shows an increase in deflection without considerable increase of load the prestressed one increases load and deflection at the same time. Evaluating the stiffness at first loading to the stiffness at ultimate load there is a ratio of 17 for the reinforced one and a ratio of 9 for the prestressed one. The irreversible deflection increases considerably for the reinforced specimen up to 7.5 mm after the last cycle before failure. In the case of the prestressed specimen the irreversible deflection amounts only to 2.5 mm. When the load cycles are omitted i.e. at monotonic loading, one can see the different behaviour even better (Fig. 11).

However, the carbon specimen shows the opposite (Fig. 8). Now, the reinforced one behaves more stiff than the prestressed one. The stiffness ratio between initial slope and slope at failure amounts to 9 for the reinforced specimen and to 18 for the prestressed one. The reason for that is once again the bad bond between carbon and concrete which is deteriorated in the prestressed element when the prestress is released. It is striking how the irreversible deflection increases for the prestressed specimen up to 7.5 mm the last cycle before failure which is amounted to 5 mm for the reinforced one. Fig. 12 gives a clearer picture under monotonic loading.

Both types of reinforcement lead to a softening behaviour in the post-peak range which could clearly be seen for Figs. 7 and 8. The carbon reinforced specimen shows this clearer than the AR glass type. Carbon is mainly pulled out whereas AR glass fractured.

<u>Crack width</u>

Crack width mirrors the deflection since the ultimate strain of concrete in tension is very low. This means that a certain deflection is caused by cracking, the question being how large the cracks are and how they are spaced. When the AR glass reinforced specimen deflected up to 8 mm there were only cracks of 0.2 mm width. Then, obviously no more cracks developed but the load increased a little and the cracks opened to a final size of 0.7 mm at 17 mm deflection. In the prestressed specimen cracks stayed at 0.1 mm until failure at 9 mm deflection. One can see at the wrinkles of the load-displacement curve that many cracks have developed.

The behaviour of the carbon reinforced specimen is opposite but also matching with the deflection. The reinforced one shows cracks with 0.1 mm width up to 8 mm deflection which coincides with the prestressed AR glass specimen. This means that the bond of carbon with concrete is very good as long it is efficient. On the contrary when the bond is disturbed by first prestressing the deflection becomes larger and also the cracks wider. The order of magnitude is the same as in the case of the AR glass reinforced specimen.

Pull-out with acoustic emission

The pull-out tests were essential to understand the different behaviour of AR glass and carbon reinforcement in the bending tests. Why were the two materials so different? The answer could be given by analysing the pull-out results. Looking to Fig. 9 one sees on the left side in the upper figure a plot of the energy and also a line of the cumulative energy vs. time. The lower plot shows the force and the slip vs. time. Combining both one can see that the first events start at the real beginning of the slip when already a force of about 75 N was reached. The first cluster of events start after one minute. There is a very minor slip and a force of 125 N. The line of cumulative energy corresponds to the slip line. When the force had reached the maximum most events occurred. The middle left figure shows the depth of the event, and the size of the point gives a relative indication of the energy of the event. There are low energetic ones in the beginning and high energetic ones at a later time. They correspond with fibre rupture. The upper right figure is a localization plot together with the projection on the x-y, x-z and y-z plane. One can clearly see that all events come from a narrow source in the prism, i.e. directly from the interface between AR glass and concrete. The places of an event are concentrated in the lower 15 mm of the roving which means that there was a good bond leading to the rupture of the reinforcing roving already with an embedment length of 15 mm. Finally the lower right plot shows the force vs. slip relation with an almost linear increase of the force and a post-peak drop due to continuous fracture of the filaments.

If this result is used to interpret Fig. 7 one has to conclude that the good bond of the AR glass causes multiple cracking in the specimen until the maximum load in the reinforcing rovings is reached which leads then to an abrupt failure. The same is true for the prestressed elements, however, with the difference that the initial strain in the element does not show up any more.

Fig. 10 is the analogue plot with respect to the carbon reinforcement. The upper left figure shows that the onset of the acoustic event clusters take already place after less than half a minute. The corresponding load is only 50 N. There is a continuous increase of slip and a continuous production of acoustic events. Also here, the line of cumulative energy corresponds well with the slip-time curve. However, there is a big difference in the sources of the events, they are scattered over the whole depth of the prism which means that slip occurs over the whole embedment length and that the 40 mm are not sufficient for a complete anchorage of the roving. The roving failed in pull-out and not in tensile rupture. The upper right plot shows how the events are distributed and the lower right figure is a plot of force-slip with a final friction type behaviour.

This behaviour means that prestressing of the specimens was doubtful. There was a visible slip of the carbon rovings into the concrete. Although the original prestress in the element was 2.5 MPa, after stress release after one day the real prestress in the concrete was smaller but not exactly known. It has to be assumed that the bond was affected by the stress release. On the other hand, the specimens which were not prestressed had a good bond. Both effects were responsible for the behaviour as depicted in Fig. 7.

Figs. 9 and 10 can be evaluated with respect to the maximum bond stress. The AR glass showed an average maximum bond stress of 225 N divided by 15 mm and the carbon showed one of 300 N divided by 40 mm. The calculated values are 15 N/mm and 7.5 N/mm resp.

CONCLUSIONS

The investigations lead to some conclusions:

- Prestressing of concrete leads to a dramatic decrease of deflection and crack width when properly executed. The AR glass specimens support this statement. The most beneficial effect is that the initial strain in the fabric is anticipated by prestressing and therefore crack width remains small.

- The carbon reinforcement showed the opposite behaviour due to early stress release and inferior bond.

- Pull-out tests together with acoustic emission monitoring and localization was a very appropriate measure to show the different bond behaviour of AR glass and carbon. AR glass failed at an effective bond length of 15 mm while carbon slipped through 40 mm.

- Evaluating bond results the unexpected behaviour of carbon reinforcement could be explained.

- To improve the bond of carbon and the long term stability of glass reinforcement it is suggested that the rovings should be impregnated or coated
by an epoxy coating.

## ACKNOWLEDGEMENT

The first author is greatly indebted to Surenda Shah for his continuous support
and interest. It was about 25 years ago when we met in Delft, Netherlands, and
from that time there was a continuing exchange of ideas and research results
with great respect to Surenda's experimental and analytical achievements.

## REFERENCES

1.  Daniel, J.I., Shah, S.P. (Eds.): *Thin-Section Fiber Reinforced Concrete and Ferrocement*. ACI SP-124, Detroit 1990

2.  Taerwe, L. (Ed.): *Non-Metallic (FRP) Reinforcement for Concrete Structures*. RILEM Proceedings 29, E & FN SPON, London 1995

3.  Japan Concrete Institute: *Non-Metallic (FRP) Reinforcement for Concrete Structures*. Proc. 3rd Intern. Symposium on Non-Metallic Reinforcement for Concrete Structures, Tokyo 1997

4.  Reinhardt, H.W.: *On the biaxial testing and strength of coated fabrics*. Experimental Mechanics 16 (1976), No. 2, pp. 71-74

5.  Helbing, A.K., Brühwiler, E.: *Eine neue Halterung für Zugversuche mit Beton-Probekörpern*. Material und Technik 15 (1987), No. 4, pp. 103-107

6.  Grosse, C.U.: WinPecker, Programme for the 3D localization of acoustic emission and the automatic determination of onset times. Manual Version 1.2a, Institute of Construction Materials, University of Stuttgart, 1999

Table 1. Concrete mix

| Component | kg/m$^3$ |
|---|---|
| Cement CEM I 42.5 R | 480 |
| Fly ash | 154 |
| Silica fume | 41 |
| Sand 0 - 0.6 mm | 460 |
| Sand 0.6 - 1.2 mm | 920 |
| Superplasticizer | 17 |
| Water (added) | 170 |

Table 2. Properties of hardened concrete

| Property | after | 1 d | 7 d | 28 d |
|---|---|---|---|---|
| Compressive strength, MPa | | 25 | 62 | 75 |
| Flexural strength, MPa | | 5 | 9 | 11.5 |
| Shrinkage, mm/m | | - | 0.5 | 0.6 |

Table 3. Properties of textile reinforcement

| Property, unit | Type | |
|---|---|---|
| | C-Sae | AR-Sae |
| Material | Carbon | AR glass |
| Structure of fabric | Biaxial 0°/90° | Biaxial 0°/90° |
| Weight, g/m² | 320 | 500 |
| Roving, tex | 1700 | 2500 |
| Area of one roving, mm² | 0.9 | 0.9 |
| Mesh size, square, mm | 10 | 10 |
| Maximum tensile force per roving, N | 1100 | 300 |

Table 4. Main results of the bending tests, mean of four results

| Reinforcement | LOP | | MOR | |
|---|---|---|---|---|
| | Stress MPa | Deflection mm | Stress MPa | Deflection mm |
| AR glass | | | | |
|    Reinforced | 7.1 | 0.34 | 14.6 | 17.9 |
|    Prestressed | 7.0 | 0.41 | 17.8 | 12.9 |
| Carbon | | | | |
|    Reinforced | 5.4 | 0.20 | 30.0 | 11.5 |
|    Prestressed | 5.8 | 0.30 | 17.7 | 18.0 |

Table 5. Crack width at various deflections

| Reinforcement | Crack width mm | Deflection mm |
|---|---|---|
| AR glass | | |
|    Reinforced | 0.2 | 8 |
| | 0.3 | 11 |
| | 0.4 | 12.5 |
| | 0.5 | 14 |
| | 0.7 | 17 (failure) |
|    Prestressed, 1.5 MPa | < 0.1 | 9 (failure) |
| Carbon | | |
|    Reinforced | 0.1 | 8 |
| | 0.2 | 10 (failure) |
|    Prestressed, 2.5 MPa | 0.1 | 4 |
| | 0.2 | 7 |
| | 0.3 | 12 |
| | 0.4 | 16 |
| | 0.6 | 17 (failure) |

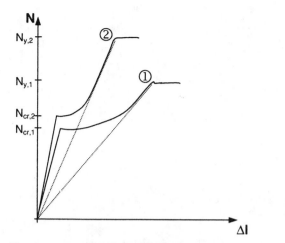

Fig. 1.   Normal force vs. elongation of a reinforced element
①  low reinforcement ratio and/or low modulus of elasticity
②  high reinforcement ratio and/or stiff reinforcing material

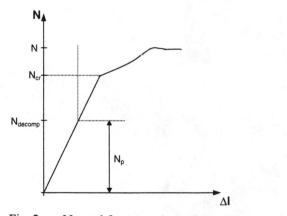

Fig. 2.   Normal force vs. elongation of a prestressed element

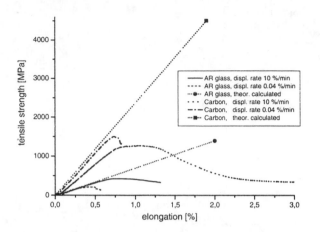

Fig. 3.    Tensile strength of rovings at different displacement rates

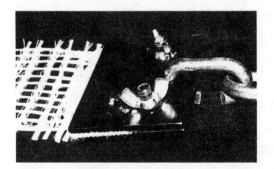

Fig. 4.    Clamping device

Fig. 5.    Prestressing frame

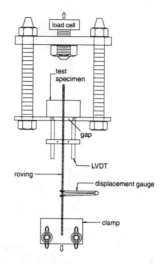

Fig. 6.    Schematic of pull-out test

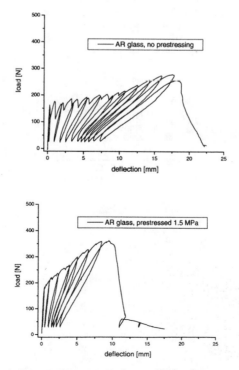

Fig. 7.    AR glass reinforced elements under cyclic load
Top: reinforced, bottom: prestressed

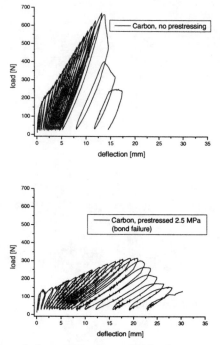

Fig. 8. Carbon reinforced elements under cyclic load
Top: reinforced, bottom: prestressed

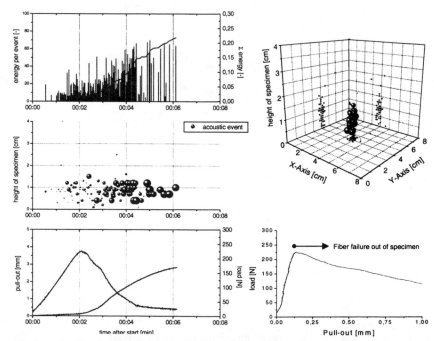

Fig. 9. Pull-out force and displacement together with acoustic emission signals (AR glass)

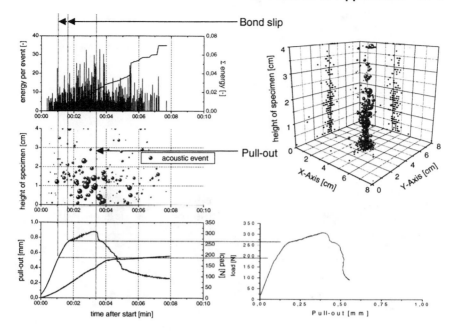

Fig. 10.   Pull-out force and displacement together with acoustic emission signals (Carbon)

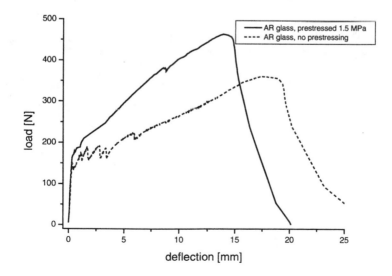

Fig. 11.   Load-deflection curve for a reinforced and a prestressed element under monotonic loading (AR glass)

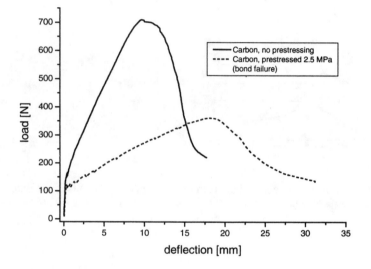

Fig. 12.   Load-deflection curve for a reinforced and a prestressed element under monotonic loading (Carbon)

# Advances in ECC Research

## by V. C. Li

Synopsis: This article reviews the recent advances in the research of Engineered Cementitious Composites (ECC), a class of microstructurally tailored fiber reinforced cementitious composites. The design basis, the processing routes, and some ECC performance characteristics in structural applications are highlighted. This article is dedicated to Professor Surendra Shah, in honor of his seminal contributions to research and education in advanced cementitious materials over the last several decades.

Keywords: composites; engineered cementitious composites(ECC); high peformance fiber reinforced cementitous composites(HPFRC); repair; retrofit; seismic structures

ACI member Victor C. Li is a Professor of Civil and Environmental Engineering. He is a fellow of the American Society of Civil Engineers and the American Society of Mechanical Engineers. His areas of interest include infrastructure engineering, fracture mechanics, micromechanics, composite design, and material/structure interactions.

## INTRODUCTION

Fiber reinforced cementitious composites (FRC) can be classified into three groups. FRC employing low fiber volume fractions (<1 %) utilize the fibers for reducing shrinkage cracking [1]. FRC with moderate fiber volume fractions (between 1% and 2%) exhibit improved mechanical properties including modulus of rupture (MOR), fracture toughness, and impact resistance. The fibers in this class of FRC could be used as secondary reinforcement in structural members, such as in partial replacement of shear steel stirrups [2,3,4], or for crack width control in structures [5,6]. In the last decade or so, a third class of FRCs, generally labeled as high performance FRC, or simply HPFRC, has been introduced. HPFRC exhibits apparent strain-hardening behavior by employing high fiber contents. These HPFRCs include SIFCON (slurry infiltrated 5-20% of steel fibers, see, e.g. [7]), SIMCON (slurry infiltrated 6% steel fiber mat, [8]), and CRC matrix (using 5-10% finer steel fibers [9]). The tensile strain capacity of HPFRC is typically about 1.5% or less.

Recently, a new kind of fiber reinforced cementitious composite, known as Engineered Cementitious Composite (ECC) has been developed. ECC exhibits tensile strain-hardening behavior with strain capacity in the range of 3-7% [10,11], yet the fiber content is typically 2% by volume or less. The ultra-high ductility is achieved by optimizing the microstructure of the composite employing micromechanical models [10] that account for the mechanical interactions between the fiber, matrix and interface. These models provide guidance to tailoring of these three phases synergistically so that high composite performance can be achieved with only a moderate amount of fibers. ECC may be regarded as an optimized HPFRC.

The differences between ECC, FRC and common HPFRC are summarized in Table 1. Micromechanical calculations [10] tend to favor fibers with diameter $d_f$ less than 50 μm, in order to achieve strain-hardening with lower fiber volume fraction $V_f$. Hence polymeric fibers, often drawn to such diameters, are preferred over steel fibers, typically in the 150-500 μm range. Steel fibers with lower $d_f$ can in principle be manufactured, although the cost becomes prohibitively high [12].

This article highlights some recent advances in ECC research, including the theoretical guidelines for ECC development, the processing routes of ECC material, and some ECC performance characteristics in structural applications. A review of the fundamental mechanical properties of ECC in tension, flexure, shear and fracture can be found in [10].

## ECC MATERIAL DESIGN BASIS

The micromechanics of tensile strain-hardening for cementitious composites reinforced with randomly oriented short fibers have been extensively studied. The requirements for steady state crack propagation [13, 14] necessary for composite strain-hardening behavior, and the micromechanics of the σ–δ relationship [15] combine to provide guidelines for the tailoring of fiber, matrix and interface in order to attain strain-hardening with the minimum amount of fibers [10].

Steady state crack propagation means that a crack extends at constant ambient tensile stress $\sigma_{ss}$, while maintaining a constant crack opening $\delta_{ss}$ (a flat crack, other than a near the crack tip region). Marshall and Cox [13] showed that this phenomenon prevails (over the typical Griffith type crack) when the condition

$$J_{tip} = \sigma_{ss}\delta_{ss} - \int_{0}^{\delta_{ss}} \sigma(\delta)d\delta \tag{1}$$

is satisfied. In Eqn. (1), $J_{tip}$ approaches the matrix toughness $K_m^2/E_m$ at small fiber content, appropriate for ECC since less than 3% fiber by volume is used. (The matrix fracture toughness $K_m$ and Young's Modulus $E_m$ are sensitive to the mix design, such as $w/c$ ratio and sand content.) The right hand side of Eqn. (1) may be interpreted as the energy supplied by external work less that dissipated by the deformation of the "inelastic springs" at the crack tip process zone opening from 0 to $\delta_{ss}$. The inelastic springs concept is a convenient means of capturing the inelastic processes of fiber deformation/breakage and interface debonding/slippage of those fibers bridging across the crack faces in the process zone. Hence Eqn. (1) expresses the energy balance during steady state crack propagation.

Figure 1 schematically illustrates this energy balance concept on a fiber bridging stress-crack opening $\sigma(\delta)$ relationship. The right hand side of Eqn. (1) is shown as the dark shaded area and is referred to as the complementary energy. Since the maximum value (hatched area) of this complementary energy $J_b'$ occurs when the shaded area extends to the peak stress $\sigma_0$ and crack opening $\delta_0$, it implies an upper limit on the matrix toughness as a condition for strain-hardening:

$$\frac{K_m^2}{E_m} \leq \sigma_o\delta_o - \int_{0}^{\delta_o} \sigma(\delta)d\delta \equiv J_b' \tag{2}$$

It is clear from Eqn. (2) that successful design of an ECC requires the tailoring of fiber, matrix and interface properties. Specifically, the fiber and interface properties control the shape of the $\sigma(\delta)$ curve and are therefore the dominant factors governing $J_b'$. Composite design for strain-hardening requires the tailoring of the fiber/matrix interface to maximize the value of $J_b'$. Similarly the matrix composition must be designed so that the value of $J_m = K_m^2/E_m$ is not excessive.

The shape of the $\sigma(\delta)$ curve and especially the rising branch associated with $J_b'$ shown in Figure 1 is related to a number of fiber/matrix interaction mechanisms. In the simplest case when fibers and matrix are in frictional contact only, the slope of the rising branch of the $\sigma(\delta)$ curve, or the stiffness of the bridges, is mainly governed by the fiber content $V_f$, diameter $d_f$, length $L_f$ and stiffness $E_f$, and the interface frictional bond $\tau_0$. In the case when chemical bond $G_d$ is present, the starting point of the $\sigma(\delta)$ is not at the origin of the plot, but is shifted upwards. This reflects the need of a certain amount of load on the fibers and interface before the interfacial chemical bond is broken. Dedonding is needed to allow for stretching of the debonded fiber segment to produce crack opening $\delta$. Thus the presence of $G_d$ typically diminishes the complementary energy $J_b'$ and is not conducive to strain-hardening. It is, however, helpful for minimizing the crack width of the multiple cracks if strain-hardening is achieved.

The peak value of the $\sigma(\delta)$ curve is mainly governed by $V_f$, $d_f$, $L_f$, $\tau_0$ in the case of simple friction pull-out. An analytic expression of $\sigma(\delta)$ and $\sigma_0$ can be found in [14]. In the case when both interface chemical bond and slip-hardening are present, an analytic expression of $\sigma(\delta)$ can be found in [16]. In the presence of $G_d$, the higher load on the fiber can lead to fiber rupture. Thus for given fiber strength $\sigma_f$, the complementary energy $J_b'$ again decreases with $G_d$.

It should be emphasized that Eqn. (2) does not prescribe a particular fiber type for ECC. Rather, it is the combination of fiber, matrix and interface parameters that creates the satisfaction of the strain-hardening condition. However, there are indeed certain characteristics of fibers which enhances $J_b'$. These include e.g., small $d_f$, high $L_f$ and low $G_d$. Any fiber with the correct profile of properties can in principle be a suitable fiber for ECC reinforcement.

Since $J_b'$ increases with $V_f$, Eqn. (2) can be used to define a critical fiber volume fraction $V_f^{crit}$ above which strain-hardening is expected to occur (when the inequality sign just holds), for given fiber, matrix and interface properties. Figure 2 plots $V_f^{crit}$ as a function of interfacial friction $\tau_0$ (while keeping all other micromechanical parameters fixed). For PE (polyethylene) fiber, the $V_f^{crit}$ continues to drop with increasing $\tau_0$ since the fiber strength is very high. In addition, the hydrophobic fiber does not have a chemical bond ($G_d = 0$) so that fiber rupture does not usually happen. (Fiber rupture can still occur if $L_f$ is excessively large. However, the material will not be processible.) For PVA fiber, the critical fiber volume fraction $V_f^{crit}$ first drops but then rises again due to fiber rupture at increasingly high $\tau_0$. Fiber rupture in the hydrophilic PVA fiber is enhanced by strong chemical bonding due to the presence of the hydroxyl group in the PVA (polyvinyl alcohol) fiber, resulting in a high $G_d$.

The concepts expressed in Eqn. (2) for strain-hardening condition have been validated by at least two studies involving ECC composites with PE and PVA fibers. Li et al [17] studied the transition from quasi-brittle to ductile behavior for a set of composites reinforced with PE fibers. They varied the sand content in the matrix which directly influenced the matrix toughness and the interface properties. Figure 3a shows the theoretical boundary between strain-hardening and quasi-brittle behavior using Eqn. (2) and the positions in the ($\tau_0$ , $J_m$) space for the four composites tested. The tensile stress-strain responses of the four composites are shown in Figure 3b. It is clear that in the region of low $J_m$ and high $\tau_0$ (with low sand content S:C = 0 - 0.5) where Eqn. (2) predicts strain-hardening, the corresponding composites (Mix I and III) revealed tensile strains of 2.27 – 5.44 %. In contrast, the composite (Mix II) with high $J_m$ (with high sand content S:C = 2) showed poor ductility of 0.20 %.

A recent study by [11] demonstrated the validity and effectiveness of using Eqn. (2) for composite constituent tailoring. For a fixed fiber volume fraction ($V_f$ = 2%) and matrix toughness ($J_m$ = 5 J/m$^2$), target values of $\tau_0$, $G_d$ and $\beta$ have been determined by Wu [18] as 1 - 2.1 MPa, 0 - 2.2 J/m$^2$, and 0 - 1.5 respectively in order to satisfy the inequality sign of Eqn. (2) for strain-hardening. These values have taken into account the variability of flaw size in the matrix [19]. The interface values were found to be successively reduced by increasing the amount of oiling agent on the fiber. When these $\tau_0$, $G_d$ parameters approached or dropped below the target values, the ductility of the composite was greatly enhanced.

Figure 4 shows the decreasing values of $\tau_0$ and $G_d$ as a function of oiling agent measured in single fiber pull-out tests, as well as the target values determined theoretically. This set of data suggests that oiling agent in excess of 0.6-0.8% is needed to achieve strain-hardening. Composite tensile tests using the same fiber and the same five levels of oiling content reveal that the ductility was enhanced from less than 1% to more than about 5% (Figure 5). The tensile strain capacity and multiple crack spacing as a function of oil content are shown in Figure 6. It is clear that the composite reaches multiple crack saturation when the oiling content reaches about 0.6-0.8% as expected.

The above discussions reveal that the design strategies for PE-ECC and PVA-ECC are very different. For the PE fiber with very high tensile strength and low bond properties, enhancing the interface bond strength leads to strain-hardening. An attempt to increase bond strength of PE fibers using plasma treatment process was demonstrated by Li et al [20], resulting in composites with tensile strain capacity of up to 7%. In contrast, for the PVA fiber with high chemical bond and moderate fiber strength, the design strategy calls for a reduction in interfacial chemical and frictional bond. In both cases, the matrix composition must be controlled to limit the toughness. These design strategies are rooted in micromechanics which provides guidelines for tailoring of material constituents for targeted composite performance.

## ECC PROCESSING

Two types of processing routes have been developed for ECC. For casting, normal casting and self-compacting casting [21-23] are available. Extrusion of ECC has also been demonstrated [24]. Spray ECC, equivalent to shotcreting, but replacing the concrete with ECC, is now being developed at the University of Michigan.

Figure 7a shows a PE-ECC pipe being extruded. The resulting pipe is very ductile under ring loading, as shown in Figure 7b. Instead of the typical brittle fractures at the four hinge zones, the PE-ECC pipe exhibits 'plastic' yielding type behavior. This figure also shows the typical FRC quasi-brittle behavior of a similarly extruded pipe containing 7% of pp-fiber.

Li and co-workers [21-23] developed self-compacting ECC via a constitutive rheological approach. In this approach, the ingredients of the mortar matrix were tailored so that high flowability is achieved, while respecting the conditions of strain-hardening for the composite as described earlier. The high flowability mortar matrix results from an optimal combination of a strong polyelectrolyte (a super-plasticizer) and a non-ionic polymer with steric action in maintaining non-aggregation of the cement particle in the dense suspension. Figure 8 shows the result of a deformability test and box test of a self-compacting PVA-ECC. The deformability $\Gamma$-value and the self-compactability L-value achieved in these tests ($\Gamma = 11.7$; $L = 0.94$) are comparable to those acceptable ($\Gamma = 8$ to $12$; $L = 0.73$ to $1.00$) for self-compacting concrete [25]. Tensile strain-hardening of these materials was reported [22-23].

## ECC PERFORMANCE CHARACTERISTICS IN STRUCTURAL APPLICATIONS

A number of investigations into the use of ECC in enhancing structural performance have been conducted in recent years. These include the repair and retrofit of pavements or bridge decks [26-28]; the retrofit of building walls to withstand strong seismic loading [29-31]; and the design of new framing systems [32]. These studies often reveal unique characteristics of ECC and R/ECC (steel reinforced ECC) in a structural context. These include high damage tolerance, resistance to shear load, energy absorption, delamination and spall resistance, and high deformability and tight crack width control for durability. These characteristics are illustrated with highlights of experimental results below.

**Damage Tolerance**: The damage tolerance of a structure is the ability for the structure to sustain load-carrying capacity even when overloaded into the inelastic range. For civil infrastructure systems, the cost of repair after a major earthquake can be enormous, as the 1995 Kobe earthquake in Japan (U.S.$95-147 billion [33]) demonstrated. Hence, damage tolerance of infrastructure is critical for both safety and economic reasons.

Parra-Montesinos and Wight [32] conducted a series of cyclic tests on the structural integrity of joints in hybrid steel-beam R/C column connections. Figure 9a,b shows the contrasts of the damage experienced by the standard connection and that experienced by a connection where the joint concrete is replaced by ECC. The corresponding load – joint deformation is shown in Figure 9c,d. The standard R/C joint suffers large crack opening, loss of bond between the axial reinforcement and the concrete, and therefore composite action, and experienced severe spalling where the steel beam bears on the concrete. In contrast, the ECC joint underwent strain-hardening with multiple micro-crack damage. The R/ECC material maintained high bond efficiency and no spalling was observed. All these were accomplished despite eliminating all the shear reinforcement used in the R/C specimen. Although the load and deflection imposed were pushed to higher levels, no repair was needed in the R/ECC connection. This experiment also demonstrated that spall failure commonly observed in high stress-concentration zones when concrete and steel elements interact can be eliminated by virtue of the damage tolerance of ECC material.

**Shear Resistance**:  In R/C elements, shear failures are typically resisted by shear reinforcements in the form of dowel bars, stirrups or hoop reinforcement. However, shear failures in short columns in bridge piers or walls between stories in buildings continue to be observed after large earthquakes.  The intrinsic brittleness of concrete makes it difficult to prevent shear failure that reflects the diagonal tensile stresses, potentially leading to catastrophic modes of collapse.

Kanda et al [29] and Fukuyama et al [34] conducted cyclic shear experiments on beam elements.  The contrast in failure modes of the R/C and R/ECC beams can be seen in Figure 10a [34]. The corresponding cyclic load-deformation curves (Figure 10b) confirm that the shear ductility of R/ECC is superior to that of R/C. For ECC, shear damage is similar to ductile tensile damage with diagonal microcracks.

**Energy Absorption**:  In seismic structures, energy absorption in hinge zones is used to dissipate the earthquake energy input.  In R/C structures, the concept of a plastic hinge is introduced by ductile yielding of the steel reinforcement in seismic detailing.  However, it is typical that only a small fraction of the steel actually undergoes yielding due to the disintegration of the surrounding brittle concrete.

Fischer and Li [35] conducted cyclic flexural experiments on cantilevered elements. The contrast in failure modes of the R/C and R/ECC beams is shown in Figure 11a,b. The corresponding cyclic load-deformation curves (Figure 11c,d) reveal that the R/ECC absorbs much higher energy than the R/C element.  This is particularly noteworthy since no hoop shear reinforcement was utilized in the R/ECC element.

**Resistance to Delamination and Spalling in Repaired Concrete Structures**:  In patch repairs, the common failure modes are spalling and/or delamination between the new and old concrete.  In bridge deck or pavements overlay repairs, reflective crack and spalling in the concrete overlays and/or delamination between the bonded overlay and the old concrete substrate are often observed.  It appears difficult to resist all modes of failure simultaneously since strengthening the interfacial bond tends to encourage spalling while enhancing the overlay strength tends to encourage delamination. As a result, the durability of concrete repair is compromised by one or the other of these failure mechanisms.

Li and co-workers [26-28] studied the resistance to delamination and spalling using ECC as the repair material. Using a specimen simulating an overlay on top of a joint together with an initial interface defect, they found that the delamination and spalling modes can be both eliminated by means of a kink-crack trapping process (Figure 12a). As the four point bend load increases, the initial interface crack extends slightly but quickly kink into the ECC overlay. The kink crack was subsequently trapped in the ECC so that further load increase forces crack extension into the interface. The kinking-trapping process then repeats itself, resulting in a succession of kink cracks in the ECC. However, spalling of the ECC was not observed since the kink crack does not propagate to the specimen surface. Delamination of the interface was also eliminated since the interface crack tip repeatedly kink into the ECC. In contrast, the specimen with a regular FRC overlay shows the expected kink-spall brittle fracture behavior. The corresponding load-deformation curves for the two specimens are shown in Fig. 12b.

**Deformability**: Under restrained drying shrinkage and/or temperature loading, concrete slabs may crack and undermine durability of the slab structure (e.g. bridge decks, parking garage slabs, factory floor slabs). Such problems have typically been dealt with by introducing joints. However the joints themselves often result in damage and are costly to maintain, especially in bridge decks. The large strain capacity of ECC can be utilized to accommodate imposed deformation on the deck by replacing standard joints with a strip of ECC material. This concept was test by Zhang et al [36].

Figure 13a shows a specimen containing a strip of ECC sandwiched between two standard concrete slabs shaped into a dog-bone for tensile loading, simulating restrained shrinkage. Even when the overall specimen has loaded to over 1.3% strain (Figure 13b), the concrete slabs experienced no cracks. Instead, the imposed deformation was localized into the ECC ductile strip which underwent tensile strain-hardening. Multiple microcrack damage is shown in the inset of Figure 13a.

**Durability of Concrete Cover**: In R/C beams serious durability problems may occur due to cracking of the concrete in the tensile zone, migration of aggressive agents through the concrete cover, corrosion of the reinforcing steel, expansion of the corroded steel bar, spall-fracturing of the concrete cover, subsequent accelerated corrosion of the reinforcing bar, and loss of load-carrying capacity of the R/C beam. Such durability problems are especially common in structures such as parking garages in cold regions where deicing salts are applied to the slabs. Maalej and Li [37] investigated the use of ECC as the cover material replacing standard concrete.

In four point bend tests loading to concrete crushing on the compression side, the standard R/C beam shows a large crack width which grows rapidly after steel yielding on the tension side. In contrast, the R/C beams with an ECC cover show crack width limited to 0.2 mm throughout the whole test range (Figure 14). Since the permeability of cracked concrete has been shown to scale with the cube of the crack width, the restriction of crack width is expected to slow down the migration of aggressive agents and therefore the corrosion rate of the re-bars, resulting in enhanced durability of the R/C structure.

## CONCLUSIONS

From the above discussions, it is clear that ECC has significant advantages in material performance over concrete and standard FRC. It also has cost and processing advantages over most common HPFRC due to its relatively low fiber content. As a result, it is plausible to envision ECC being applied to new structures such as moment frames [32] and R/C structures with ECC cover [36]. Applications of ECC to structural repair in the form of patch repair or as overlays [26-28] are plausible. Self-compacting processing [21-23] and spray processing will make these repair applications even more attractive. From the energy absorption point of view, the use of ECC in seismic retrofitting in building [29] and especially in hospitals [31] may be logical directions. Many of these application possibilities will require further investigations, but the research work that has been carried out lays a sound foundation for the use of ECC in enhancing the performance of infrastructure systems.

Further materials development of ECC may be expected in the areas of adaptation to lower cost material ingredients, adaptation of material ingredients and mixing methods to achieve desirable fresh properties for various processing routes, the addition of special functionalities to the hardened material performance, and standardization of material for easy adoption by industrial concerns. For structural applications, there is a strong need for standardized testing and design guidelines. The Japanese Concrete Institute has recently formed a Committee to investigate these issues. Novel structural systems with smart responses to loading and that take advantage of the unique characteristics of ECC may be expected. A truly high performance concrete material should embody strength, ductility, durability, cost-effective and easy and flexible processing routes. ECC represents an attempt towards this ideal.

## REFERENCES

[1]   Balaguru, P., and S. Shah, Fiber Reinforced Cement Composites, McGraw Hill, 1992.

[2]   Batson, G., Jenkins, E. Spatney, R., "Steel fibers as shear reinforcement in beams", *ACI Journal*, 69, 10, 640-644, 1972.

[3]   Sharma, A. K., "Shear strength of steel fiber reinforced concrete beams", *ACI proceedings*, 83, 4, 624-628, 1986.

[4]   Swamy, R.N., Bahia, H.M., "The effectiveness of steel fibers as shear reinforcement", *Concrete International*, 35-40, 1985.

[5]   Stang, H., Aarre, T. "Evaluation of crack width in FRC with conventional reinforcement", Cement & Concrete Composite, 14, 2, 143-154, 1992.

[6]   Stang, H., Li, V.C., Krenchel, H. "Design and structural applications of stress-crack width relations in FRC", *RILEM J. Materials and Structures*, 1993.

[7]   Naaman, A.E., and Homrich, J.R., "Tensile stress-strain properties of SIFCON" ACI Materials J., 86, 3, 244-251, 1989.

[8]   Krstulovic-Opara, N., and H. Toutanji, "Infrastructural repair and retrofit with HPFRCCs," in High Performance Fiber Reinforced Cement Composites 2, Eds. Naaman, A.E., and H.W. Reinhardt, E&FN Spon, 423-439, 1996.

[9]   Bache, H.H., Introduction to Compact Reinforced Composites, Nordic Concrete Research, Publication N6, 19-33, 1987.

[10] Li, V.C., "Engineered Cementitious Composites – Tailored Composites Through Micromechanical Modeling," in *Fiber Reinforced Concrete: Present and the Future*. Eds. N. Banthia et al, CSCE, Montreal, 64-97, 1998.

[11] Li, V.C., S. Wang, and C. Wu, "Tensile Strain-Hardening Behavior of PVA-ECC", Submitted, *ACI J. of Materials*, Jan., 2001.

[12] Bakaert Corp., Belgium, private communication, 2000.

[13] Marshall, D. and Cox, B.N. "A J-integral Method for Calculating Steady-State Matrix Cracking Stress in Composites," Mechanics of Materials 7, 127-133, 1988.

[14] Li, V.C., "Post-Crack Scaling Relations for Fiber Reinforced Cementitious Composites", ASCE *J. Materials in Civil Engineering*, 4(1), 41-57, 1992.

[15] Li, V.C. and Leung, C.K.Y., Steady State And Multiple Cracking Of Short Random Fiber Composites, ASCE J. Engineering Mech., 118(11) 1992, 2246-2264.

[16] Lin, Z., T. Kanda and V.C. Li, "On Interface Property Characterization and Performance of Fiber Reinforced Cementitious Composites," *J. Concrete Science and Engineering, RILEM,* 1, 173-184, 1999.

[17] Li, V.C., D.K. Mishra and H.C. Wu, "Matrix Design for Pseudo Strain-Hardening Fiber Reinforced Cementitious Composites," RILEM *J. Materials and Structures*, 28(183), 586-595, 1995.

[18] Wu, C., "Micromechanical Tailoring of PVA-ECC for Structural Applications," PhD Thesis, Department of Civil and Environmental Engineering, University of Michigan, Ann Arbor, USA, Jan., 2001.

[19] Kanda, T. and V.C. Li, "A New Micromechanics Design Theory for Pseudo Strain Hardening Cementitious Composite," *ASCE J. of Engineering Mechanics*, 125(4), 373-381, 1999.

[20] Li, V.C., H.C. Wu, and Y.W. Chan, "Effect of Plasma Treatment of Polyethylene Fibers on Interface and Cementitious Composite Properties," *J. of Amer. Ceramics Soc.*, 79(3), 700-704, 1996.

[21] Li, V.C., H.J. Kong, and Y.W. Chan, "Development of Self-Compacting Engineered Cementitious Composites," in Proceedings, International Workshop on Self-Compacting Concrete, Kochi, Japan, 46-59, 1998.

[22] Li, V.C., J. Kong, and S. Bike, ""High Performance Fiber Reinforced Concrete Materials," in Proc. of High Performance Concrete – Workability, Strength, and Durability, 71-86, Eds. C.K.Y. Leung et al, China, 2000.

[23] Li, V.C., J. Kong, and S. Bike, "Constitutive Rheological Design For Development Of Self-Compacting Engineered Cementitious Composites", In Proc., 2nd Int'l Workshop on Self-Compacting Concrete, Tokyo, Japan, 2001.

[24] Stang, H. and V.C. Li, "Extrusion of ECC-Material", in Proc. Of High Performance Fiber Reinforced Cement Composites 3 (HPFRCC 3), Ed. H. Reinhardt and A. Naaman, Chapman & Hull, pp. 203-212, 1999.

[25] Ozawa, K., Sakata, N., and Okamura, H., "Evaluation of Self-Compactability of Fresh Concrete Using the Funnel Test", *JSCE Concrete library*, **25**, 59-75, 1995.

[26] Lim, Y.M. and V.C. Li, "Durable Repair of Aged Infrastructures Using Trapping Mechanism of Engineered Cementitious Composites" *J. Cement and Concrete Composites*, 19(4) 373-385, 1997.

384  Li

[27]   Kamada, T. and V.C. Li, "The Effects of Surface Preparation on the Fracture Behavior of ECC/Concrete Repair System," *J. of Cement and Concrete Composites*, 22, 6, 423-431, 2000.

[28]   Zhang, J. and V.C. Li, "Monotonic and Fatigue Performance in Bending of Fiber Reinforced Engineered Cementitious Composite in Overlay System," Accepted, *J. of Cement and Concrete Research*, 2001.

[29]   Kanda, T., S. Watanabe and V. C. Li, "Application of Pseudo Strain Hardening Cementitious Composites to Shear Resistant Structural Elements", in *Fracture Mechanics of Concrete Structures* Proc. FRAMCOS-3, AEDIFICATIO Publishers, D-79104 Freiburg, Germany, 1477-1490, 1998.

[30]   Kabele, P., V.C. Li, H. Horii, T. Kanda and S. Takeuchi, "Use of BMC for Ductile Structural Members," in Proc. of 5th Int'l Symp. on *Brittle Matrix Composites* (BMC-5), Warsaw, Poland, 579-588, 1997.

[31]   Kesner, K.E., and S. L. Billington, "Investigation of Ductile Cement-Based Composites for Seismic Strengthening and Retrofit," in Fracture Mechanics of Concrete Structures, de Bost et al (eds), A.A. Balkema, Netherlands, 65-72, 2001.

[32]   Parra-Montesinos, G.J., and J.K. Wight, "Seismic Response of Exterior RC Column-to-Steel Beam Connections," ASCE J. Structural Engineering, 1113-1121, 2000.

[33]   EQE International, "The January 17, 1995 Kobe Earthquake", An EQE Summary Report, April 1995.

[34]   Fukuyama, H., Y. Matsuzaki, K. Nakano, and Y. Sato, "Structural Performance of Beam Elements with PVA-ECC," in Proc. Of High Performance Fiber Reinforced Cement Composites 3 (HPFRCC 3), Ed. Reinhardt and A. Naaman, Chapman & Hull, pp. 531-542, 1999.

[35]   Fischer, G., and V.C. Li, "Influence of Matrix Ductility on the Tension-Stiffening Behavior of Steel Reinforced Engineered Cementitious Composites (ECC)," Accepted, *ACI J. of Structures*, Jan., 2001.

[36]   Zhang, J., V.C. Li, A. Nowak and S. Wang, "Introducing Ductile Strip for Durability Enhancement of Concrete Slabs," Accepted, *ASCE J. of Materials in Civil Engineering,* April, 2001.

[37]   Maalej, M., and V.C. Li, "Introduction of Strain Hardening Engineered Cementitious Composites in the Design of Reinforced Concrete Flexural Members for Improved Durability," *ACI Structural J.,* 92(2), 167-176, 1995.

Table 1: Comparison between FRC, common HPFRC and ECC

| | FRC | Common HPFRC | ECC |
|---|---|---|---|
| **Composite Design Methodology** | NA | Use high $V_f$ | Micromechanics based, minimize $V_f$ for cost and processability |
| **Fiber** | Any type, $V_f$ usually < 2%; $d_f$ (steel) ~ 500 μm | Mostly steel, $V_f$ usually > 5%; $d_f$ ~ 150 μm | Tailored, polymer fibers most suitable; $V_f$ usually < 2%; $d_f$ < 50 μm |
| **Matrix** | Coarse aggregates used | Fine aggregates used | Controlled for matrix toughness and initial flaw size; fine sand used |
| **Interface** | Not controlled | Not controlled | $G_d$ and $\tau_o$ controlled |
| **Tensile behavior** | Strain-softening | Strain-hardening | Strain-hardening |
| **Tensile strain capacity** | 0.1% | < 1.5% | >3%; 8% demonstrated |
| **Crack width** | Unlimited | Typically several hundred μm, unlimited for $\varepsilon$ >1.5% | Typically < 100 μm during strain-hardening |
| **Processing** | Self-compaction demonstrated; Extrudability demonstrated | Self-compaction impossible due to high $V_f$, often requires high frequency vibration (e.g. in CRC); Extrudability demonstrated | Self-compaction demonstrated; Extrudability demonstrated |

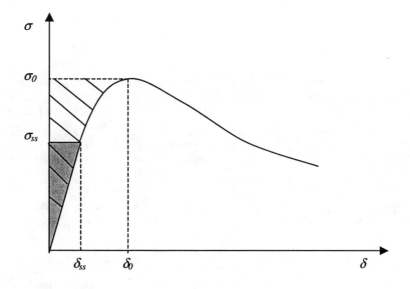

Figure 1: Schematic illustration of energy balance concept on a fiber bridging stress-crack opening $\sigma(\delta)$ plot

(a)

(b)

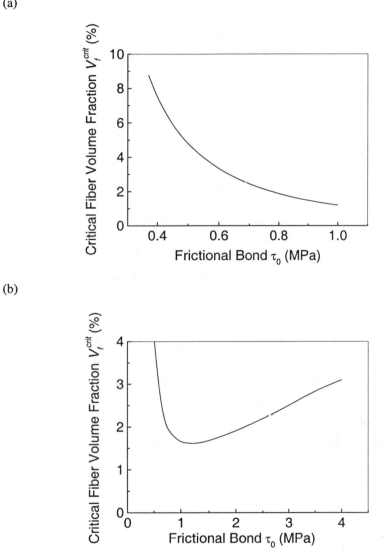

Figure 2: $V_f^{crit}$ illustrated as a function of interfacial friction $\tau_0$ (a) for PE-ECC; (b) for PVA-ECC.  All other dependent parameters fixed.

(a)

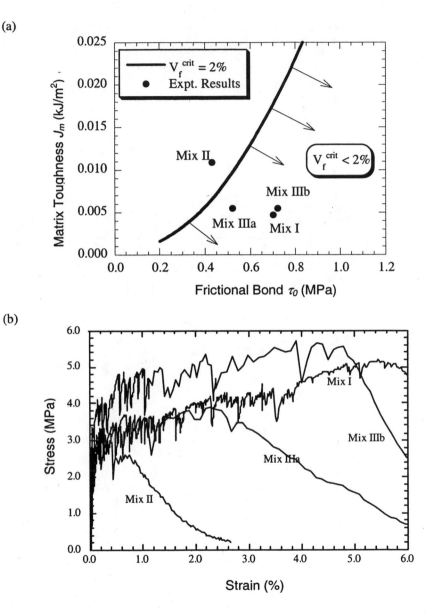

(b)

Figure 3: (a) Theoretical boundary between strain-hardening and quasi-brittle behavior using Eqn. (2) and the positions in the $(J_m, \tau_0)$ space for four composites tested. Shaded region indicates strain-hardening achievable with $V_f < 2\%$. (b) Experimentally determined tensile stress-strain curves for the four composites shown in (a).

(a)

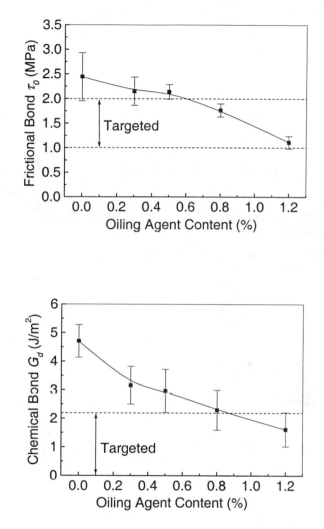

(b)

Figure 4: Measured (a) $\tau_0$ and (b) $G_d$ as a function of oiling content.  Also shown are the target values determined theoretically.

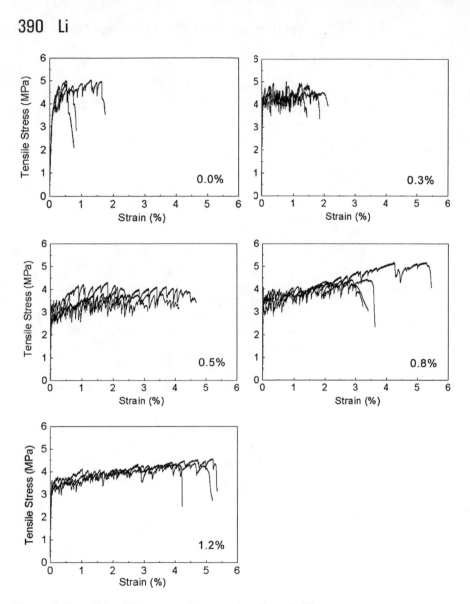

Figure 5: Ductility of the composites as a function of oiling content.

(a)

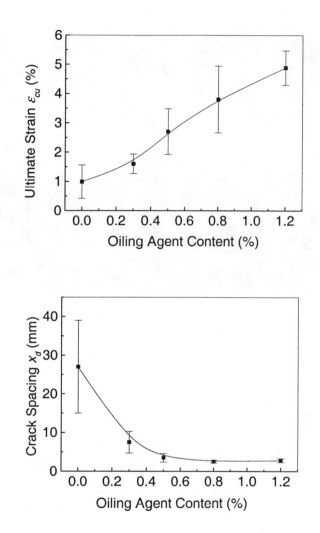

(b)

Figure 6: (a) Tensile strain capacity and (b) multiple crack spacing as a function of oiling content.

(a)

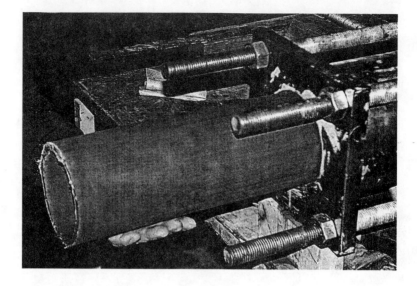

(b)

(c)

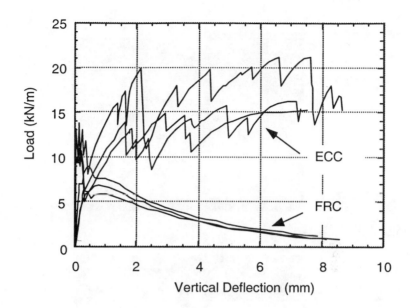

Figure 7: (a) PE-ECC pipe being extruded, (b) Ductile deformation of pipe under ring load, (c) Load-deflection curves for PE-ECC. Also shown for contrast are curves for PP-FRC pipe.

(a)

(b)

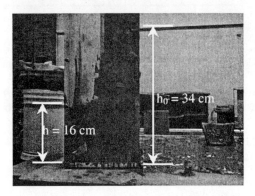

Figure 8: Result of (a) a deformability test of a self-compacting PVA-ECC ($\Gamma$ = 11.7), and (b) a box test of a self-compacting PVA-ECC (L = 0.94).

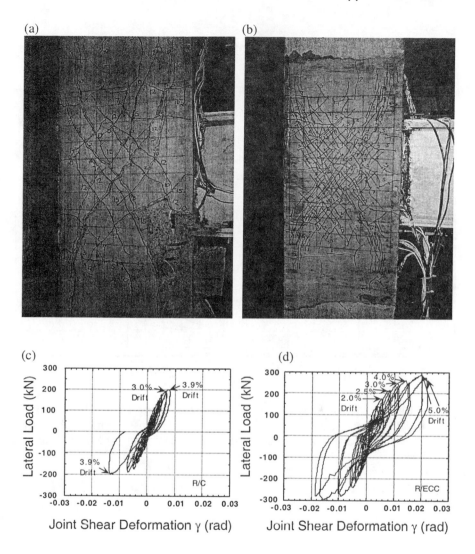

Figure 9: Damage of joint panel in RCS connection subjected to cyclic loading for (a) standard R/C joint; and (b) R/ECC joint. The corresponding load – joint deformation hysteresis loop for (c) standard R/C joint; and (d) R/ECC joint are also shown.

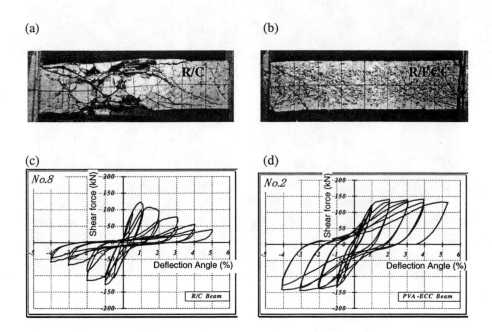

Figure 10: Damage of beam element subjected to shear cyclic loading for (a) standard R/C beam; and (b) R/ECC beam. The corresponding load – beam deformation hysteresis loops for (c) standard R/C beam; and (d) R/ECC beam are also shown.

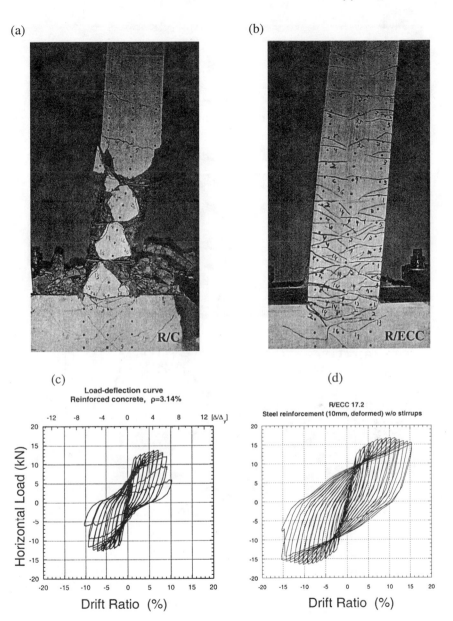

Figure 11: Damage of beam element subjected to flexural cyclic loading for (a) standard R/C beam; and (b) R/ECC beam.  The corresponding load – beam deformation hysteresis loops for (c) standard R/C beam; and (d) R/ECC beam are also shown.

(a)

(b)

(c)

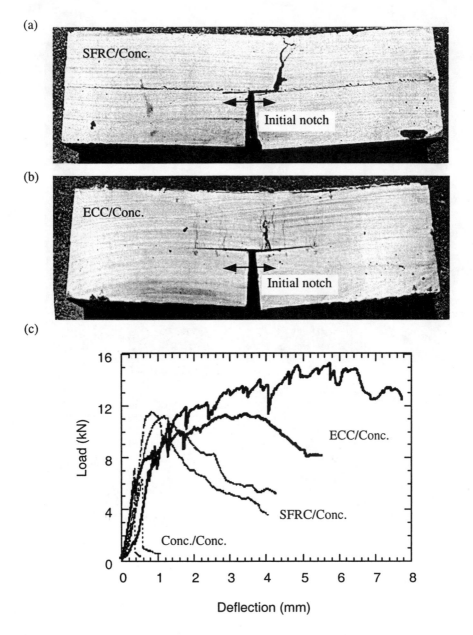

Figure 12: Damage of overlay system subjected to four point bend loading for (a) FRC overlay; and (b) ECC overlay.  The corresponding load – deflection curves for standard concrete, FRC, and ECC overlay systems are shown in (c).

(a)

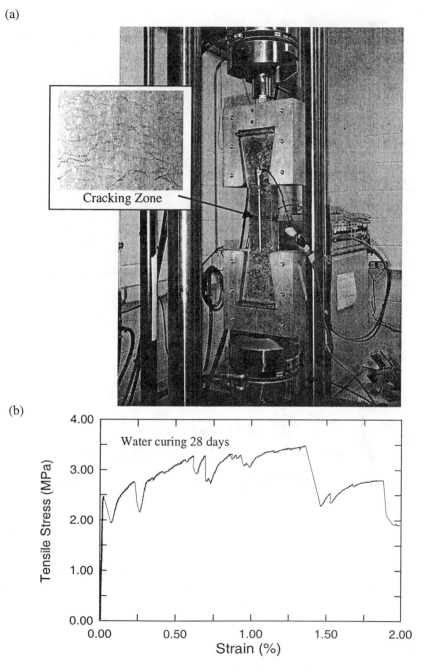

(b)

Figure 13: (a) Ductile strip specimen with ECC sandwiched between two standard concrete slabs subjected to simulated restrained shrinkage tensile loading. The corresponding load-deformation curve is shown in (b).

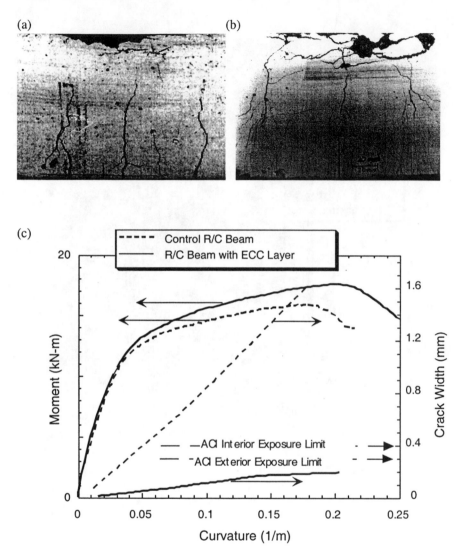

Figure 14: Damage of R/C beam subjected to four point bend loading for (a) standard concrete cover; and (b) ECC cover. The corresponding load – beam deflection curves and crack width in the cover are shown in (c).

# PVA Polymer Modified Glass Fiber Reinforced Cementitious Composites

## by Z. Li, A. C. P. Liu, and C. K. Y. Leung

Synopsis: This report presents an experimental investigation on the effect of polyvinyl alcohol (PVA) polymer modified alkali resistance (AR)-glass fiber reinforced cementitious composites (GFRC). The performance of the alkali resistance glass fibers in blended cementitious matrix was evaluated by a series of tensile and flexural tests. Specimens with different amounts of PVA polymer were tested. Specimens under dry air and moist curing environment were compared. The results indicate that the addition of 2% (weight) PVA polymer shows improvement in strength, toughness, ductility, and deflection. SEM studies indicated that PVA forms a thin film covering on the glass fibre surface, which inhibits the nucleation of calcium hydroxide crystals on the glass fibres surface and to enhance the formation of C-S-H during the hydration. The presence of PVA shows the ability to reduce the embrittlement of the GFRC and changes the failure mode from brittle to ductile. EDAX results shows that the Ca/Si ratio in the fiber interface of the PVA modified specimens is greatly reduced.

Keywords: ductility; EDAX; embrittlement; fiber interface; glass fiber-reinforced cementitious composites; polyvinyl alcohol (PVA); SEM; toughness

Authors:

Zongjin Li is an associate professor of civil engineering at Hong Kong University of Science and Technology.

Alvin C.P. Liu is an MPhil graduate student in the Department of Civil Engineering at Hong Kong University of Science and Technology.

Christopher K.Y. Leung is an associate professor of civil engineering at Hong Kong University of Science and Technology.

## INTRODUCTION

Glass fibers are incorporated in Portland cement matrices to produce composites with substantially increased in strength and ductility. Because of durability problem associated with aging under wet condition, use of such composites is limited currently to non-structural purposes. Under such severe conditions the GFRCC loses part of its strength and most of its ductility (toughness). It becomes effectively as brittle as an unreinforced cement matrix.

The embrittlement of GFRCC has been attributed to chemical attack on the glass resulting from the alkaline cement environment. Such attack may result in etching of the glass surface and lead to loss of strength of the glass fiber. Development of high zirconia glass fibers (alkali resistance glass fibers) with better resistance to alkaline environment has resulted in improved aging characteristics [1,2]. Despite the use of alkali-resistant glass fibers and pozzolanic materials, the durability problems still exist; their use resulting in a slowing down of the rate of strength and ductility loss, but not in preventing such loss. In part this is due to the fact that the durability problem is not only a response to chemical degradation of that glass, but also is associated with the deposition and growth of calcium hydroxide in the space between the individual filaments of the glass fiber, cementing them together and reducing the flexibility of the glass filaments in the strand. It has been proposed that such loss of flexibility leads to embrittlement since the fracture process involves the development of significant local flexural stresses.

If both mechanisms can cause embrittlement, failure of the composite will be initiated at a local zone where either flaw exist on the surface of glass fibers, or the deposition of dense hydration products between the filaments is such as to give rise to high local stress concentrations that cannot be relieved by relative motion of the individual filaments.

## EXPERIMENTAL DETAILS

### Materials and Mix Proportion

Type I Portland cement, two kinds of silica sand (600-300μm and 150-90μm in diameter), ground blast-furnace slag, alkaline resistant glass fibers and polyvinyl alcohol (PVA) polymer were used as basic materials. The PVA polymer (specific gravity is 1.23) used was 86-89% hydrolyzed poly(vinyl acetate) in powder form. Its molar mass was 100000-205000 number averages and the degree of polymerisation was 2000-4200. The AR glass fibers had a length of 12mm with an average diameter of 8μm, and an average tensile strength and elastic modulus of 3600MPa and 70GPa. Four batches with different amounts of PVA polymer were prepared. The detail of the mixing proportion by weight is listed in Table 1.

Prior to mixing, the PVA powder was put into the hot water and kept stirring until all the PVA powder was dissolved. After the PVA powder dissolved, kept the PVA solution in room temperature until it completely cooled down. Before the extrusion, slurry was prepared by coating the glass fibers with the cementitious powder and PVA solution.

### Specimen Preparation

The GFRCC specimens were manufactured by the extrusion technique [3, 4, 5]. The mixed dough-like fresh fiber-reinforced cementitious composite is extruded from a single screw vacuum extruder. The advantages of the extrusion technique include: mass-production capability, the ability to produce complicated and varied shapes, and the capability of improving the material properties. This improvement of properties can be attributed to the achievement of low porosity and good interfacial bond under high shear and compressive stress during the extrusion process. With properly designed dies and controlled mix, fibers can be aligned in the load-bearing direction during the die forming process. In the present study, a thin-sheet die with a cross sectional area of 300mm × 6mm was used. After extrusion, the fresh extrudates were cured under a plastic cover for 28 days under condition of 22°C and 76% R.H. for bending test. For tensile test, the specimens were cured under the accelerated aging of one day in wet storage at 80°C in order to evaluate the long-term performance of PVA modified GFRC.

### Flexural Test

In accordance with ASTM C947-89, a four-point bending test was performed on a closed-loop servo-hydraulically controlled test system (MTS 810) with a load capacity of 250kN. Under the midpoint of the specimen, two LVDTs were used to measure the midpoint deflection of the specimen. The average deflection was used as a feedback signal to form a close loop control. The loading rate of the crosshead was set at 0.3mm/min.

Tensile Test

For tensile tests, the specimens were obtained by cutting the extruded board into 75mm × 250mm × 6mm plates where its fiber orientation is along the tensile force direction. Each end of a specimen was glued onto two aluminum plates which were fixed to loading fixtures by pins. The loading fixtures were connected to MTS hydraulic grips. On each side of a specimen, a LVDT was attached to measure the deformation of the specimen (Fig.1). These two LVDTs were connected to the digital controller of a MTS machine through two AC conditioners. The average displacement of these two LVDTs was used as a feedback signal to form a closed loop control. To cover the cracks at the boundary of the aluminum plate and the test portion of the specimen, the span of the LVDT was slightly extended beyond the boundary. To ensure proper alignment of the LVDT and friction-free movement of the electric core, pre-load checking was performed.

## TEST RESULTS AND DISCUSSION

Flexural Test Results

Flexural test curves of the dry air cured is shown in Fig.2. It can be seen from the figure that with the increase of PVA addition, the flexural strength, ultimate deflection, ductility, and toughness are increased. It clearly demonstrates that PVA powder has a significant effect to improve the mechanical properties of GFRC. Within the dosage of 2%, the higher the addition of PVA, the better the enhancement of the mechanical properties.

Tensile Test Results

Tensile curve of the control and PVA modified GFRC specimens are shown in Fig. 3. With the 2% addition of PVA, great improvement in ultimate tensile strain, ductility, and toughness is achieved. As shown in the Figure, the specimen without PVA addition failed in a brittle mode, without any strain hardening response. During the test, the specimens failed suddenly without any warning when the stress peaks were reached. The tensile strain corresponding to the peak stresses of this aged GFRC sheet is only around $3.0 \times 10^{-4}$. Comparing this to the brittle failure mode of aged GFRC sheet, the tensile strain and toughness is significantly increased for the PVA modified GFRC sheet. A strain plateau appeared and the failure mode changed. The ratio of the ultimate strain to the BOP (bend over point) strain was close to five. These macro-structural tests suggest that the aging effect of GFRC can be prevented by the addition of small amounts of PVA powder.

## MICROSTRUCTURAL ANALYSIS

The research work of Robertson *et al.* [6, 7, 8] has shown that the addition of PVA to a cement matrix enhanced the bond strength and the pullout work was more than doubled. The postpeak drop in load was also less drastic for the PVA-modified material than without PVA. This ductile behaviour is probably the result of changes in the inorganic materials at the interface and the presence of bulk polymer. PVA is molecularly dispersed in the water during mixing, each PVA molecule can diffuse through the water and attach onto hydrophilic surface site. The result is an increased concentration of PVA near the fiber surfaces, and the polymer may even coat the surface with a monomolecular film. PVA has been found to inhibit the nucleation of CH crystals around the fibres and to enhance the formation of C-S-H during the hydration [6, 8].

SEM examination disclosed that after 28 days dry air curing, the GFRCC exhibited relatively ductile behaviour with fibers pulled out as shown in Fig. 4. Fractured surfaces of the batch T1 and T2 samples were observed with a polarizing optical microscope at a magnification of 500x. Typical micrographs of short fibers are shown in Fig. 5 and Fig. 6. In the optical micrograph without PVA powder (Fig. 5), a smooth surface of the glass fiber is seen. No etch pits and local damages were found. This observation is consistent with the SEM images by Bentur and Diamond [9]. When PVA powder is added, the observable glass fiber surfaces were completely different from those without the PVA powder addition. A typical glass fiber from a fractured specimen (batch T2) is shown in Fig.6. It can be seen that polymer depositions attach to the glass fiber surface. As addressed by Bijen [1], the surface energy of the glass fiber filament is able to attract a water-soluble polymer molecule and thus lead to an accumulation of such polymer on its surface. This process can be accelerated by the slurry coating adopted in this study. As a result, this water-soluble PVA forms a thin polymer film covering on the glass fiber surface to prevent the formation and accumulation of hydration products, especially the brittle calcium hydroxide. In the mean time, this ductile polymer film cover modifies the bond between the fibers and matrix such that the stress transition between fibers and the matrix is smoother and the stress localization is reduced. As shown in the macrostructural tensile test, the tensile strain and toughness are considerably increased (Fig. 3).

In order to further demonstrate the presence of PVA polymer at the fiber interface, EDAX was used to investigate the chemical components at the fibre-matrix interface and in the bulk matrix. The atomic ratios of calcium to silicon taken from the small region ($5\mu m$ by $10\mu m$) of the GFRC in the glass fibre interface and in the bulk cementitious matrix are given in Table 2. Ca/Si ratios were essentially the same for the bulk matrix in case of with and without PVA. Before or even after the accelerated aging, the Ca/Si ratio in the interface of PVA modified GFRC was smaller than that of the normal GFRC.

## CONCLUSIONS

1. Addition of PVA polymer can improve flexural strength, ultimate deflection, ductility, and toughness of GFRC. With the increment of PVA powder, the results are more amplified.

2. For the tensile performance, the PVA modified GFRC shows a great improvement in tensile strain, ductility, and toughness.

3. From the optical observations, the PVA modified polymer seemed to form a thin film covering on the glass fiber surface, which prevents the formation and accumulation of hydration products. Hence, the PVA polymer has the ability to reduce the embrittlement of the GFRC and change the failure mode from brittle to relatively ductile.

4. According to EDAX results, the Ca/Si ratios were essentially the same for the bulk matrix in case of with and without PVA. Before or even after the accelerated aging, the Ca/Si ratio in the interface of PVA modified GFRCC was smaller than that of the normal GFRCC.

## ACKNOWLEDGEMENT

The financial support from GRC under grant CRC 98/01.EG04 and the technical support from the MCPF at HKUST are greatly acknowledged.

## REFERENCES

1.  Bijen, J. (1983). "Durability of Some Glass Fiber Reinforced Cement Composites", *ACI Materials Journal*, **80**(4), pp305-311.

2.  Hayashi, M., Sato, S. and Fujii, H. (1985). "Some Ways to Improve Durability of GFRC", *Proceedings, Durability of Glass Fiber Reinforced Concrete Symposium*, Prestressed Concrete Institute, Chicago, pp270-284.

3.  Li, Z. and Mu, B. (1998). "Application of Extrusion for Manufacture of Short Fiber Reinforced Cementitious Composite," *Journal of Materials in Civil Engineering*, **10**(1), pp2-4.

4.  Shao, Y. and Shah, S. P. (1996). "High Performance Fiber-Cement Composites by Extrusion Processing.", *Materials for the New Millennium*, Edited by Ken P. Chong, **2**, pp251-260.

5.  Shao, Y., Marikunte, S. and Shah, S. P. (1995). "Extruded Fiber-Reinforced Composites.", *Concrete International*, **17**(4), pp48-52.

6.  Chu, T.J., Robertson, R.E., Najm, H and Naaman, A.E. (1994). "Effect of Poly(vinyl alcohol) on Fiber Cement Interfaces. Part II: Microstructures", *Advanced Cement Based Materials*, **1**(3), pp122-130.

7.  Kim, J.H., Robertson, R.E. and Naaman, A.E. (1999). "Structure and Properties of Poly(vinyl alcohol)-Modified Mortar and Concrete", *Cement and Concrete Research*, **29**, pp407-415.

8.  Najm, H., Naaman, A.E., Chu, T.J. and Robertson, R.E. (1994). "Effects of Poly(vinyl alcohol) on Fiber Cement Interfaces. Part I: Bond Stress-Slip Response", *Advanced Cement Based Materials*, **1**(3), pp115-121.

9.  Bentur, A. and Diamond, S. (1987). "Aging and Microstructure of Glass Fiber Cement Composites", *Durability of Building Materials*, **4**(3), pp201-226.

Table 1  Mix proportion.

| Batch | Slag/C | GF | PVA/C | w/c | Test type |
|-------|--------|-----|-------|------|-----------|
| B1 | 1 | 2% | 0% | 0.26 | |
| B2 | 1 | 2% | 0.5% | 0.26 | |
| B3 | 1 | 2% | 1% | 0.26 | Four point bending |
| B4 | 1 | 2% | 2% | 0.26 | |
| T1 | 1 | 2% | 0% | 0.26 | |
| T4 | 1 | 2% | 1.5% | 0.26 | Uniaxial tension |

Table 2  Atomic ratios of calcium to silicon from selected regions of the
GFRCC obtained from EDAX.

|  | Ca/Si Ratio | |
| --- | --- | --- |
| Type of Specimen | Fibre Interface | Bulk Matrix |
| T1 (0% PVA) | 0.4 | 3.6 |
| T2 (2% PVA) | 0.2 | 3.4 |

Fig. 1 Tension test setup.

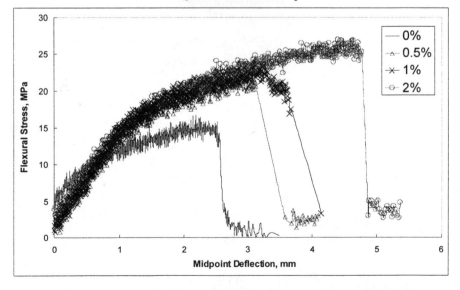

Fig. 2  Flexure curves of GFRC with different amount of PVA after 28 days
dry air curing.

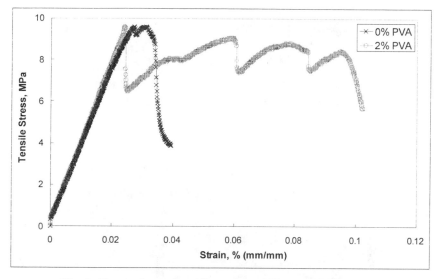

Fig. 3  Comparison of tensile curves of PVA modified GFRC.

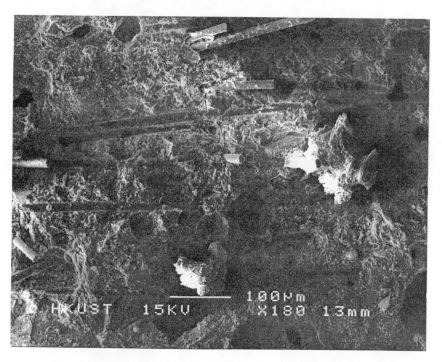

Fig. 4  Failure pattern of the GFRC with PVA powder after
bending test.

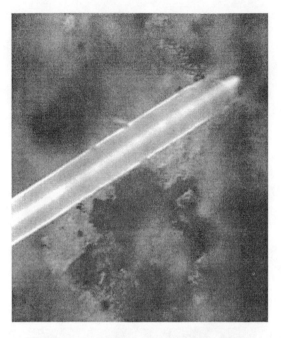

Fig. 5  Optical observation for fiber without
PVA powder addition.

Fig. 6  Optical observation for fiber with
PVA powder addition.

# Fiber Reinforced Cement Based Composites Under Drop Weight Impact Loading: Test Equipment & Material Influences

## by N. Banthia and V. Bindiganavile

Synopsis: Impact resistance of concrete remains to date one of its least understood properties. There are no standardized test methods and no meaningful comparisons can be made between results from various labs. This paper describes some of the issues at hand, and examines the response of plain and fiber reinforced concrete to impact loading through a multi-pronged investigation using instrumented drop weight impact tests. The study involved understanding the influence of machine specific parameters such as hammer mass and drop-height, specimen specific parameters such as specimen geometry and size and, several material related parameters including the type of fiber reinforcement. Among the fibers investigated, particular attempt was made at comparing steel and polymeric fibers under varying rates of load application.

It was found that machine parameters such as the hammer mass and drop-height greatly influence the apparent resistance of concrete to impact and its apparent sensitivity to stress-rate. Both flexural strength and toughness factors were seen to exhibit a size effect under impact loading.

In the case of fiber reinforced concrete, both plain and fiber reinforced concrete exhibited an increase in strength with stress-rate. In the case of polypropylene fiber, however, due to the viscoelastic nature of the fiber material, a much greater improvement in fracture energy absorption under impact loading was noted. This effect was so pronounced, that under very high stress-rates, the energy absorption capability of polypropylene fiber reinforced concrete exceeded that of steel fiber reinforced concrete. The paper discusses the relevance of these data in designing structures under impact and blast loading, and identifies areas of further research.

Keywords: drop-weight test; fiber reinforced concrete; impact; polymeric fibers; size effect; steel fibers

N. Banthia, is a Professor of Civil Engineering in the University of British Columbia, Vancouver, Canada.

V. Bindiganavile, is a Graduate Research Assistant at the Department of Civil Engineering in the University of British Columbia, Vancouver, Canada.

## INTRODUCTION

Cement based materials are quasi-brittle and are known to exhibit a highly stress-rate sensitive behavior (1,2). In structures that are subjected to impact forces, this causes concern in two ways: first the brittleness may result in a catastrophic failure without warning and second, the properties of concrete during such events may be very different from those measured in standardized quasi-static tests. Unfortunately, there are no standardized tests available for testing concrete under impact loading and there is significant confusion as to what constitutes an appropriate test. A number of tests have been proposed including the charpy pendulum test, the split hopkinson pressure bar test and the drop weight impact test, but clearly all these tests generate data based on different loading configurations and stress-rates and invoke different energy dissipation mechanisms and losses. Thus, even for the simplest of cement-based materials, such as plain concrete, the data from even the fully instrumented versions of these tests cannot be compared. Need exists to adopt a specific test, understand the influence of various variables and generate standards that allow for a rational determination of impact properties. In this study, the drop weight impact test, which is believed by most to be the simplest of all and one providing the most meaningful data, was critically studied for a number of test parameters.

Fiber reinforcement is highly recommended for situations involving impact and fatigue loading, but the rate sensitivity of the fiber materials such as steel (3) and polymers (4) must also be considered as they directly affect the ultimate composite properties. One may divide the load-displacement response of fiber reinforced concrete (FRC) into two qualitative phases: first, the pre-peak response during which only the relatively weaker and more brittle cement-based matrix undergoes tensile micro-fracturing and second, the post-peak response where fibers bridge fully coalesced matrix cracks and undergo bond-slip (5) processes that consume energy. Clearly one must consider both regions of response if highly optimized FRCs are desired. While its enhanced resistance to impact is well known, there is only limited data in the literature related to impact performance of FRC (6,7). This is particularly true in the case of polymeric fibers (8,9). It is believed that introducing fibers significantly reduces the crack velocity, which in turn leads to increased energy dissipation at higher rates (10).

This paper examines three major issues related to impact loading on concrete and fiber reinforced concrete. First, within the context of drop weight impact tests, a number of machine parameters were examined including capacity size (150 J-15,000 J) and drop heights (1.2 m – 2.5 m). It was found that the machine parameters strongly influence the observed material response to impact. Second, the influence of specimen size on the impact response was investigated. Finally, a comprehensive test program was launched where steel and polymer fibers with widely different constitutive properties were compared as reinforcement in concrete under impact loading.

## EXPERIMENTAL PROGRAM

Three instrumented drop-weight impact machines with varying hammer masses and capacities were employed. These included the small machine with a mass of 12 kg (capacity of 150 Nm); the medium machine with a mass of 60 kg (capacity of 1,500 Nm) and the large machine with a mass of 580 kg (capacity of 15,000 Nm). As can be seen, these three machines offered a two orders of magnitude variation in the incident potential energy (150-15,000 J). A detailed description of the instrumentation and data acquisition is given in Banthia et *al.* (7). The data acquisition was at the rate of $1.25 \times 10^5$ Hz for the two larger machines and $2 \times 10^5$ Hz for the small machine. For the study on machine effects, two broad impact criteria were employed: first, that of an equal hammer-potential energy and second, that of an equal hammer-velocity (or drop height), Table 1. Plain and fiber reinforced concrete beams (100 mm x 100 mm x 350 mm) were cast and cured for 28 days prior to testing. Fiber reinforcement was in the form of flat-end steel fiber (30 mm long; 0.7 mm dia) and crimped polypropylene fiber[1] (30 mm long). Companion quasi-static tests were performed in a 150 kN floor mounted Instron test system.

In the other series, the stress-rate sensitivity of plain and fiber reinforced concrete was investigated using the medium capacity machine at four different drop-heights of 200 mm, 500 mm, 750 mm and 1000 mm. Accordingly, the hammer potential energy was   120 J, 300 J, 450 J and 600 J respectively. The effect of specimen size on its rate-sensitivity was investigated through four different beam sizes for plain concrete and three sizes for the fiber reinforced mixes as shown in Table 2. Six beams were tested under each loading condition.

---

[1] The deformation, in mm is given by $y = A\sin(2\pi x/L)$ where $A = 0.5$ mm; $L = 3.9$ mm and $x$ is measured along the fiber length.

## RESULTS AND DISCUSSION

Analysis of Test Data

As has been previously described (11), during impact, significant specimen accelerations occur, and an appreciable part of the applied load is manifested as inertial load, which needs to be accounted for. By performing a dynamic analysis of the system, the following expression for the inertial load $P_i$ (t) may be obtained (11).

$$P_i (t) = \rho A (d^2u/dt^2)\{ 1/3 + 8(ov)^3/3l^2 \} \qquad \qquad ...(1)$$

Where, u (t)   =   mid-span deflection of the beam at time 't'
$\rho, A$   =   density and cross-sectional area of the beam
l   =   clear span of the beam
$ov$   =   length of overhang on each side of the supports.

Machine Effects

Figure 1 shows the load vs. displacement results under impact loading for plain concrete when tested under the conditions of equal potential energy, and Figure 2 shows the same under the conditions of equal drop height. The quasi-static response is also shown for comparison. In nearly all cases under impact, beams failed by fracturing in to two halves. The exception being those beams tested on the smallest machine under the equal drop height regime (Figure 2).

As can be seen from Figures 1 and 2, for drop-weight impact testing of concrete, parameters such as the hammer mass and drop height have considerable influence on the apparent concrete strength and toughness.

As may be seen from Figure 3, for the same potential energy, since the larger machines incorporated heavier hammers, their drop heights were lower and this resulted in an impact pulse of longer duration. Therefore, with heavier machines, the impact pulse was 'flatter' and spread over the period of impact in comparison with those from the smaller machines. Previous studies (12) have reported that for a given hammer mass, the drop height affects only the amplitude of the impulse. While the exact influence of the hammer mass on either the amplitude or the duration of the impulse could not be clearly identified in this study, the duration of the impulse was found to vary inversely with the drop height (or hammer velocity). For identical drop heights (or hammer impact velocity), the pulse duration was nearly identical with all machines (Figure 4).

Regardless of machine type, the fracture energy of a plain concrete beam was found to be dependent on the induced stress rate and the machine influence diminished at higher rates (Figure 5).

It is clear from Figure 6 that under the equal potential energy regime of testing, the stress rates induced are vastly dissimilar (over an order of magnitude) for different machines. This is perhaps because both the hammer mass and the drop height are varied under this regime. On the other hand, when the impact velocity is maintained between the various machines by using an identical drop height, the difference between the results from two machines having vastly different capacities is significantly reduced. Thus from a comparative standpoint, data from different labs where two machines of different capacities are used can be compared only when velocities involved are identical. If potential energies are kept constant, then the larger machine will yield lower values. Standardized testing therefore must involve maintaining a fixed drop height with a relatively large tolerance on the hammer mass.

Load-time plots for the medium and large machines under identical drop heights are shown in Figure 7. Since the beams tested in the small machine under the same drop height did not fracture, these results are not shown. The raw 'tup' load data were corrected for inertia and these are also shown in Figure 7. The true stressing loads (obtained after subtracting the inertial loads from the measured raw loads) are plotted in Figure 8. As may be seen, the inertial loads form about 10 % of the raw loads.

Notice in Figure 8 that two widely different machines can produce an almost identical stressing load history if the drop height is maintained. Once again, this illustrates that the hammer mass is less critical than the hammer drop height, and a variety of dynamic load cases may be simulated in a laboratory with a single drop-weight machine with a sufficiently large hammer mass. By varying the drop height, the desired stress rate may be induced and appropriately corrected for inertial forces. With the help of data shown in Figure 5, the fracture energy for a cement based composite under similar stress rates may be conveniently estimated.

Stress-rate sensitivity

Stress rate sensitivity may be obtained from the formulation given by Nadeau et al. (13). This is shown graphically in Figure 9 for plain oncrete, SFRC and PFRC. It is significant to note that based on the parameter 'N', plain concrete appears to be more sensitive to stress rate than polypropylene fiber reinforced concrete and steel fiber reinforced concrete.

## Size effect on plain concrete during impact

Size effects in quasi-static flexural testing of concrete have been previously reported (14). Unfortunately, no data exists to support similar trends under impact loading. Further, the issue of size effects under impact loading is made more difficult by the fact that maintaining an identical stress-rate during impact tests on specimens of varying sizes is nearly impossible. For identical drop heights, larger beams experience lower stress rates, and hence what one notices is a combined influence of size and stress-rate, rather than of just the size alone. This is shown in Figure 10.

Notice in Figure 10 that smaller beams depict higher apparent strength due not only to their smaller size but also due to having experienced a higher stress rate. Clearly, one must normalize the test data with respect to constant stress rate in order to observe the true influence of specimen size. In this study, this normalization was carried out on the basis of the stress rate sensitivity plot determined earlier (Figure 9). Figure 11 shows the effect of specimen size alone on flexural strength of plain concrete under impact where the influence of stress rates has been eliminated.

Bazant developed the Size Effect Equation in mathematical terms (15), and the same is given as:

$$\sigma_r = B.f'_t(1+D/D_0)^{-1/2} \qquad \qquad ...(2)$$

where,

$\sigma_r$       = flexural strength of the specimen under 3-point loading for impact and 4-point loading for quasi-static case respectively;

$f'_{ti}, f'_{ts}$   = tensile strength of the material under impact and quasi-static case respectively;

$D$        = characteristic specimen dimension (here, the depth of the beam)

$B, D_0$    = empirical constants depending on specimen geometry and loading configuration

The data obtained in this study is fitted using the Bazant's Size Effect Equation and the same is shown in Figure 12. Results from a previous study (16) under quasi-static, 4-point loading are fitted with the Bazant's Size Effect Equation in Figure 13. A good match can be noticed, however, a different set of constants appear to operate for impact loading while defining size effects as per Bazant's model.

Impact response of FRC

As indicated previously, the medium capacity machine was used in order to investigate the response of fiber reinforced concrete to impact loads, where the height of hammer drop was varied. Figures 14 and 15 show the flexural load-displacement plots under variable drop heights for polypropylene and steel fiber reinforced concrete, respectively. Note a clear increase in the peak loads with an increase in the height of hammer drop (or the stress-rate). The point of greater interest, however, is the influence of stress-rate on the energy absorption capability or flexural toughness.

The flexural toughness of the fiber reinforced specimens tested in this program reveals an interesting trend. As expected, steel fiber reinforcement provided a higher toughness than the polypropylene fiber reinforcement under quasi-static rates. Under impact loading however, as the stress rate was increased by increasing the hammer drop height, steel fiber reinforced concrete depicted a continual decrease in its energy absorption capacity. Polypropylene fiber reinforced concrete on the other hand, demonstrated an increase in the energy absorption capacity with an increase in the stress rate, and in fact, surpassed the toughness of the steel fiber reinforced concrete at very high rates (Figure 16).

Clearly, this 'switch' between the steel and polypropylene fibers under varying stress rate arises from their different constitutive responses, stress rate sensitivities and bonding characteristics. Studies on strain rate sensitivity of steel (3) and polypropylene (4) indicate that while steel undergoes an increase of 20 % in strength in the range of strain rate $10^{-4}$ to $10^{2}$, polypropylene experiences an increase of 300 % in the same range. The modulus also follows a similar trend, and polypropylene depicts a much stiffer response under impact loading. These factors all support the use of polypropylene fiber reinforcement for structures subjected to impact loading.

Size effect on FRC Toughness under impact

In the case of fiber reinforced concrete, an increase in the specimen size resulted in a decrease in the flexural toughness as shown in Figure 17.

This is contrary to the findings of Chen (16) who under quasi-static loading did not find a significant size effect on flexural toughness factor based on JSCE values for SFRC beams. One possible reason for this could be the increasingly linear elastic behaviour of the fracture process under impact compared to the non-linear, micro-fracturing response of FRC under quasi-static loading. Admittedly, the effect of size was investigated at only one impact rate. In order to establish the nature of the size-effect under impact, a few more rates should be studied. This forms a part of the ongoing research program.

## CONCLUSIONS

The following conclusions may be drawn from this study:
1.  For cement-based materials, the measured impact response is highly dependent on the characteristics of the drop-weight impact machine used for testing. The pulse duration was found to depend upon the drop-height, with greater drop-heights leading to shorter pulses. Results appear to be far less sensitive to the mass of the hammer than to the drop-height. This observation forms a useful basis for standardizing impact testing of plain and fiber reinforced concrete. Results from two different machines with varying hammer masses can be compared if the drop-heights were identical.
2.  Crimped polypropylene fiber is less effective than steel fiber at quasi-static rates of loading. However, at higher stress rates, it performed better than the steel fiber. This 'switch' in the behaviour of FRC is attributed to the greater strain rate sensitivity of polypropylene vis á vis steel.
3.  Specimen size effect appears in impact loading similar to those previously reported for quasi-static loading. In addition, there is a significant influence of size on the flexural toughness factor under impact loading, which is generally not seen in quasi-static loading.

## ACKNOWLEDGEMENT

The work reported here was supported in part by the Natural Science and Engineering Research Council of Canada.

## REFERENCE:

1.  Reinhardt, H.W., *'Strain rate effects on the tensile strength of concrete as predicted by thermodynamic and fracture mechanic models'*, Cement based composites: Strain rate effects on fracture (Eds. Mindess S. and Shah, S.P.), vol. 64, Pittsburgh, 1985, pp. 1-12.

2.  Ross, C.A., *'Review of strain rate effects in materials'*, Structures under extreme loading conditions, ASME Pressure Vessels and Piping Conference, Orlando, FL, 1997, pp.255-262.

3.  Malvar, L.J., *'Review of static and dynamic properties of steel reinforcing bars'*, ACI Materials Journal, vol. 95, no.5, Sept-Oct 1998, pp. 609-616.

4.  Kawahara, W.A., Totten, J.T. and Korellis, J.S., *'Effects of temperature and strain rate on the nonlinear compressive mechanical behaviour of polypropylene'*, SANDIA Report: SAND89-8233; UC-13, 1989, pp.3-39.

5.  Banthia, N. and Trottier, J-F., *'Concrete reinforced with deformed steel fibers—Parts 1 & 2: Bond-Slip Mechanisms & Toughness Characterization'*, ACI Materials Journal, vol. 91 no. 5 Sept-Oct 1994 pp. 435-459 & vol. 92. no. 2, Mar-Apr 1995, pp. 146-154.

6.  Gopalaratnam, V.S. and Shah, S.P., *'Properties of steel fiber reinforced concrete subjected to impact loading'*, ACI Journal Proceedings, vol. 83, no. 1 Jan-Feb 1986, pp. 117-126.

7.  Banthia, N., Mindess, S. and Bentur, A., *'Impact behaviour of concrete beams'*, Materials & Structures, 20, 1987, pp.293-302.

8.  Gupta, P., Banthia, N. and Yan, C., *'Fiber reinforced wet-mix shotcrete under impact'*, ASCE Journal of Materials in Civil Engineering, vol. 12, no. 1, 2000, pp.81-91.

9.  Mindess, S. and Yan, C., *'Perforation of plain and fiber reinforced concrete subjected to low velocity impact loading'*, Cement and Concrete Res., vol. 23, no. 1, 1993, pp. 83-92.

10. Mindess, S., Banthia, N., Ritter, A. and Skalny, J.P., *'Crack development in cementitious materials under impact loading'*, Mat. Res. Soc. Symp. Boston, vol. 64, Cement based composites: Strain rate effects on fracture, (Eds. Mindess, & Shah), 1985, pp. 217-224.

11. Banthia, N.P., Mindess, S. and Bentur, A., *'Energy balance in instrumented impact tests on plain concrete beams'*, SEM/RILEM International conference on Fracture of Concrete and Rock, Houston, June, 1987, pp.26-36.

12. Wang, N., *'Resistance of concrete railroad ties to impact loading'*, Ph. D. Thesis, The University of British Columbia, Vancouver, Canada, 1996.

13. Nadeau, J.S., Bennet, R. and Fuller, E.R. (Jr), *'An explanation for the rate-of-loading and the duration-of-load effects in wood in terms of fracture mechanics'*, Journal of Materials Science, vol. 17, 1982, pp. 2831-2840.

14. Wright, P.J.F., *'The effect of the method of test on the flexural strength of concrete'*, Magazine of concrete research, no. 11, 1952, pp. 67-76.

15. Bazant, Z.P. and Planas, J., *'Fracture and size effect in concrete and quasi-brittle materials'*, CRC Press, 1998.

16. Chen, L., *'Flexural toughness of fiber reinforced concrete'*, Ph D. Thesis, The University of British Columbia, Vancouver, Canada, 1995.

17. JSCE SF 4, *'Method of tests for flexural strength and flexural toughness of steel fiber reinforced concrete'*, Concrete library of JSCE No.3, June 1984.

Table 1  Test Program to study Machine Effects

| Equal Potential Energy | | |
|---|---|---|
| Machine Type | Drop Height (mm) | Potential energy (J) |
| Large | 20 | 120 |
| Medium | 200 | 120 |
| Small | 1000 | 120 |
| Equal Drop Height | | |
| Machine Type | Drop Height (mm) | Potential Energy (J) |
| Large | 200 | 1200 |
| Medium | 200 | 120 |
| Small | 200 | 24 |

Table 2 Test Program to study the 'Size Effect'

| Beam Dimension | Plain Concrete | Steel FRC | Polypropylene FRC | Span/Depth |
|---|---|---|---|---|
| 25 x 25 x 350 | √ | — | — | 75/25 |
| 50 x 50 x 450 | √ | √ | √ | 150/50 |
| 100 x 100 x 350 | √ | √ | √ | 300/100 |
| 150 x 150 x 480 | √ | √ | √ | 450/150 |

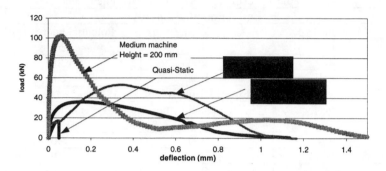

Figure 1 Impact performance of plain concrete under equal potential energy

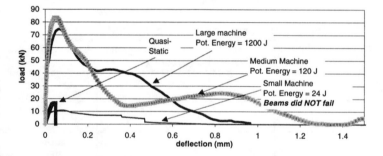

Figure 2 Flexural response of plain concrete to impact through equal drop height

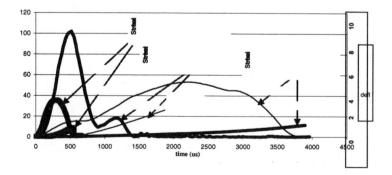

Figure 3 Impact performance of plain concrete under equal potential energy
(time histories)

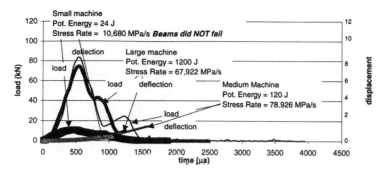

Figure 4 Impact performance of plain concrete under drop height (time histories)

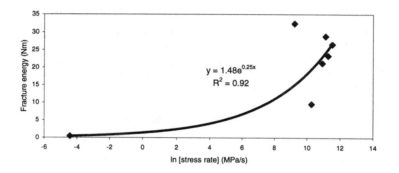

Figure 5 Effect of stress rate on fracture energy of plain concrete under different machines

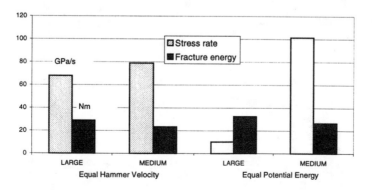

Figure 6 Induced stress rate and resulting fracture energy from impact testing under different regimes

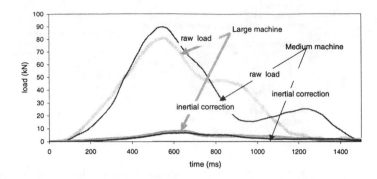

Figure 7 Raw load and corresponding inertial forces for impact under equal drop height regime

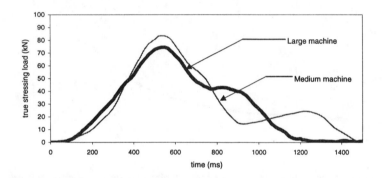

Figure 8 True bending load-pulse for different machines under equal drop heights

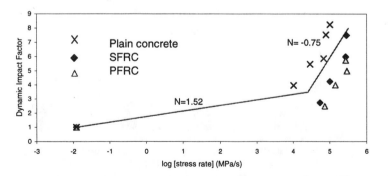

Figure 9 Stress rate sensitivity of plain and fiber reinforced concrete under flexure

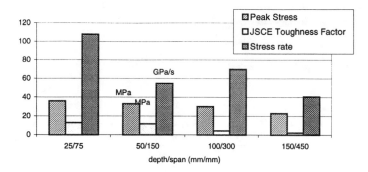

Figure 10 Size effect in plain concrete beams under impact

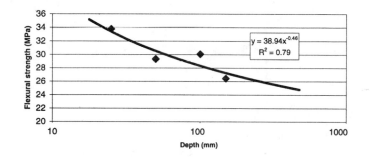

Figure 11 Size effect on the flexural strength for plain concrete under impact

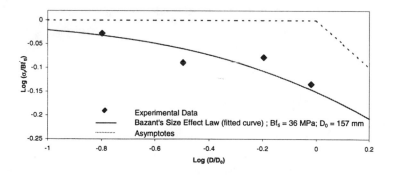

Figure 12 Bazant's size effect model fitted for data from impact loading

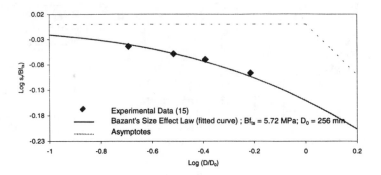

Figure 13 Bazant's size effect model fitted for data from quasistatic loading

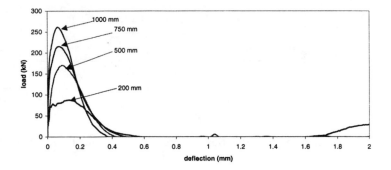

Figure 14 Impact performance of PFRC under flexure

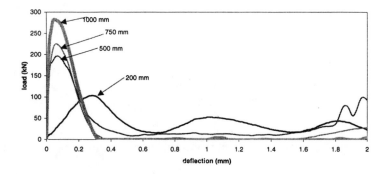

Figure 15 Impact response of SFRC under flexure

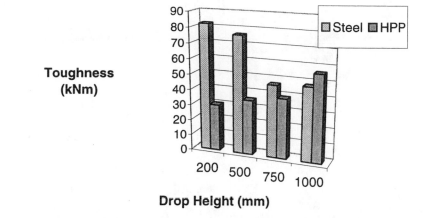

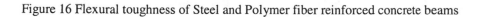

Figure 16 Flexural toughness of Steel and Polymer fiber reinforced concrete beams

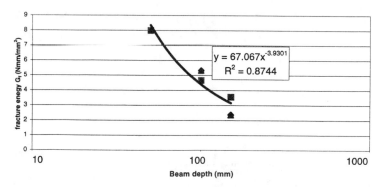

Figure 17 Size effect on flexural toughness of FRC under impact loading

# Electrical Resistance of Carbon Fiber Cement Composites Under Compression

## by F. Reza, G. B. Batson, J. A. Yamamuro, and J. S. Lee

Synopsis: The incorporation of a small volume of carbon fibers into a concrete mixture produces a strong, durable concrete and at the same time lends the product a smart material property. Durability is enhanced because of increased crack resistance and the carbon fiber's high resistance to wear, heat and corrosion. The smart property is a damage-sensing ability. This intrinsic capability can be tapped by using simple electrical resistance techniques. There is the potential for these techniques to be used as nondestructive testing methods to assess the integrity of the composite. The results of some fundamental investigations on the bulk electrical properties of carbon fiber cement composites (CFCC) under compressive loading are presented. Well-defined patterns are exhibited in the electrical resistance behavior that can be correlated with the stress-strain behavior. The resistance behavior was evaluated for various fiber volume contents. The effects of taking resistance measurements both parallel and perpendicular to the axis of loading were investigated.

Keywords: carbon fibers; cement; composites; electrical resistance; fiber-reinforced concrete; non-destructive testing; smart materials

ACI member **Farhad Reza** is Assistant Professor of Civil Engineering, Department of Civil Engineering, Ohio Northern University, Ada, Ohio. He is a member of ACI Committee 544.

ACI member **Gordon B. Batson** is Professor Emeritus, Civil and Environmental Engineering Department, Clarkson University, Potsdam, New York and is a member of ACI committees 544 and 549.

**Jerry A. Yamamuro** is Assistant Professor in the Department of Civil and Environmental Engineering at the University of Delaware, Newark, Delaware.

**Jong S. Lee** is an Associate Professor in the Civil Engineering Department at Hanyang University, Korea.

## INTRODUCTION

The addition of small amounts of fiber-reinforcement is one alternative for providing a concrete with a high strength as demanded by industry, but with significantly improved durability. Carbon fiber-reinforced concrete in particular has been shown to have excellent durability properties due to the fiber's high resistance to wear, heat and corrosion (1).

In addition, the ability of a material to serve as a smart material that could provide information as to the level of internal stresses when measured by simple techniques would be invaluable. Recent studies indicated that carbon fiber-reinforced composites might have the ability to serve as smart materials (2), (3). In this paper, the results of electrical resistance measurements taken parallel and perpendicular to the axis of compression loading of cubes of carbon fiber reinforced mortar (CFRM) are presented. A correlation between the resistance measurements and internal conditions is then hypothesized.

## EXPERIMENTAL SETUP

The mixture proportions used for the CFRM specimens are shown in Table 1. Carboxy methylcellulose was added at a dosage of 0.001 g per $cm^3$ of the mixing water to aid in the dispersion of the carbon fibers. The use of a high quantity of silica fume and high-range water-reducing admixture is recommended for carbon fiber-reinforced concrete (1). Since silica fume acts as a cement substitute, appropriate constituents were proportioned according to the total cementitious material content (that is, portland cement plus silica fume) rather than just the cement content.

The cement was Type III portland cement. The silica fume, which was added in dry form, met ASTM 1240 requirements. The sand was ASTM C778 graded Ottawa sand. The high-range water-reducing admixture met the specifications of ASTM C494 Type A & F. The fibers used in this study were

RK 10 carbon fibers.   These were isotropic polyacrylonitrile (PAN)-based fibers of 8-$\mu$m diameter and 6-mm length with 1-2 % water-soluble sizing.

The cubical specimens were made in standard brass split-molds of 50.8 mm size.  The molds ensured that five out of the six faces turned out flat and square.  Two of these flat ends were used to apply the loads.  Specimens were continuously stored in a moist-curing room until the time of testing.

The compression testing was performed on a 250-kN capacity closed loop control servo-hydraulic system.  A stainless steel plate was threaded onto the load cell and another stainless steel plate was rigidly coupled by a threaded rod to the actuator.  Ceramic plates were placed on the loaded ends of the cubical specimens to distribute the load and they also served to electrically isolate the specimens from the loading frame.

The electrical resistance parallel to the applied loading was measured using a four-ring electrode configuration.  Four thin strips of conductive silver paint were applied to the specimen and then copper wire was tightly wound around the cube and in contact with the silver paint to serve as the electrodes.  In this method, current is passed through the outer two electrodes and the potential drop is measured across the inner electrodes.

The electrical resistance perpendicular to the applied load was measured using a parallel plate electrode configuration.  Silver paint was applied to opposite parallel faces of a cube and thin copper plates held in contact with the silver paint by electrical tape served as the electrodes.

The four-ring configuration could not be used to measure resistance perpendicular to the loading direction.  This is because the electrodes would create an uneven loading surface.  The two-plate configuration could not be used for the case of measurement parallel to the loading direction.  This is because two-electrode measurements are sensitive to contact pressure of the electrodes (4) and would have changed during loading.  Thus it was necessary to utilize the two different configurations.

Resistance measurements parallel and perpendicular to the compression loading were made with an HP–4284A inductance, capacitance and resistance (LCR) meter.  The LCR meter separated the complex impedance into a resistance and reactance.  The resistance was measured with an AC test signal of 100 Hz frequency.  A personal computer was utilized to simultaneously collect resistance measurements from the LCR meter, compression load from the load cell, and displacements from the actuator.  Figure 1 shows the electrode arrangements for the two loading cases.

The compression tests were controlled by setting the actuator rate of displacement to a constant rate of 0.001 mm/sec.  The actuator displacement consisted of the displacement of the specimen and the displacement of the testing system(loading train).  This problem was largely eliminated when corrections were made for the stiffness of the testing system (including hydraulics and mechanical components). The load-displacement relationship of the system was determined by loading all of the test fixtures without the specimen in between.  The test setup was modeled as a set of springs in series and the displacement induced in the specimen was found by subtracting the displacement of the loading system from the actuator displacement. This is depicted schematically in Figure 2.

## TEST RESULTS OF CUBE SPECIMENS

In order to investigate the differences between resistance measurements taken in a direction perpendicular to the direction of loading and measurements taken parallel to the loading direction, several cubical specimens were prepared with 0.4% and 1.0% carbon fibers by volume. A four-ring configuration of the electrodes was used for resistance measurement parallel to the load and a parallel-plate electrode configuration was used for resistance measurements perpendicular to the load as shown in Figure 1.   It should be noted that the resistance values for the parallel-plate method may be higher than the true resistance.  The four-ring method also produces slightly larger values than the true resistance if the electrode spacing is extremely close as was required in this instance, but this does not affect the ability of either method to detect resistance changes due to internal modifications or damage of the carbon fiber reinforced concrete. For the reasons mentioned and also because of possible anisotropy, electrical resistivity is not reported but rather a normalized change in resistance with respect to the resistance at a zero stress state is utilized.   A detailed discussion on electrode configurations is presented in (4).

The compressive strengths of the cube specimens at 240 days of age are given in Table 2.

Figures 3a and 3b show the stress/strain and resistance changes from the tests of cubical specimens containing 0.4% and 1.0% by volume of fibers for measurements made parallel to the loading direction.  The stress values were normalized with respect to the ultimate stress and the resistance values were normalized with respect to the initial resistance. The resistance curve mirrors the stress-strain curve in many aspects. During the initial phase of the test, the electrical resistance decreases.  This may be attributed to the decreased contact resistance between fibers as they are pushed into closer proximity and also due to the simple fact that the conduction length is being decreased.  It should be noted that these results are in contrast to those obtained by (2) who found the electrical resistance (using DC) to increase during this stage.  When the stress has reached approximately 85% of the ultimate stress, the electrical resistance gradually begins to increase.  This phenomenon is attributed to the onset of significant microcracking.  As microcracks form, the bridging fibers are now in tension.  As the fibers are elongated, the overall resistance increases.  Shortly after attaining the peak stress, the microcracks coalesce into larger cracks.  As the cracks propagate, fiber pullout occurs which breaks the conduction network causing significant increases in the electrical resistance behavior.

Figures 4a and 4b show the stress-strain and resistance changes from the tests of cubical specimens containing 0.4% and 1.0% by volume of fibers for measurements made perpendicular to the loading direction. There is no observed change in the resistance until the stress value reaches approximately 80% of the ultimate stress.   In this case, there is certainly no decrease in conduction length, and if Poisson's ratio effects are assumed to be minimal, there is no significant increase in the conduction length. However, as cracks propagate vertically there is a significant decrease in the conduction paths available for the current to flow from one side to the other.   Hence, the observed changes in resistance after the peak of the stress-strain curve are much larger when measured perpendicular to the loading than when measured parallel to the loading. This occurs as fibers are either pulled out or broken.   In theory, if the specimen were to split into two or more pieces, severing all conduction paths, the resistance would approach infinity.

## SUMMARY AND CONCLUSIONS

Well-defined patterns are identifiable in the electrical resistance behavior of carbon fiber cement composites during compressive loading. When resistance is measured parallel to the direction of loading, there is a decrease in resistance during elastic compression due to reduction in conduction length and decrease in contact resistance between fibers. An increase in resistance occurs when substantial microcracking is initiated.  Finally, there is a substantial increase in the resistance when large cracks propagate.

When measurements are taken in a direction perpendicular to the direction of the applied compressive load, no indication of internal elastic stress is discernible through observation of the electrical resistance, however the resistance is much more sensitive to the propagation of large cracks.

The potential exists for utilizing the carbon fiber reinforced cement composites as structural materials and the continuous monitoring of the electrical resistance of critical parts of the structure could provide indications as to the state of internal stress as well as the structure's physical integrity.

## ACKNOWLEDGEMENTS

This work was supported by the National Science Foundation under Grant No. CMS-9522726.  Ken Chong was the Program Director.  The authors also gratefully acknowledge the support of the following companies for their donation of materials to the research program - Globe Metallurgical Inc. of Niagara Falls, NY who supplied the silica fume and Master Builders Technologies of Cleveland, OH who supplied the high range water-reducing admixture.

## REFERENCES

1.  Banthia, N., 1994, "Carbon Fiber Cements: Structure, Performance, Applications and Research Needs," *Fiber Reinforced Concrete Developments and Innovations*, J.I. Daniel and S.P. Shah, eds, American Concrete Institute, Farmington Hills, Mich., pp. 91-120.

2.  Chen, P., and Chung, D. D. L., 1996, "Carbon Fiber Reinforced Concrete as an Intrinsically Smart Concrete for Damage Assessment during Static and Dynamic Loading," *ACI Materials Journal*, V. 93, No. 4, pp. 341-350.

3.  Lee, J. S., and Batson, G. B., 1996, "Electrical Tagging of Fiber Reinforced Cement Composites," *Proceedings of the 4th Materials Engineering Conference*, ASCE, Washington D.C., pp. 887-896.

4.  Reza, F., Batson, G. B., Yamamuro, J. A., and Lee, J. S., 2001, "Volume Electrical Resistivity of Carbon Fiber Cement Composites," *ACI Materials Journal*, V. 98, No. 1, pp. 25-35.

## LIST OF NOTATIONS

| | |
|---|---|
| $\delta_{actuator}$ | Displacement of the actuator |
| $\delta_{system}$ | Displacement of components of the testing system |
| $\delta_{specimen}$ | Displacement induced in the specimen |
| P | Load |
| R | Electrical Resistance |
| $R_o$ | Electrical Resistance at the start of the test (zero stress state) |

Table 1 Mixture Proportions of Mortar for Cubical Specimens

| Constituent | Content per unit volume of mortar (kg/m$^3$) | Ratio by weight to total cementitious material |
|---|---|---|
| Cement | 477 | 0.86 |
| Silica Fume | 72 | 0.14 |
| Water | 439 | 0.80 |
| Sand | 875 | 1.60 |
| High-Range Water Reducer | 0 - 3 | 0.000 - 0.005 |
| Methylcellulose | 1 kg/m$^3$ of water | NA |
| Carbon Fibers | 0-1% vol. of mortar | NA |

Table 2 Compressive Strength of Cube Specimens

| Fiber content (volume percentage) | Compressive strength (MPa) |
|---|---|
| 0.0% | 32.7 33.1 |
| 0.4% | 32.8 32.0 |
| 1.0% | 30.5 34.0 |

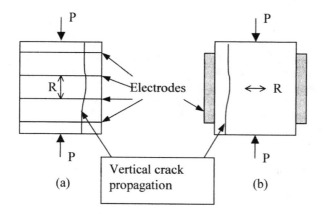

*Figure 1 Changes in electrical resistance measured (a) parallel and (b) perpendicular to the direction of loading.*

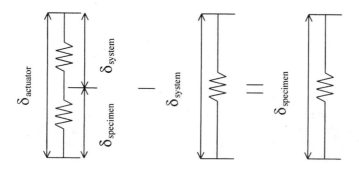

*Figure 2 Correction in global strain measurement made by compensating for the stiffness of the testing system.*

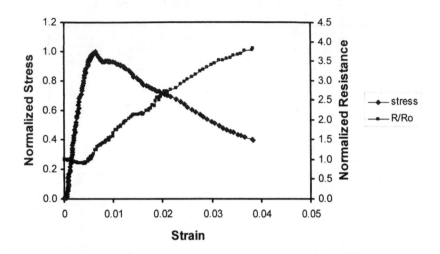

*Figure 3a-Typical Stress-Strain and Resistance behavior for cubical specimen. The resistance measurement was made parallel to the loading direction. The specimen contained 0.4% by volume of fibers and was 240 days old. The normalizing value for stress was the ultimate stress of 32.8 MPa and the value for resistance was the initial resistance of 10.27 Ω*

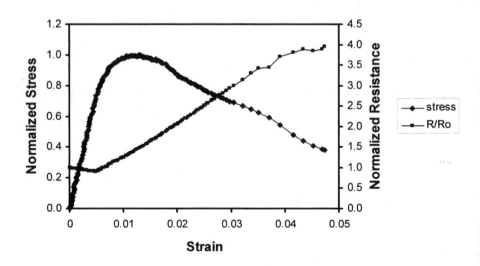

*Figure 3b -Stress-Strain and Resistance behavior for cubical specimen. The resistance measurement was made parallel to the loading direction. The specimen contained 1.0% by volume of fibers and was 240 days old. The normalizing value for stress was the ultimate stress of 30.5 MPa and the value for resistance was the initial resistance of 4.96 Ω*

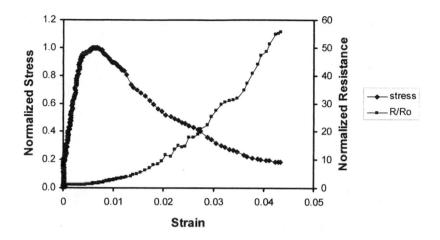

*Figure 4a-Typical Stress-Strain and Resistance behavior for cubical specimen. The resistance measurement was made perpendicular to the loading direction. The specimen contained 0.4% by volume of fibers and was 240 days old. The normalizing value for stress was the ultimate stress of 32.0 MPa and the value for resistance was the initial resistance of 13.87 $\Omega$*

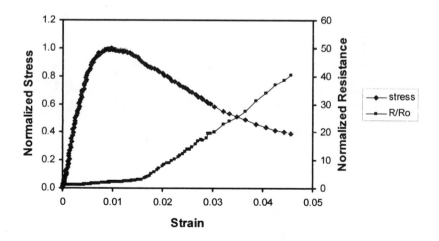

*Figure 4b-Stress-Strain and Resistance behavior for cubical specimen. The resistance measurement was made perpendicular to the loading direction. The specimen contained 1.0% by volume of fibers and was 240 days old. The normalizing value for stress was the ultimate stress of 34.0 MPa and the value for resistance was the initial resistance of 5.98 $\Omega$*

# Strength, Toughness and Durability of Extruded Cement Boards with Unbleached Kraft Pulp

## by Y. Shao and S. Moras

Synopsis:
    The use of extrusion technology for the production of cement boards with unbleached kraft pulps is evaluated in this paper. Cement boards reinforced by both hardwood and softwood pulp of different percentages were fabricated using an auger type lab extruder. The moisture content, water absorption and density of all batches were measured. The flexural response of the cement boards was used to investigate the strength, the toughness, the anisotropy, the age effect and the resistance to natural weathering as well as to freeze-thaw cycling. With a relative ease of manufacture and a much cleaner production, extrusion was found to be a suitable means for making cement boards with up to 8% pulp by weight. Higher pulp content increased the toughness of the material but didn't enhance the flexural strength appreciably due to a higher water content required for extrudability. Anisotropic behavior and age effect were observed. The extruded products exhibited good resistance to weathering and freeze-thaw cycling. Hardwood pulps, cheaper and more available than the softwood ones, were found to be more suitable for extrusion production in terms of extrudability, finished surface and long term mechanical properties.

Keywords: age effect; anisotropy; extrusion; fiber-cement board; freeze-thaw cycling; hardwood pulp; natural weathering; softwood pulp

ACI member Yixin Shao is an assistant professor of Department of Civil Engineering and Applied Mechanics at McGill University. His research interests include fiber reinforced cementitious materials, fiber reinforced polymeric materials, processing and building product development.

Shylesh Moras is a graduate research assistant of Department of Civil Engineering and Applied Mechanics at McGill University. His research is related to the processing of fiber cement composite building products.

## INTRODUCTION

In North American, the fiber reinforced cement boards are primarily used in the single family house market for three major applications: (1) siding, (2) roofing and (3) backboard. It is reported that the fibercement siding is the fastest growing building material in the US market in the 90's. It shares about 7-10% siding market currently and will grow to 25-30% in year 2005[1]. The replacement of conventional plywood and organic bonded particleboard by fiber-cement products has been facilitated by a higher fire resistance, moisture resistance and better durability. Also, problems commonly related to wood like rot and insect attack are altogether eliminated. However, as environmental concerns grow and timber supplies become constrained, a further consideration should be taken that fiber-cement composite has only about 10 percent wood content compared to 90 percent or more in timber products.

Hatschek process is the major production method presently used in North America in making fiber-cement flat sheets for sidings and roofings. It works on the principle similar to paper production [2-3]. Extrusion process was attempted to make similar thin sheets as well as hollow panels [4]. This paper evaluates the effect of extrusion process on wood pulp-cement composites, and compares the hardwood pulp with the softwood for extruded products, since hardwood pulps are cheaper and more available. Six batches were extruded - three containing 2, 4 and 8% by weight hardwood pulp and three 2, 4 and 8% by weight softwood pulp. Following the extrusion of these batches, mechanical and physical tests were performed, including the longitudinal and transverse properties, the age effect, the natural weathering effect and the resistance to freeze-thaw cycling.

### EXTRUSION OF CEMENT BOARD WITH KRAFT PULP

Table 1 shows the physical properties of the unbleached kraft pulp used in this project. Softwood pulps had longer fiber length and smaller diameter when compared to the hardwood, both of which were obtained as a wet pulp with a 30% consistency (30% solids). Kraft pulps were removed from their parent wood by a chemical process. This kraft process produced low yield (approximately 50%), ribbon-like delignified fibers which are chemically and physically compatible with cement. The other materials used in making the composites were Type 10 Portland cement, silica fume, methylcellulose and water. Silica fume was used primarily as a dispersing aid to help disperse the pulps, and methylcellulose as a processing aid to modify the viscosity of the mix. Table 2 shows the mixture proportion for all the six batches. For each of the hardwood and softwood batch, the cement content was reduced with the increase of fiber content. In effect, the fibers were replacing the cement and this was required in order to provide a certain consistency in the quality of the material. The increase of fiber content in a batch required more water to keep the same viscosity for extrusion. A flow diagram of the extrusion manufacture process is shown in Figure 1. The extruded cement board had a dimension of 6.3 mm thick and 76 mm wide. It was observed that the hardwood pulp required less water to achieve the suitable viscosity and produced a much smoother surface of the finished product, owing probably to the shorter fiber length. The cement board was cut to 305 mm long and cured in a chamber with 80% relative humidity.

## EXPERIMENTAL PROCEDURE

Flexural Strength Tests

Three-point bending tests with a span of 101.6 mm were carried out to evaluate the longitudinal flexural strength and toughness of the extruded composites at 7 days and 180 days. The loading rate for all tests was set at 0.012 mm/s. The flexural toughness of thin sheets was defined as the total area under the entire load-deflection curve. To compare the longitudinal (in extrusion direction) and transverse properties of an extruded product, three-point bending tests of 63.5 mm span were also conducted with specimens cut in both longitudinal and transverse directions. A shorter span length was used in comparison study because the extruded samples were only 76 mm wide. At least five samples were tested and averaged for each batch. The 180-day specimens were used as a control to study the effects of aging, weathering, and freeze-thaw cycling.

Natural Weathering Tests

To study the weathering resistance of the extruded pulp-cement composites, a weathering panel was constructed. Extruded sheets of 305 mm long were selected from the softwood and hardwood batches and screwed tightly on to a wood panel at both ends. This weathering panel was subsequently placed in an outdoor environment for a period of 5 months from November to March in the Montreal area. The highest temperature for this period was around 15°C (early December) and the lowest around –27°C (middle of January). Approximately 20 cycles was observed between the maximum and minimum temperatures. At the end of March, the flexural strength test with a span of 101.6 mm was carried out at the age of 180 ± 5 days. The results were compared to that of control specimens at the same age but moist cured in laboratory conditions.

Freeze Thaw Tests

The freeze thaw tests were conducted to evaluate the degradation of the extruded composites on exposure to repeated cycles of freezing and thawing. The procedure followed the guidelines set out in ASTM C 666 and ASTM C 1185. Specimens from each batch were tested at the age of 180 days. The cycle temperature variation was between +10°C and -20°C over a period of 2 hours. Both the freezing and thawing were done in water. The specimens were tested at 300 cycles. The flexural strengths of the cycled specimens were evaluated at a span of 101.6 mm and averaged with at least three sample tests. This was done to investigate the possible strength deterioration due to the cycling damage. The results were compared to that of uncycled control specimens of the same age, which remained in moist curing conditions. The strength retained was defined as the ratio of the flexural strength of the cycled specimen to that of the uncycled control specimen.

Physical Property Tests

Moisture content, water absorption and density were measured following the procedures outlined in ASTM C 1185. Three specimens from each batch, 180 ± 5 days old and measuring 127 mm by 76 mm by 6.3 mm, were selected. The specimen size was not standard and limited by the width of extrudate of 76 mm. They were conditioned in an oven to dry out the moisture. Subsequently all the samples were left in the air at room temperature for a period of 60 days. At the end of 60 days, the initial mass of the samples were measured and they were dried out in an oven at 102 ± 2°C. Mass readings were taken again at 2 hour intervals until two consecutive readings differed by less than 1%. The final oven dry mass was recorded. The moisture content is defined as the difference between the oven dry mass and the initial mass and reflects the ability of the composites to absorb moisture from the environment. Once the specimens cooled down, they were submerged in water at 25°C for a 48 hour period. After this 48-hour period, the saturated surface dry (SSD) mass was recorded. The water absorption was obtained from the difference between the oven dry mass and the saturated surface dry mass. Finally, Archimedes principle was used to obtain the volume of the specimens for density calculations. The density was calculated based on the oven dry mass.

**RESULTS**

Flexural Response at 7 days

The typical flexural stress vs. deflection curves for the six batches at 7 days are shown in Figure 2. A slight increase in the flexural strength was observed with increasing fiber content in both hardwood and softwood composites. However, this strength increase was not proportional to the increasing fiber content. The toughness nevertheless improved considerably as the fiber content increased. At 7 days, for a given fiber content, the softwood composites generally performed better than the hardwood specimens, displaying higher values of flexural strength and toughness.

To investigate the effect of the extrusion process on transverse properties, three-point bending tests were carried out in both longitudinal and transverse directions at 7 days using a 63.5 mm span. The results are summarized in Figure 3. The strength and toughness of the cement boards in transverse direction were reduced when compared with that in longitudinal direction. This difference was more obvious at higher fiber contents and indicative of possible fiber alignment in the direction of extrusion. The softwood cement board at 8% pulp content exhibited the best performance in strength gain, toughness gain and transverse property retention. The reduction in hardwood cement board was proportional to the pulp content, suggesting that more fibers were possibly aligned in extrusion direction. Fracture surfaces of cement board were examined under SEM. Typical pictures of hardwood and softwood pulps on fracture surface are shown in Figures 4a and 4b. The fractured fibers featured a flat and hollow ribbon-like structure. The pullout fiber length was generally short. The SEM pictures also showed the hollow softwood fiber had thicker wall than the hardwood. Significant damage on the surface of the pulled-out hardwood fibers was observed. A close-up of hardwood fiber on fracture surface is displayed in Figure 5. The hollowed structure is apparent. The polished cross section of cement board was also examined under SEM (Figure 6). The pulps (white spots) were likely dispersed and a large number of distributed air voids (black spots) was noted. The dispersion of the pulp in cement board was mostly attributed to the kneading process and the air voids were due to the use of methyl cellulose polymer[5]. The distributed and disconnected air voids had a maximum diameter smaller than 0.1 mm and a spacing less than 1 mm. The entrained air was expected to improve the resistance to freeze-thaw cycling.

Effect of Age

Age effect was studied by comparing the performance of cement board tested at 7 days and 180 days cured under the same conditions. The results are shown in Figure 7. There was a prevalent trend with age in which the specimens exhibited a more brittle type of behavior, displaying higher values in flexural strength and a gradual loss in toughness. For the softwood composites, the trend of increasing flexural strength and toughness with increasing fiber content was quite noticeable. For the hardwood composites, a more or less constant flexural strength of 25 MPa was obtained for all three fiber percentages at 180 days. It was noticed that the hardwood composites had an appreciable strength gain from 7 to 180 days with a good retention of toughness, when compared to softwood. Obviously, aging had reduced toughness more significantly in softwood composites than in hardwood cement boards.

Effect of Natural Weathering

Natural weathering tests were conducted to investigate the combined effects of freezing-thawing and wetting-drying cycles on the mechanical properties of extruded sheets. There was significant frost action during the period of test along with plenty of rain, snow, and sun. Figure 8 summarizes the results for the flexural tests of composites that had been exposed to natural weathering, along with the unexposed control specimens left in the lab. There was a decrease in flexural strength for all composites under exposed conditions. The reduction in flexural strength for the softwood specimens was between 1-5% and that for the hardwood specimens was between 4-25%. For the hardwood composites, the reduction was more significant at the lower fiber fraction ratio.

Effect of Freeze-Thaw Cycling

The flexural strengths of cement board subject to 300 freeze-thaw cycles are shown in Figure 9, together with the uncycled specimens tested at the same age as reference. The percent of retained strengths after 300 cycles are also labeled. The severe freeze-thaw cycles reduced strengths of all composites. The reduction was proportional to water to cement ratio (see Table 2). The higher the water/cement ratio used, the more the reduction was observed. High pulp ratio required more water to keep the workability, so did the softwood pulp when compared with hardwood. Therefore, the hardwood composites had better frost resistance. However, the retained strengths of all extruded boards were over 75% of the control after 300 cycles between +10 $^0$C and –20 $^0$C.

Physical Properties

Moisture content, water absorption and density for all batches are shown in Table 3 together with standard deviations. For both the hardwood and softwood composites, an increase in moisture content was observed with increasing fiber content, indicating that moisture absorption of the material from its environment was proportional to the amount of the wood fibers used. Water absorption represents the capacity of the material to absorb water when in contact with water. Relatively, the hardwood composites had lower water absorption with an average of 15%, compared to the softwood batches averaged at 21%. There were no clear correlations linking water absorption to the fiber content, the water/cement ratio or the methylcellulose/cement ratio, for both composites. If the standard deviations were considered, the percent absorption was close to a constant for each hardwood and softwood composite. It implied that the porosity generated in each composite was also close to a constant, independent from W/C ratio, fiber ratio or methylcellulose/cement ratio. As expected, increasing fiber content had reduced the density of the composites.

Also, for a given fiber fraction ratio, the hardwood samples displayed slightly higher densities when compared to the softwood samples.

## CONCLUSIONS

The hardwood pulps had a better extrudability with cement matrix and smoother finished surface than the softwood. At 7 days, the flexural performance of the softwood composites proved to be superior, displaying higher values for strength and toughness. However, at 180 days, the hardwood specimens exhibited much improved flexural strength and toughness. In general, longer curing time led to a higher strength and a lower toughness.

The extrusion process was found to have an effect of aligning the fibers in the direction of extrusion. This was inferred from higher values of flexural strength (17-86%) and toughness (20-94%) tested in the extrusion direction.

Higher pulp content had a significant effect on the toughness of the material. However, if the increase in pulps required an increase in water to keep the workability, there was no accompanying significant increase in flexural strength. The natural weathering exposure decreased the flexural strength for all composites. So did the freeze-thaw cycling. The retained strengths were however all above 75%. It might be the air voids entrained by methyl cellulose that contribute to the improved frost resistance. The moisture absorption of the composites from the environment was found to increase, and the density to decrease, with increasing pulp content. However, the water absorption was not proportional to the amount of pulps used. Instead, water absorption for each composite was close to a constant if the standard deviations were considered. Relatively, the softwood composites absorbed more water than their hardwood counterparts.

## ACKNOWLEDGMENT

Financial supports from the Natural Science and Engineering Research Council (NSERC) of Canada and from Le Fonds pour la Formation de Chercheurs et l'Aide à la Recherche (FCAR) of Quebec are gratefully acknowledged.

## REFERENCES

1. Kurpiel, F.T. (1997). Diffusion of Cellulose Fiber-Cement Siding and Roofing into North America. Proceedings of *Inorganic-Bonded Wood and Fiber Composite Materials*, A.A. Moslemi (Ed.), Vol. 5. pp. 41-44

2. Coutts, R. S. P. (1987). Air cured woodpulp fiber/cement mortars. *Composites*, Vol. 18, No. 4, pp. 325-328.

3. Soroushian, P., Marikunte S., and Won J. (1995). Statistical Evaluation of Mechanical and Physical Properties of Cellulose Fiber Reinforced Cement Composites. *ACI Materials Journal*, Vol. 92, No. 2, pp. 172-180.

4. Shah, S. P., Peled, A., Deford, D. (1998), Extrusion technology for the production of fiber-cement composites, Proceedings of *Inorganic-Bonded Wood and Fiber Composite Materials*, A.A. Moslemi (Ed.), Vol. 6. Pp261-277.

5. Shao, Y., Qiu, J. and Shah, S. P. (2001), Microstructure of Extruded Cement-Bonded Fiberboard. *Cement and Concrete Research*, Vol. 31, pp 1153-1161.

Table 1: Physical Properties of Hardwood and Softwood Pulp

| Fiber Type (Species) | Fiber Length, mm | Fiber Diameter, µm | Wood Density, kg/m$^3$ |
|---|---|---|---|
| Hardwood (Aspen) | 1.0 - 2.0 | 25 - 45 | 400 - 480 |
| Softwood (Black Spruce) | 3.5 - 4.5 | 10 - 40 | 432 - 609 |

Table 2: Mixture Proportions for Hardwood and Softwoods Composites

| % by weight | Hardwood | | | Softwood | | |
|---|---|---|---|---|---|---|
| | 2% HW2 | 4% HW4 | 8% HW8 | 2% SW2 | 4% SW4 | 8% SW8 |
| cement | 67.0 | 64.0 | 57.0 | 69.0 | 60.0 | 53.0 |
| silica fume | 10.0 | 10.0 | 9.5 | 5.5 | 10.0 | 10.0 |
| fibers | 2.0 | 4.0 | 8.0 | 2.0 | 4.0 | 8.0 |
| methylcellulose | 1.5 | 1.5 | 1.5 | 1.5 | 1.5 | 1.5 |
| water | 19.5 | 20.5 | 24.0 | 22.0 | 24.5 | 27.5 |
| W/C | 0.29 | 0.32 | 0.42 | 0.32 | 0.41 | 0.52 |

Table 3: Physical Properties of Extruded Cement Board

| HW2 | HW4 | HW8 | SW2 | SW4 | SW8 |
|---|---|---|---|---|---|
| Moisture content (%) | | | | | |
| 3.69 ± 0.43 | 4.14 ± 0.87 | 5.24 ± 0.24 | 4.41 ± 0.01 | 5.60 ± 0.03 | 5.81 ± 0.62 |
| Water absorption (%) | | | | | |
| 14.56 ± 0.43 | 13.36 ± 0.72 | 17.19 ± 1.05 | 18.24 ± 0.65 | 24.58 ± 3.98 | 20.23 ± 2.98 |
| Density (g/cc) | | | | | |
| 1.83 ± 0.04 | 1.68 ± 0.01 | 1.39 ± 0.01 | 1.75 ± 0.02 | 1.56 ± 0.05 | 1.38 ± 0.03 |

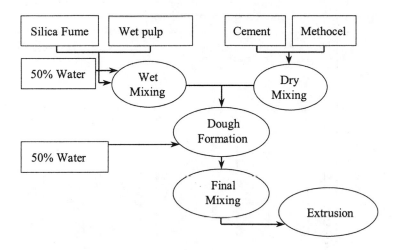

Figure 1: Flow diagram of extrusion process

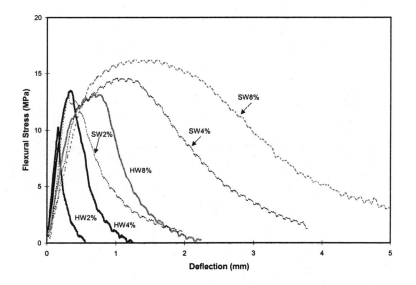

Figure 2: Typical flexural stress – deflection curves of extruded cement boards
(7 days, span = 101.6 mm)

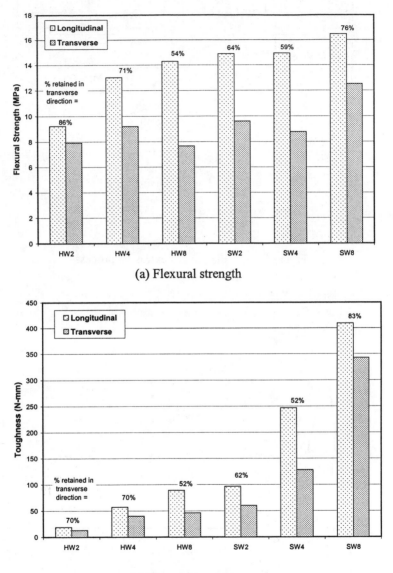

(a) Flexural strength

(b) Flexural toughness

Figure 3: Comparison of longitudinal and transverse flexural response
(7 days, span = 63.5 mm)

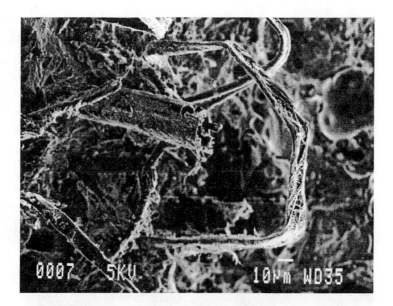

(a) Hardwood pulp cement board

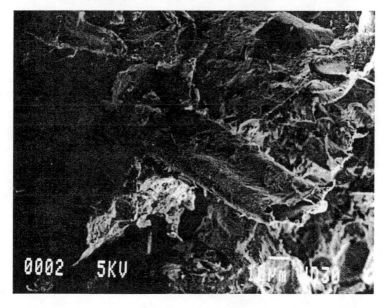

(b) Softwood pulp cement board

Figure 4: Fracture surface of extruded cement boards

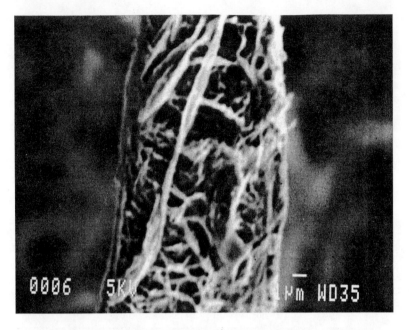

Figure 5: Close-up of hardwood pulp fiber on fracture surface

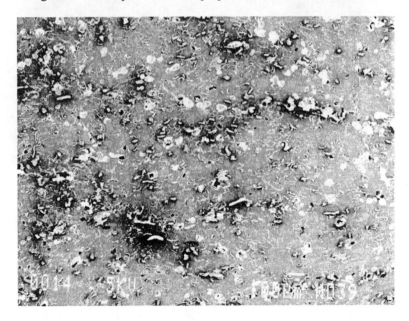

Figure 6: Polished cross-section showing distributed pulps and entrained air

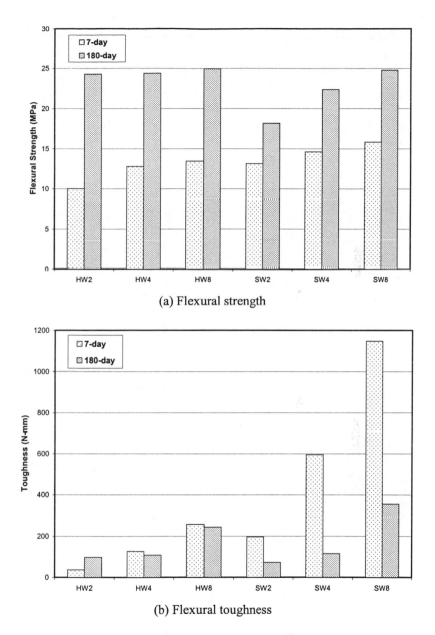

(a) Flexural strength

(b) Flexural toughness

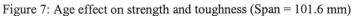

Figure 7: Age effect on strength and toughness (Span = 101.6 mm)

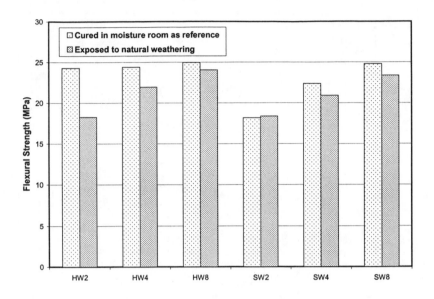

Figure 8: Effect of exposure to natural weathering

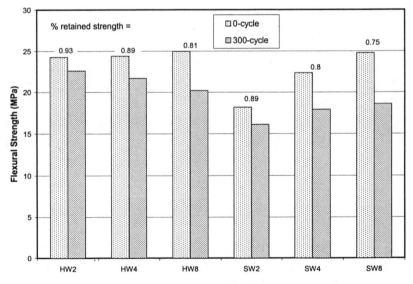

Figure 9: Strength retained after 300 freeze-thaw cycles

# Development and Seismic Behavior of High-Performance Composite Frames

## by N. Krstulovic-Opara and V. Kilar

Synopsis: High-Performance Fiber Reinforced Concrete (HPFRC) exhibits features particularly desirable for increasing earthquake resistance such as high tensile strength and ductility. However, since HPFRC is substantially different from conventional materials, using existing design and construction procedures does not lead to the most cost-effective solutions. To address this issue, this paper presents a way of selectively using Slurry Infiltrated Mat Concrete (SIMCON), Slurry Infiltrated Fiber Concrete (SIFCON), and High Strength - Lightweight Aggregate Fiber Reinforced Concrete (HS-LWA FRC) to construct partially precast High Performance Composite Frames (HPCFs). The objective is to improve cost effectiveness by simplifying both construction and post-earthquake repair, increasing the construction speed, lowering seismically induced forces and increasing overall seismic resistance. This is achieved by combining SIMCON stay-in-place formwork, and precast, replaceable SIFCON fuses with cast-in-place HS-LWA FRC.

First, the paper presents experimentally evaluated behavior of HPCF members. Next the seismic response of a four story HPCF building is investigated analytically and compared to that of a seismically designed R.C. frame. HPCF reached an overall good seismic response. As compared to the reference building, it exhibited slower strength and stiffness degradation, lower top displacements and story drifts. Overall building damage was lower for the HPCF building, and even under 33% higher seismic excitations the HPCF building had higher seismic resistance than the reference R.C. building.

Keywords: composite members; experimental investigation; fiber reinforced concrete; high performance construction materials; metal fibers; non-linear seismic analysis; reinforced concrete; seismic resistant design; slurry infiltrated fiber cocrete (SIFICON); slurry infiltrated mat concrete (SIMCON); slurries

**Neven Krstulovic-Opara** is an Assistant Professor of Civil Eng. at North Carolina State Univ. He received his M.Sc. from Imperial College, London, and his Ph.D. from Carnegie Mellon Univ. His research interests are in the areas of fracture mechancis, "smart" and high-performance fiber reinforced composite materials and structures. He is a member of ACI 544, 446, 348 and 325.

**Vojko Kilar** is an Assistant Professor in the Faculty of Architecture, Univ. of Ljubljana, Slovenia. He received his M.S. and Ph.D. degrees from the University of Ljubljana, Slovenia. His research interest is in the area of earthquake engineering, and the development of computer programs for linear and non-linear analysis of concrete structures under seismic excitations.

## INTRODUCTION

Major progress has been made in recent decades in the development of Fiber Reinforced Concrete (FRC), and High-Performance FRC (HPFRC). However, progress has been slower in the cost-effective structural use of these advanced materials. The goal of the presented research was to reconsider existing procedures for design and construction of R.C. frames, and develop an alternative approach better suited for cost-effective, seismically resistant use of FRCs and HPFRCs. Such a frame, termed herein **High-Performance Composite Frame** (HPCF), is made by selectively using: High Strength - Lightweight Aggregate FRC (HS-LWA FRC), Slurry Infiltrated Mat Concrete (SIMCON) and Slurry Infiltrated Fiber Concrete (SIFCON). The proposed HPCF consists of three main elements: (1) a composite beam member made with a precast SIMCON "jacket" and cast-in-place HS-LWA FRC core, (2) a composite HS-LWA FRC filled steel tube column member, and (3) a precast SIFCON "fuse" connecting the two, as shown in Figure 1. Material properties are summarized first, followed by a brief overview of the HPCF member response and results of a more detailed analysis of the system response.

## BACKGROUND ON FRC

FRC is made by adding fibers to concrete. Based on the level of improved behavior, FRCs are classified as conventional FRCs or HPFRCs. Practically, conventional FRCs are made with up to 2 % volume fraction of discontinuous fibers, which primarily increases its ductility and energy dissipation capacity, leading to a large improvement in seismic performance: improved concrete confinement, higher shear and moment strength, higher stiffness and slower stiffness degradation, higher ductility and energy dissipation capacity. The improvements can be so large that shear reinforcement can be eliminated (e.g., [1]). On the other hand, HPFRCs exhibit strain-hardening behavior which significantly increases tensile and compressive strengths, ductility and energy dissipation capacity. HPFRCs used in this research are SIMCON [4-9] and SIFCON (e.g., see [10]). Both are made by infiltrating pre-placed continuous steel fiber-mats, as shown in Figure 3 a, or discontinuous fibers, respectively, with a cement-based slurry.

## High Strength - Lightweight Aggregate FRC

Low fiber volume fraction FRC used in this research was made by adding short steel fibers to otherwise extremely brittle high strength – lightweight aggregate (HS-LWA) concrete. The high strength and low weight leads to lower seismically induced forces. However, due to their extremely brittle failure [2,3], HS-LWA concretes are not used in seismic resistant design. To overcome this limitation, 1.5 % and 2 % volume fraction of steel fibers was added to HS-LWA concrete used in column and beam members, respectively.    Average compressive strengths were 107.8 MPa (15,636 psi) and 80.7 MPa (11,700 psi), for 1.5 % and 2.0 % fiber volume fraction, respectively, as shown in Figure 2

## HPFRCs:  SIMCON and SIFCON

Due to its fiber-mat configuration, SIMCON exhibits high strengths and ductilities, as shown in Figure 3 b [5], and is also well suited for casting thin retrofit jackets or stay-in-place formwork members shown in Figure 3 c.  This markedly increases member's flexural and shear capacities, ductility and overall seismic response [6,8,11-13]. Average 28-day tensile and compressive strengths of 6.4 % fiber volume fraction SIMCON used in this research were 15.9 MPa (2,300 psi) and 68.9 MPa (10,000 psi), respectively [5,9].  In the case of SIFCON, a higher volume fraction of short fibers is needed to achieve behavior similar to that of SIMCON [4].  However, SIFCON is better suited for manufacturing "thick" members, such as "fuse" zones [14-16].

## EXPERIMENTAL INVESTIGATION

The HPCF consists of three members: (1) a "hybrid" beam with stay-in-place SIMCON formwork and cast-in-place HS-LWA FRC core, (2) a CFT column with cast-in-place HS-LWA FRC core, and (3) a precast SIFCON fuse. All three members are connected on site through bolting, allowing the speed of construction per story similar to that of prefabricated steel frames. Monolithic construction, required for good seismic resistance, is achieved by casting-in-place HS-LWA FRC member "cores." Finally, since the use of FRC eliminates secondary reinforcement [17,18]: (1) problems of reinforcement congestion are eliminated, and (2) speed of construction is increased.  A more detailed description of the experimentally determined member response is provided next.

**Beam Member:** No conventional reinforcement was used in 40.6 cm (16 in.) high and 2.5 cm (1 in.) thick formwork walls.  Use of SIMCON eliminated the need for stirrups, while its high strength and toughness permitted direct bolting of shear studs and connection bolts into the formwork walls, as shown in Figure 3 c.  The section "core" was cast-in-place, 2% fiber volume fraction, 80.7 Mpa (11,700 psi) HS-LWA FRC, with the unit weight of 19.6 kN/m$^3$ ( 123 lb/ft$^3$). The beam specimen was tested under cantilever-type reverse cyclic loading.

**SIFCON Fuse** provides an energy "sink" under severe seismic excitations by forming a plastic hinge [14-16].  To both increase the construction speed and simplify replacement of damaged fuses after an earthquake, precast "fuses" were developed. The effect of reinforcement layout and SIFCON properties on fuse behavior was evaluated and optimized [19]. A representative response, exhibiting a curvature ductility of 6.9, is shown in Figure 4. Measured displacement ductilities of 4.2 [19] compare well with the values of 4.5 and 6, reported for cast-in-place SIFCON and conventional FRC fuses, respectively [14-16]. It should be noted that the reported displacement ductility of 6 [14], was calculated as the ratio between (**a**) the maximum "usable" displacement, and (**b**) the displacement that is 30 % higher than the displacement reached at 75 % of the calculated yield load. If the same definition is used in this research, a value of 6.4 is obtained [19]. Furthermore, Abdou et al. [15,16] designed fuses so that yielding of the connection between the fuse and precast members provides a significant contribution to overall energy absorption. On the contrary, since in the present research it was anticipated that the fuses should be replaceable after the damage occurred, the goal was to minimize or eliminate yielding of the interface sections. This goal was successfully achieved and the main source of energy absorption was the zone in the middle of the fuse.

**Column Member:** If adequately confined, HS-LWA concrete is ideally suited for column members [20,21]. To maximize confinement HS-LWA FRC was encased in a steel tube, as shown in Figure 5. Under a static-reversed cyclic loading specimen exhibited stable response, hysteretic loops were wide and a comparatively high displacement ductility of 7.7 was reached, as shown in Figure 5 c. Specimen failure initiated by the "elephant foot" buckling of the steel tube, followed by formation of steel cracks first observed in the 31st cycle. Extension of these cracks led to the final failure of the specimen.

## ANALYTICAL INVESTIGATION

The goal of the anaytical investigation was to both (**1**) evaluate the effect of the fuse length on the beam behavior, and (**2**) evaluate the seismic behavior of the HPCF by comparing it to the behavior of a reference R.C. frame building.

The reference R.C. (i.e., "Ispra") structure, shown in Figures 6 and 7 [22], was designed in accordance with the seismic requirements of the Eurocode 2 and 8 for High Ductility structures and was tested under pseudodynamic loading in the European Laboratory for Structural Assessment [22]. Additional dead load from floor finishing and partitions of 2.0 $kN/m^2$, live load of 2.0 $kN/m^2$, peak ground acceleration of 0.3g, soil type B, importance factor of 1 and behavior factor, $q = 5$, were used. The same building dimensions were used for HPCF. Thus, the HPCF beam, fuse and column cross-sectional dimensions were scaled up by 19%, as shown in Figure 8 and summarized in Table 2, while the column diameter was scaled to 0.386 m (15.2 in). Experimentally obtained member properties were scaled up following the "true model" scaling laws [23], as shown in Figure 9 and sumarized in Table 1. Resulting elastic moduli of the SIMCON beam member, fuse and column were equal to $E_b = 4.15 \cdot 10^7$ $kN/m^2$ (6.02 * $10^6$ psi), $E_f = 1.16 \cdot 10^7$ $kN/m^2$ (1.68 * $10^6$ psi), and $E_c = 5.78 \cdot 10^7$ $kN/m^2$

(8.38 * $10^6$ psi), respectively.  First 20 cm (7.87 in.) of the fuse length facing the column (idimension $L_r$ in Figure 8) were modeled as infinitely rigid.  Use of HS-LWA FRC decreased by 10% the beam mass as compared to the R.C. beams.  The effect of floor slabs was not considered.

## Behavior of the Hybrid-Beam Members

Behavior of the entire "hybrid" (i.e., HPC) beam, fuse, and the reference R.C. (i.e., "Ispra") beam, are compared in Figures 10 and 11.  The HPC beam exhibited much higher strength and ductility than the R.C. beam, as shown in Figure 10 a.  Only when the percent of reinforcement of the R.C. beam is increased to 3%, do both beams reach similar strengths.  However, curvature ductility of the R.C. beam decreases further as can be seen by comparing Figures 10 a and b.  Comparing the response of the HPC beam to that of only the fuse, shown in Figure 10 a, indicates that even though the fuse yields first, some beam yielding still takes place at higher deformation levels.  This can be clearly observed from Figure 11 a, where points "3" and "4" denote fuse yielding, while point "5" marks beam yielding.  Increasing the fuse length, $L_f$, by 50 % eliminates beam yielding, as shown in Figure 11 b, in which case point "d" marks fuse yielding.  However, changing the fuse length has an inverse effect on stiffness, as shown in Figure 11 b.  Another alternative is to either add additional longitudinal reinforcement to the beam top, or consider the effect of slab reinforcement [27].  In the following analytical investigation it was assumed that three 14 mm diameter bars were added to the beam top, as shown in Figure 8, section b-b, while the fuse length of 0.876 m (34.5 in.) was kept constant.

## Behavior of HPCF Building

**Building Member Modelling:**   Non-linear seismic analysis was performed using the CANNY-E computer program [24].  The damping matrix was assumed to be proportional to the instantaneous stiffness matrix.  The assigned damping ratio was 2%.  Both beam and column members were modelled using a non-linear uniaxial-bending model that considered the effect of elastic shear deformation.   Axial and torsional beam deformations, as well as torsional column deformations were not considered.  All inelastic flexural deformation was lumped at the element ends using two non-linear "rotational"-springs, connected to the joint through a rigid zone.  Moment-rotation response of these springs was obtained using experimentally determined moment-curvature relationships.   An asymmetrical moment distribution was assumed along the length of the element.  A tri-linear skeleton curve was used to represent the cross-sectional member behavior.  While the hysteretic behaviour followed the Takeda rules, pinching effect was also considered.  The best correlation with the experimental results of the reference R.C. building were obtained when small values of the unloading stiffness were used (e.g., unloading stiffness in each cycle was reduced by 50% over its value in the preceding unloading cycle).  A more detailed discussion of the effect of different model parameters on the response of the reference R.C. building has been presented elsewhere [25].

**Seismic Input:** An artificial "S7" accelerogram, derived from the 1976 Friuli earthquake record and shown in Figure 12 a, was used for testing the reference R.C. building in the European Laboratory for Structural Assessment [22]. The response to this accelerogram fits that given by Eurocode EC8 for soil profile B at 5% damping, as shown in Figure 12 b. The pseudodynamic tests of the R.C. building were performed using accelerogram S7 scaled by 1.5. Thus the analytical investigation was performed using the same scaling. Additionaly, the behavior of the HPCF building under 2 times the S7 accelerogram was also considered. The loading was oriented in the X direction, as defined in Figure 6. Periods of both buildings are summarized in Table 5. Lower mass and slower stiffness degradation of the HPCF building resulted in smaller periods.

**Results of Non-linear Dynamic Analysis:** Responses of the R.C. and HPCF buildings exposed to 1.5 times the S7 accelerogram, as well as the HPCF building exposed to 2 times the S7 accelerogram, are shown in Figures 13 to 17. The global damage index, presented in Figure 15, gives an overall estimate of the building damage [24]. The damage index is calculated per each non-linear spring by linear combination of normalised deformation and energy dissipation [26]. Rotational ductility factors, shown in Figure 18, are defined as a ratio between yield rotation and ultimate rotation reached during non-linear dynamic analysis. Values equal or greater than 1.0 indicate member yielding.

Under the same seismic excitations HPCF building experiences much smaller damage, as shown in Figure 15. Only after the seismic record was increased by 33 % did the overall damage equal that of the reference building exposed to 1.5 times the S7 accelerogram. A decreased amount of damage results in slower decrease in building stiffness, and hence higher base shear at lower top displacements, as shown in Figures 14 and 13. Even when an HPCF building is exposed to 33 % higher seismic accelerations, its displacements are up to 2 times smaller than those of the reference building, as shown in Figure 14. Under the same excitations, the maximum story drifts are 2.5 times larger in the reference building, as shown in Figure 16. Finally, under 1.5 times the S7 accelerogram, all the damage in the HPCF building occurs only within the fuse zones, while SIMCON/HS-LWA FRC beam members behave fully elastically. Only when the accelerogram is increased to 2 times the S7 accelerogram is the onset of yielding recorded at the bottom of the columns.

## CONCLUSIONS

Behavior of the tested beam member demonstrated the feasibility of manufacturing SIMCON stay-in-place beam formwork that was sufficiently strong, stiff and "sturdy" to be easily handled. Tests show that shear keys or bolts for connecting beam formwork to the precast fuse, can be effectively used. Precast SIFCON fuses behaved well. Measured displacement ductilities of 4.2 [19] compare well with the values reported by other authors [14-16]. Since damaged fuses should be replaceable after the earthquake, yielding of the end zones was minimized, and the main sources of energy absorption were all other fuse zones.

The inability of the confined HS-LWA concrete columns to engage lateral confinement before the onset of column failure is the key reason why these concretes have not been successfully used in seismic-resistant design. This problem was resolved by adding 1.5% of steel fibers. As a result, HS-LWA concrete columns exhibited a very ductile seismic response, reaching high displacement ductility of 7.7, as shown in Figure 5. Even higher ductilities could be anticipated if the premature ripping of the steel tube is prevented.

Good seismic properties of tested members translate to good seismic response of the HPCF building. In this case, as long as the issue of possible SIMCON/HS-LWA FRC beam damage is properly addressed, all the damage occurs only in the replaceable fuse members. The HPCF building exhibits slower strength and stiffness deterioration, lower top displacement and story drifts than the reference R.C. building. The observed plastic mechanisms of HPCF are favorable, and even under 2 times the S7 accelerogram the building behaved well, indicating higher seismic resistance than that achieved using conventional R.C. design.

Finally, it should be pointed out that while behavior of all frame members was experimentally evaluated, behavior of connections between these members was not tested. Instead, connections were assumed to be "sufficiently strong" not to fail under seismic excitations. Since proper detailing of these connections and evaluation of their performance under seismic excitations was not fully addressed in the presented project, this critical issue should be explored in a greater detail in the future investigations.

## Acknowledgements

This investigation is partially supported by NSF grants CMS-9632443 with Dr. S. C. Liu as Program Director as well as Ms. C. Dudka as International-Program Director. The authors are grateful to Professors J. Hanson, M. Leming and P. Zia of NCSU, S. Ahmad of American Concrete Institute, Research Associates Dr. J. M. Shannag and Dr. J. Brzezicki, Mr. J. Atkinson and Mr. B. Dunleavy from NCSU, graduate students J. Punchin, B. Brezac, E. Dogan, B. Wood, P. Thiedeman and J. Becchio, who assisted in the presented work.

## REFERENCES

1. Katzensteiner, B., Mindess, S., Filiatrault, A., Nathan, N. D., Banthia, N., "Use of Steel Fiber Concrete in Seismic Design," RILEM Symposium, *Fiber Reinforced Cement and Concrete*, 1992, pp. 613 - 628.
2. Martinez, S., Nilson, A. H., Slate, F. O., "Spirally Reinforced High-Strength Concrete Columns," ACI Journal, *Proceedings*, 1984 Sep, pp. 431 - 442.
3. Bjerkeli, L., Tomaszewicz, A., Jensen, J. J., "Deformation Properties and Ductility of High-Strength Concrete," Proceedings of the *Second International Symposium on Utilization of High-Strength Concrete*, University of California, Berkeley, 1990 May 20, pp. 215 - 238.

4. Hackman, L. E., Farrell, M. B., Dunham, O. O., "Slurry Infiltrated Mat Concrete (SIMCON)," *Concrete International*, December 1992, pp. 53-56.
5. Krstulovic-Opara, N., Malak, S., "Tensile Behavior of SIMCON," ACI *Materials Journal*, January – Feb., 1997, pp. 39 - 46.
6. Krstulovic-Opara, N., Dogan, E., Uang, C. -M, Haghayeghi, A., "Flexural Behavior of Composite R.C. - SIMCON Beams," ACI *Structural Journal*, September-October 1997, pp. 502 - 512.
7. Krstulovic-Opara, N., Malak, S., "Micromechanical Tensile Behavior of Slurry Infiltrated Fiber Mat Concrete (SIMCON)", ACI *Materials Journal*, September - October 1997, Vol. 94, No. 5, pp. 373 - 384.
8. Krstulovic-Opara, N., Al-Shannag, M. J., "Slurry Infiltrated Mat Concrete (SIMCON) - Based Shear Retrofit of Reinforced Concrete Members," ACI *Structural Journal*, January - February 1999, pp. 105 - 114.
9. Krstulovic-Opara, N., Al-Shannag, M. J., "Compressive Behavior of SIMCON," ACI *Materials Journal*, May - June 1999, pp. 367 - 377.
10. Naaman, A. E., "SIFCON: Tailored Properties for Structural Performance," *HPFRC Composites*, E & FN Spon, 1992, pp. 18-38.
11. Dogan, E., *Retrofit of Non-Ductile Reinforced Concrete Frames Using High Performance Fiber Reinforced Composites*, Ph.D. Thesis, NCSU, 1998.
12. Dogan, E., Hill, H., Krstulovic-Opara, N., "Suggested Design Guidelines for Seismic Retrofit With SIMCON & SIFCON," *HPFRC in Infrastructural Repair and Retrofit*, ACI SP-185, 1999, pp. 207-248.
13. Krstulovic-Opara, N., LaFave, J., Dogan, E., Uang, C. - M., "Seismic Retrofit With Discontinuous SIMCON Jackets," *HPFRC in Infrastructural Repair and Retrofit*, ACI Special Publication SP-185, 1999, pp. 141-185.
14. Vasconez, R. M., Naaman, A. E., Wight, J. K., *Behavior of Fiber Reinforced Connections for Precast Frames Under Reversed Cyclic Loading*, Report Number UMCEE 97-2, The University of Michigan, 1997.
15. Abdou, H., Naaman, A., Wight, J., *Cyclic Response of Reinforced Concrete Connections Using Cast-in-Place SIFCON Matrix*, Report No. UMCE 88-8, Department of Civil Eng., University of Michigan, 1988.
16. Soubra, K. S., Wight, J., Naaman, A., *Fiber Reinforced Concrete Joints for Precast Construction in Seismic Areas*, Report No. UMCEE 92-2, Department of Civil and Env. Eng., The University of Michigan, 1992.
17. Craig, R., Mahhader, S., Patel, C., Viteri, M., Kertesz, C., "Behavior of Joints Using Reinforced Fibrous Concrete," ACI SP - 81, 1984, pp. 125-167.
18. Sood, V., Gupta, S., "Behavior of Steel Fibrous Concrete Beam-Column Connections," ACI SP - 105, 1987, pp. 437 - 474.
19. Wood, B. T., Use of Slurry Infiltrated Fiber Concrete (SIFCON) in Hinge Regions for Earthquake Resistant Concrete Moment Frames, Ph.D. Thesis, Department of Civil Engineering, North Carolina State University, 2000.
20. Shah, S., Ahmad, S., *High Performance Concretes: Properties and Applications*, McGraw Hill, 1994.
21. El-Dash, K., *Ductility of Lightweight High Strength Columns*, Ph.D. Thesis, North Carolina State University, Raleigh, 1995.
22. Negro, P., Verzelletti, G., Magonette, G.E., and Pinto, A.V., *Tests on a Four-Storey Full-Scale R/C Frame Designed According to Eurocodes 8 and 2*, Preliminary Report, European Laboratory for Structural Assessment, Italy, 1994.

23. Sabnis, G. M., Harris, H. G., White, R. N., and Mirza, M. S., "Structural Modeling and Experimental Techniques," *Civil Engineering Series*, Prentice Hall, Inc., 1983.

24. Li, K.N., *CANNY-E: Three-Dimensional Non-linear Dynamic Structural Analysis Computer Program Package*, Canny Consultants Pte Ltd., 1997.

25. Faella G., Kilar V., and Magliulo G., "Overstrength factors for 3D R/C Buildings Subjected to Bidirectional Ground Motion," *Proceedings of the 12th World Conference on Earthquake Engineering*, Auckland, New Zealand, 2000.

26. Park, Y.J. and Ang, A. H-S., "Mechanistic Seismic Damage Model for R.C.," *Journal of Structural Engineering, ASCE*, Vol. 111, No. ST4, pp. 722-739, 1985.

27. Pantazopoulou, S. J., and French, C. W., "Slab Participation in Practical Earthquake Design of Reinforced Concrete Frames," ACI *Structural Journal*, July – August 2001, pp. 479 – 489.

Table 1: Characteristic moment-curvature points used in non-linear analysis of HPC beams.

|  | Fuse | | Beam (bottom in tension) | | Beam (top in tension) | | Column | |
|---|---|---|---|---|---|---|---|---|
|  | Curv. (1/m) | Moment (kNm) | Curv. (1/m) | Moment (kNm) | Curv. (1/m) | Moment (kNm) | Curv. (1/m) | Moment (kNm) |
| **Crack** | 0.0077 | 255 | 0.0012 | 142 | 0.0008 | 95 | 0.0084 | 524 |
| **Yield** | 0.012 | 327 | 0.009 | 355 | 0.009 | 228 | 0.012 | 610 |
| **Max.** | 0.059 | 392 | 0.033 | 399 | 0.033 | 302 | 0.113 | 712 |
| **Failure** | 0.288 | 416 | 0.150 | 399 | 0.150 | 302 | 0.182 | 690 |

Table 2: Dimensions of the modelde beam, as defined in Figure 8.

| Variable | $L_r$ | $L_f$ | $L_b$ | $H$ | $H_c$ | $B$ | $B_c$ |
|---|---|---|---|---|---|---|---|
| Value | 0.2 m | 0.876 m | 3.028 m | 0.483 m | 0.423 m | 0.3 m | 0.242 m |

Table 3 a: Beam – end displacements in terms of the overall beam length, $L$. Point numbers refere to points in Figure 8.

| Point Number | Event | HPC Beam ($L_f * 1.0$) | HPC Beam with additional reinforcement ($L_f * 1.0$) |
|---|---|---|---|
| 1 | Beam cracking at section C | 0.17% $L$ | 0.17% $L$ |
| 2 | Beam cracking at section D | 0.24% $L$ | 0.24% $L$ |
| 3 | Fuse yielding at section E | 0.65% $L$ | 0.65% $L$ |
| 4 | Fuse yielding at section B | 0.74% $L$ | 0.74% $L$ |
| 5 | Fuse yielding at section C | **1.04%** $L$ | **3.52%** $L$ |

Table 3 b:  Beam – end displacements in terms of the overall beam length, $L$. Letters refer to points in Figure 11 a.  Beam sections are defined in Figure 8.

| Point | Event | Reference "Ispra" Beam |
|-------|-------|------------------------|
| a | Beam cracking at sections A and F | 0.07% $L$ |
| b | Beam yielding at section F | 0.94% $L$ |
| c | Beam yielding at section A | 1.07% $L$ |

Table 4:  Beam – end displacements at different loading stages in terms of the overall beam length, $L$.  Point numbers and letters refer to points in Figure 11 b. Section letters refer to sections defined in Figure 8.

| HPC beam ($L_f$*0.5) | | | HPC beam ($L_f$*1.5) | | |
|-------|-------|-----------|-------|-------|-----------|
| Point | Event | Top displ. | Point | Event | Top displ. |
| 1 | Beam cracking at section C | 0.11% $L$ | a | Beam cracking at section C | 0.30% $L$ |
| 2 | Beam cracking at section D | 0.13% $L$ | b | Beam cracking at section D | 0.43% $L$ |
| 3 | Beam yielding at section C | **0.48%** $L$ | c | Fuse yielding at section E | 0.78% $L$ |
| 4 | Fuse yielding at section E | 0.56% $L$ | d | Fuse yielding at section B) | 0.80% $L$ |
| 5 | Beam yielding at section D | **0.73%** $L$ | | | |

Table 5:  Building periods before and after dynamic step-by-step analysis. Accelerogram S7 scaled 1.5 times was used.

| | R.C. Building | | HPCF Building | | |
|----|---------|------------------------------|---------|------------------------------|------------------------------|
| | Elastic | After nonlinear dyn. analysis | Elastic | After nonlinear dyn. analysis | After nonlinear dyn. analysis |
| X1 | 0.56 sec. | 1.24 sec. | 0.55 sec. | 0.60 sec. | 0.67 sec. |
| X2 | 0.18 sec. | 0.36 sec. | 0.17 sec. | 0.18 sec. | 0.19 sec. |
| X3 | 0.10 sec. | 0.20 sec. | 0.09 sec. | 0.10 sec. | 0.10 sec. |

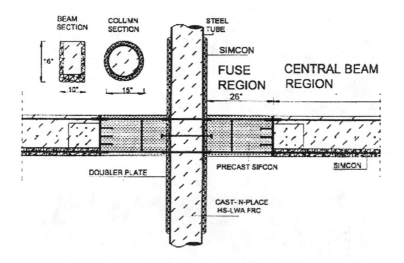

Figure 1:  Layout of the HPCF beam - column connection.

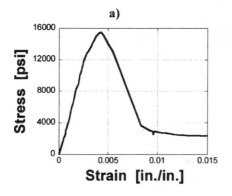

Figure 2:   Compressive response of HS-LWA concrete and HS-LWA FRC made with (a) 1.5 % and (b) 2 % fiber volume fraction.  (1,000 psi = 6.9 MPa)

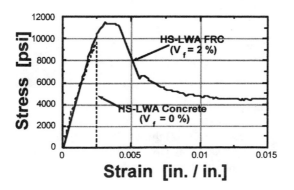

a)

b)

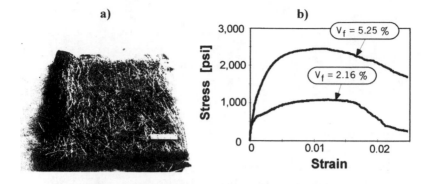

c)

Figure 3:    (a) Continuous fiber-mats used in manufacturing of SIMCON (courtesy of Ribbon Technology Corp., Inc., Ohio).  (b) Behavior of SIMCON in direct tension.  (c) The 2.5 cm (1 in.) thick 1.52 m (5 ft) long, 40.6 cm (16 in.) high and 25.4 cm (10 in.) wide SIMCON stay-in-place formwork.  Shear studs that were bolted directly into the formwork sides are clearly visible.  (1,000 psi = 6.9 MPa)

**a)**

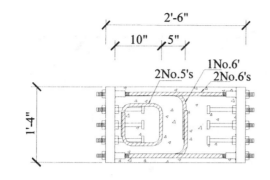

Figure 4:   (a) Layout and (b) load-displacement response of a representative SIFCON fuse specimen [19]. Specimen width was 10 in. (1 in. = 2.54 cm, 1,000 lb – in = 1.13 kNm)

**b)**

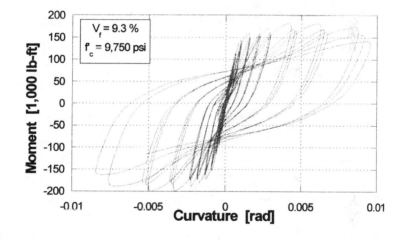

## GENERAL LAYOUT

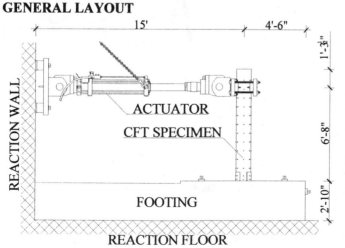

Figure 5 a: Test setup for evaluation of HS-LWA FRC filled CFT specimen behavior. (1 in. = 2.54 cm, 1,000 lb = 4.45 kN)

**b)**

# HORIZONTAL SECTION

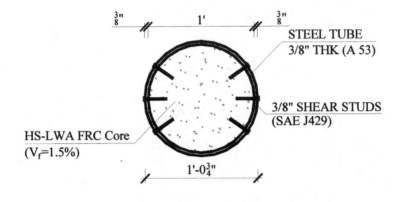

**c)**

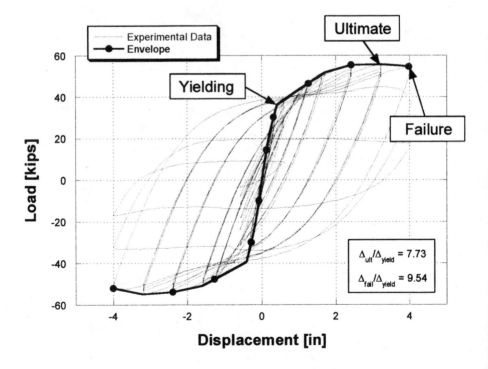

Figure 5 b: Specimen cross-section, and **(c)** load-displacement of HS-LWA FRC filled CFT specimen.  (1 in. = 2.54 cm, 1,000 lb = 4.45 kN)

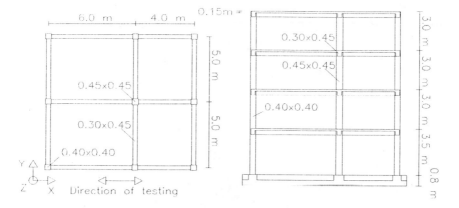

Figure 6:  Reference R.C. ("Ispra") building [22].  All dimensions are in meters. (1 m = 39.4 in. = 3.28 ft)

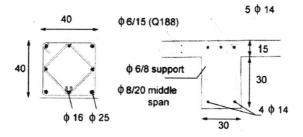

Figure 7:   Layout of corner columns and beam sections of the reference R.C. ("Ispra") building.  (1 cm = 0.39 in)

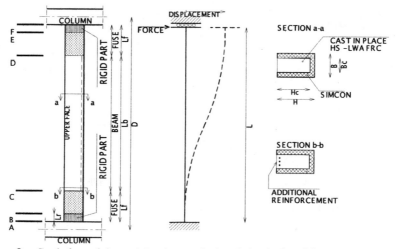

Figure 8:   Scaled model used in the analysis of the isolated beam.  Specific beam dimensions are listed in Table 2.

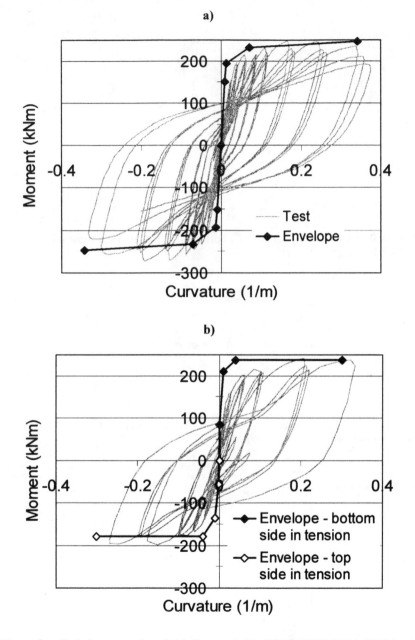

Figure 9:  Scaled test results of (**a**) fuse and (**b**) SIMCON/HS-LWA FRC beam member response under reversed cyclic loading, and moment-curvature envelopes used in nonlinear analysis.

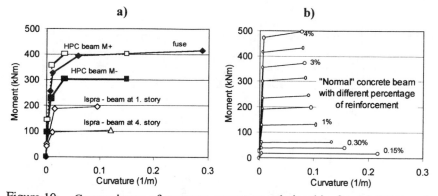

Figure 10:   Comparisson of moment curvature relationship for (a) HPC and reference R.C. "Ispra" beams, as well as for (b) reference R.C. beams with different procentage of reinforcement (symmetrically positioned reinforcement, top=bottom, C25/30).  (1 m = 39.4 in. = 3.28 ft, 1 kNm = 885 lb-in)

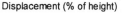
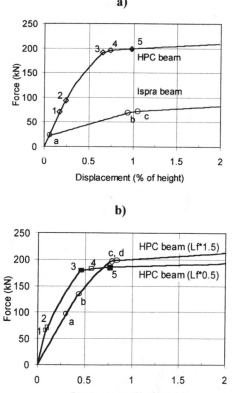

Figure 11:   Force-displacement relationship for HPC and Ispra beam.  Specific displacement values are listed in Tables 3 a, b, and 4.  (1 kN = 224.7 lb)

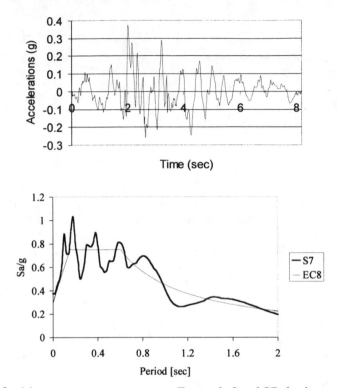

Figure 12:  **(a)** Artificial accelerogram S7. **(b)** Eurocode 8 and S7 elastic spectra.

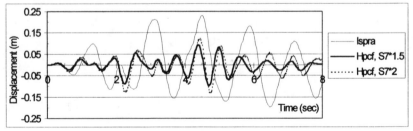

Figure 13:  Top displacement time history.  (0.1 m = 3.94 in)

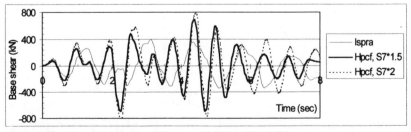

Figure 14:  Base shear time history.  (1 kN = 224.7 lb)

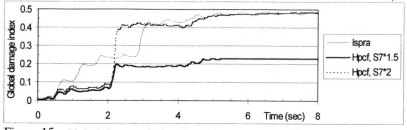

Figure 15:  Global damage index time history.

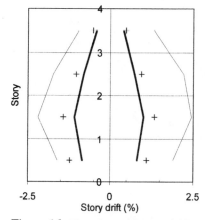

Figure 16: Envelope of story drifts
for the reference and HPCF building.

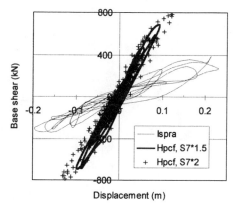

Displacement (m)

Figure 17:  Base shear – top displacement
relationship.  (1 kN = 224.7 lb, 1 m = 3.28 ft)

a)

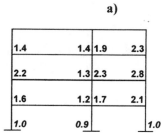

b)

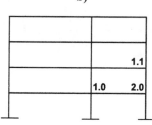

c)

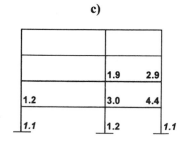

Figure 18:  Rotational ductility factors
for (a) the reference ("Ispra") building
subjected to 1.5 times S7
accelerogram, and the HPCF building
subjected to (b) 1.5 and (c) 2 times S7
accelerogram.  Factors equal or bigger
than 1.0 indicate plastic damage.

# THE FUTURE OF RESEARCH AND EDUCATION IN CONCRETE

# Applications of Fracture Mechanics to Concrete: Where Do We Go From Here?

## by S. Mindess

Synopsis: Fracture mechanics concepts were first applied to concrete in 1959, and since then several thousand papers have been published on this topic. Professor Shah himself has contributed many papers in this area, dating back to 1971. The first attempts at utilizing fracture mechanics in concrete research dealt with linear elastic fracture mechanics. However, it soon became apparent that this was insufficient to characterize a heterogeneous, non-linear elastic material such as concrete. Thus, a variety of non-linear fracture mechanics models were developed to try to better describe the fracture and failure of concrete. Unfortunately, despite a great deal of research, both theoretical and experimental, fracture mechanics concepts are still seldom used in the design of concrete structures, particularly in North America. For instance, they are not even mentioned in the current ACI 318 - *Building Code Requirements for Structural Concrete,* or in the Canadian CSA Standard A23.3 - *Design of Concrete Structures.* It is the purpose of this paper to review briefly the current position of fracture mechanics in concrete applications, and then to look to possible future developments.

Keywords: fracture mechanics; linear elastic fracture mechanics; non-linear fracture mechanics; reinforced concrete

S. Mindess has been a Professor of Civil Engineering at the University of British Columbia, Canada, since 1969. He is the author or co-author of over 250 technical publications, dealing primarily with the cracking and fracture of concrete and fiber reinforced concrete, and with the properties of concrete under impact loading. He is a Fellow of the American Concrete Institute, of the American Ceramic Society, and of RILEM.

## INTRODUCTION

The cracking and fracture of concrete have been studied seriously since at least 1928, when Richart, Brandtzaeg and Brown (1) published their first investigations into the development of cracks in concrete under load. The first direct application of fracture mechanics concepts to concrete was by Neville (2) in 1959, who attempted to relate the dependence of concrete strength on size to the distribution of Griffith flaws. The first experimental study to measure the linear elastic fracture parameters of concrete was carried out by Kaplan (3) in 1961. Since that time, several thousand papers dealing with the applications of fracture mechanics to concrete have been published. There is now general agreement that a fracture mechanics approach is necessary to analyze or predict the crack patterns that develop in concrete due to the application of stress. This is because crack extension requires a certain amount of energy (the fracture energy), and thus an energy criterion, which is fundamental to fracture mechanics, should be used.

The question then remains: Why are fracture mechanics concepts so little used in the analysis and design of concrete structures, particularly in North America? Why are these concepts not even mentioned in the two principal North American design codes, ACI 318, *Building Code Requirements for Structural Concrete,* (despite the fact that there has been, since 1985, an active ACI Committee 446 - Fracture Mechanics) and the Canadian Standard *A23.3, Design of Concrete Structures?* In what follows, the current use of fracture mechanics in concrete design, and possible future developments, will be discussed.

There is, unfortunately, not enough space available here to describe in detail the many contributions of Professor Shah to fracture mechanics. Suffice it to say that his first paper specifically applying fracture mechanics to concrete was published in 1971 (4), and since then he (and his students) have carried out countless studies, both experimental and analytical, on the applications of fracture mechanics to cementitious materials, only a few of which will be referred to below. We are all in his debt for the many insights he has provided.

FRACTURE MECHANICS OF CONCRETE:
CURRENT STATUS

The earliest studies of the application of fracture mechanics to concrete dealt primarily with Linear Elastic Fracture Mechanics (LEFM). This work has been reviewed in detail by Mindess (5-8). In his pioneering study, Kaplan (3) concluded that "the Griffith concept of a critical strain-energy release rate being a condition for rapid crack propagation and consequent fracture, is applicable to concrete." This view was generally accepted at the time, and led to numerous experimental determinations of $K_c$ and/or $G_c$. However, a number of problems with these measurements soon became apparent:

- There was great variability in the values of $K_c$ (or $G_c$) obtained by different investigators.
- The experimental results were strongly dependent upon the specimen size and geometry.
- It was found that there was an increased energy demand as crack growth progressed.
- There was a realization that some sort of "fracture process zone" (or zone of diffuse microcracking) developed around the apparent crack tip as it propagated, creating a much larger fracture surface than would be calculated from the specimen dimensions.

This led Kesler *et al.* (9) to conclude that "the concepts of linear elastic fracture mechanics are not directly applicable to cement pastes, mortars and concretes," and this is the view that now generally prevails.

That LEFM does not apply directly to concrete is not, of course, surprising. LEFM can be used only for very *brittle* materials, for which there is little or no non-linear (inelastic) behaviour up to the onset of fracture. To model the behaviour of concrete, however, a detailed knowledge of the post-peak $\sigma$-$\varepsilon$ and/or $\sigma$-$\delta$ relationships, and of the size and properties of the fracture process zone, must also be known. One fact becomes readily apparent – concrete failure exhibits a strong size effect. This may be seen in Fig. 1(10), in which the relationship between stress and deflection, as a function of specimen size, is shown schematically. Fig. 2 (10) shows the relationship between the logarithm of strength and the logarithm of specimen size; for large sizes (such as dams), LEFM applies, while for small sizes (typical of laboratory specimens), a strength criterion applies Most real structural elements lie between these two sizes. As well, as shown in Fig. 3 (10), small structures appear to be more "ductile" than large structures.

Thus, only for extremely large concrete structures, in which the cracks are very large, and the size of the fracture process zone is small compared to the overall dimensions of the structure, can LEFM be applied in a sensible fashion. For instance, Chappell and Ingraffea (11) were able to use LEFM to analyze the Fontana gravity dam, while Linsbauer *et al.* (12, 13) used LEFM to simulate the cracking in the Kölnbrein arch dam in Austria.

As a result, for about the last thirty years, the principal work on the application of fracture mechanics to concrete has been in the development of appropriate non-linear fracture mechanics (NLFM) models. (Regrettably, over that period, a considerable effort still continued, and continues to this day, on applications of LEFM to small concrete specimens). A number of NLFM models have been proposed, all of which give reasonably good descriptions of concrete behaviour if applied properly. Some of these are described very briefly below.

Perhaps the first NLFM model to receive extensive development was the *fictitious crack-model* of Hillerborg and his co-workers (14). This is a *cohesive crack-model,* in which it is assumed that some stress can still be transferred across a crack near its tip, even as the crack extends. That is, there is a zone of microcracking around the crack tip. Using this model, both crack initiation and crack growth can be predicted. Hillerborg (15) later used this model to develop a method for determining the fracture energy ($G_f$) of concrete, which became the basis of a RILEM test recommendation (16).

Other cohesive crack models have also been developed, all of which simulate in one way or another the nonlinear behaviour of the concrete near the apparent crack tip. These models vary from each other in the way in which the crack closing mechanisms are defined physically: microcracking and aggregate interlock in concrete, fibres bridging across the crack in fibre reinforced concrete, and so on. As well, the models differ in the way in which the (post-peak) strain softening is defined. Some investigators have modeled the curves as being bilinear (e.g., Petersson (17), Wittmann *et al.* (18)). Others have used smooth continuous curves (e.g., Gopalaratnam and Shah (19), Planas and Elices (20)). However, there is still no generally accepted mathematical formulation of strain softening, although the curves referred to above are quite close to each other, and all provide a reasonable fit with the available experimental data.

A different approach to modeling the non-linear behaviour of concrete is through the use of *equivalent crack models,* such as the Jenq and Shah (21, 22) *two-parameter model*. In this model, failure (or unstable crack growth) is defined in terms of a critical crack tip opening displacement ($CTOD_c$) and the critical stress intensity factor, $K_{Ic}$, at the equivalent LEFM crack tip, as shown in Fig. 4 (10). This model was later extended to handle impact loading (23) and mixed mode loading (24). The two parameters may be obtained experimentally as described in a RILEM recommendation (25).

Still another approach has been the use of the size effect law proposed by Bazant and his co-workers (26, 27) to determine the fracture parameters of concrete (28). This technique permits the prediction of the load-displacement behaviour of a cracked specimen or a structure.

### FRACTURE MECHANICS IN CONCRETE STRUCTURAL DESIGN

It is clear that, particularly for structures subjected to dynamic loading, the energy required to propagate a crack must be taken into account in the design. The ductility (or energy absorption capability) of a structure must also be included in the analysis. Finally, the size effect must be considered in proper structural design. Unfortunately, current strength-based design procedures do none of these in an objective way

From the brief account above, it may be seen that we are now at the stage at which we understand reasonably well the physical mechanisms by which fracture takes place in concrete, and we have at our disposal the analytical tools of NLFM to permit us to predict both material and structural behaviour. We also have methods for determining the relevant fracture mechanics parameters. Why, then, do we not use fracture mechanics in North America in the design of concrete structures? Indeed, the author has been unable to find a single North American textbook on the design of concrete structures in which the phrase "fracture mechanics" is even mentioned!

It is not that researchers have not examined the practical use of fracture mechanics in design. If one were to examine, some of the publications dealing with fracture mechanics applications to reinforced concrete design (eg., 29, 30) one would find papers dealing with flexure, shear, torsion, punching shear, bond with reinforcing steel, anchorage to concrete, pipes, dams, joints, slabs, cyclic loading, and so on. There is even a formal introduction of fracture mechanics data in the CEB-FIP Model Code (31), though the code itself remains strength based. Why is there resistance to using this information?

There appear to be a number of reasons for this state of affairs:

1.  For fracture mechanics concepts to be incorporated into structural analysis and design, there must be some unambiguous way of measuring fracture parameters such as $K_{Ic}$ or $G_{Ic}$. While a number of techniques have been proposed for such determinations, the different techniques often lead to quite different results. As yet, there is no single standardized test in North America for the measurement of facture parameters. [The CEB-FIP Model code (31) does not give an empirical equation for $G_{Ic}$, as

$$G_{Ic}=a_d f_c^{0.7} \tag{1}$$

where $a_d$ is a coefficient that depends upon the maximum aggregate size, and $f_c$ is the concrete compressive strength]. Until some standard tests are developed, and achieve general acceptance, there will be no rational way of incorporating fracture mechanics into concrete design.

2.  In North America, Civil Engineering undergraduate programs rarely even mention the science of fracture mechanics, except perhaps for a couple of introductory lectures in a materials science course. Even at the graduate level, fracture mechanics, if it is taught at all, is taught in terms of being a *research* tool, not a *design* tool. Thus, before we can hope to introduce fracture mechanics concepts into design, we must first educate our students in both the theory and application of both LEFM and NLFM, just as we currently teach them limit state design.

3.  North American civil engineers are entirely fixated (obsessed?) by design based entirely on strength considerations; the concrete compressive strength is deemed to be the prime indicator of when concrete will fail. It is currently

difficult, if not impossible, to convince them to look at energy considerations in addition to strength.

4.  The current North American design codes are, quite properly, conservative documents. Since they are, in effect, consensus documents, they are difficult to change in any fundamental way. In any event, the current strength-based theories for the design of reinforced and prestressed concrete structure (while semi-empirical) are rational, powerful, and relatively easy to use. Fracture mechanics will be successfully introduced into concrete design codes only if it can be shown to be as easy to apply as the existing approach, and if it can be used to solve problems that cannot be handled properly by the existing methods. For instance, while crack widths and crack spacing in ordinary reinforced concrete structures can be predicted well using the conventional approach, crack propagation and crack patterns still cannot be properly predicted.

5.  As an academic community, we have tended to focus on our own individual fracture mechanics research topics. We have not tried very hard either to force more fracture mechanics into our respective curricula, nor have we been effective in introducing these concepts to the design community. The various reports of ACI Committee 446 - Fracture Mechanics appear to be largely ignored outside of the committee itself

I must admit, of course, that the current strength based design methods have, on the whole, been successful. They produce structures that are both safe and economical. However, our challenge now is to show that there are still problems that can be better handled by the fracture mechanics approach. Can this be considered fundamental research? Probably not. Nonetheless, we should not shrink from the task of carrying out enough experimental work of an applied nature to convince our colleagues that fracture mechanics does have a place in concrete design.

## REFERENCES

1.  Richart, F. E., Brandtzaeg, A. and Brown, R. L., "A Study of the Failure of Concrete under Combined Compressive Stresses," *Bulletin No. 185,* Engineering Experiment Station, University of Illinois, 1928.

2.  Neville, A. M., "Some Aspects of the Strength of Concrete," *Civil Engineering* (London), Part 1: Vol. 54, 1959, pp. 1153-1156; Part II: Vol. 54, 1959, pp. 1308-1310; Part III: Vol. 54, 1959, pp. 1435-1439.

3.  Kaplan, M. F., "Crack Propagation and the Fracture of Concrete," *Journal of the American Concrete Institute, Vol. 58,* Nov. 1961, pp. 591-610.

4.  Shah, S. P. and McGarry, F. J., "Griffith Fracture Criterion and Concrete," *Journal of the Engineering Mechanics Division,* ASCE, Vol. 97, EM6, 1971, pp. 1663-1676.

5. Mindess, S., "The Application of Fracture Mechanics to Cement and Concrete: A Historical Review," pp. 1-30 in F. H. Wittmann, ed., *Fracture Mechanics of Concrete,* Elsevier, Amsterdam, 1983.

6. Mindess, S., "The Cracking and Fracture of Concrete: An Annotated Bibliography, 1928-1981," pp. 539-661 in F. H. Wittmann, ed., *Fracture Mechanics of Concrete,* Elsevier, Amsterdam, 1983.

7. Mindess, S., "The Cracking and Fracture of Concrete: An Annotated Bibliography, 1982-1985," pp. 627-699 in F. H. Wittmann, ed., *Fracture Toughness and Fracture Energy of concrete,* Elsevier, Amsterdam, 1986.

S. Mindess, S., " Fracture Toughness Testing of Cement and Concrete," pp. 67-110 in A. Carpinteri and A. R. Ingraffea, eds., *Fracture Mechanics of Concrete,* Martinus Nijhoff Publishers, The Hague, 1984.

9. Kesler, C. E., Naus, D. J. and Lott J. L., "Fracture Mechanics - Its Applicability to Concrete," pp. 113-124, Vol. IV in *Proceedings of the International Conference on Mechanical Behaviour of Materials,* Kyoto, 1971. The Society of Materials Science, Japan, 1972.

10. ACI Committee 446, Fracture Mechanics of Concrete: Concepts, Models and Determination of Material Properties, ACI 446.IR-91, American Concrete Institute, Farmington Hills, Michigan, 1991.

11. Chappell, J. F. and Ingraffea, A. R., *A Fracture Mechanics Investigation of the Cracking of Fontana Dam,* Report 81-7, School of Civil and Environmental Engineering, Cornell University, Ithaca, New York, 1981.

12. Linsbauer, H. N., Ingraffea, A. R., Rossmanith, H. P. and Wawryznik, P. A., "Simulation of Cracking in Large Arch Dam: Part I," ASCE *Journal of Structural Engineering,* Vol. 115, No. 7, 1989, pp. 1599-1615.

13. Linsbauer, H. N., Ingraffea, A. R., Rossmanith, H. P. and Wawryznik, P. A., "Simulation of Cracking in Large Arch Dam: Part II," ASCE *Journal of Structural Engineering,* Vol. 115, No. 7,1989, pp. 1615-1630.

14. Hillerborg, A., Modéer, M. and Petersson, P. E., "Analysis of Crack Formation and Crack Growth in Concrete by Means of Fracture Mechanics and Finite Elements," *Cement and Concrete Research,* Vol. 6, No. 6, 1976, pp. 773-782.

15. Hillerborg, A., "The Theoretical Basis of a Method to Determine the Fracture Energy $G_F$ of Concrete," *Materials and Structures (RILEM),* Vol. 18, 1985, pp. 291-296.

16. RILEM, "Determination of the Fracture Energy of Mortar and Concrete by Means of Three-Point Bend Tests on Notched Beams," [RILEM Draft Recommendation, TC 50-FMC Fracture Mechanics of Concrete], *Materials and Structures (RILEM),* Vol. 18,1985, pp. 285-290.

17. Petersson. P. E., *Crack- Growth and Development of Fracture Zone in Plain Concrete and Similar Materials.* Report No. TVBM-1006, Division of Building materials, Lund Institute of Technology, Sweden, 1981.

18. Wittinann, F. H., Rokugo, K, Brühwiler, E., Mihashi. H. and Simonin, P., "Fracture energy and Strain Softening of Concrete as Determined by Means of Compact Tension Specimens," *Materials and Structures (RILEM),* Vol. 21, 1988, pp. 21-32.

19. Gopalaratnam, V. S., and Shah, S. P., "Softening Response of Plain Concrete in Direct Tension," *Journal of the American Concrete Institute,* Vol. 82, No. 3, March 1985, pp. 310-323.

20. Planas, J., and Elices, M., "Towards a Measure of $G_F$. An Analysis of Experimental Results," pp. 381-390 in F. H. Wittmann, ed., *Fracture Toughness and fracture Energy of Concrete,* Elsevier, Amsterdam, 1986.

21. Jenq, Y.S. and Shah, S. P., "A Fracture Toughness Criterion for Concrete," *Engineering Fracture Mechanics,* Vol. 21, No. 5, 1985, pp. 1055-1069.

22. Jenq, Y. S., and Shah, S. P., "Two Parameter Fracture Model for Concrete, ASCE *Journal of Engineering Mechanics,* Vol. 111, N.. 10, 1985, pp. 1227- 1241.

23. John, R. and Shah, S. P., "Fracture of Concrete Subjected to Impact Loading," *Cement, Concrete and Aggregates,* Vol. 8, No. 1, 1986, pp. 24-32.

24. John, R. and Shah, S. P., "Mixed Mode fracture of Concrete Subjected to Impact Loading," ASCE *Journal of Structural Engineering,* Vol. 116, No. 3, 1990, pp. 585-602.

25. RILEM, Determination of the Fracture Parameters ($K^s_{IC}$ and $CTOD_c$) of Plain Concrete Using Three-Point Bend Tests on Notched Beams," (RILEM Draft Recommendation, TC 89-FMT Fracture Mechanics of Concrete - Test Methods), *Materials and Structures (RILEM),* Vol. 23, 1990, pp. 457-460.

26. Bazant, Z. P., "Size Effect in Blunt Fracture. Concrete, Rock, Metal," ASCE *Journal of Engineering Mechanics,* Vol. 110, 1984, pp. 518-535.

27. Bazant, Z. P. and Planas, J., *Fracture and Size Effect in Concrete and Other Quasibrittle Materials,* CRC Press, Boca Raton, Florida, 1998.

28.   RILEM, "Size-Effect Method for Determining Fracture energy and Process Zone Size of Concrete," (RILEM Draft Recommendation, TC 89-FMT Fracture Mechanics of Concrete - Test Methods), *Materials and Structures (RILEM),* Vol. 23, 1990, pp. 461-465.

29.   Gerstle, W. and Bazant, Z. P. (eds.), *Concrete Design Based on Fracture Mechanics,* ACI SP-134, American Concrete Institute, Farmington Hills, Michigan, 1992.

30.   Elfgren, L. (ed.), *Fracture Mechanics of Concrete Structures,* Report of RILEM TC 90-FMA Fracture Mechanics of Concrete - Applications, Chapman and Hall, London, 1989.

31.   CEB, "CEB-FIP Model Code 1990, final draft," pp. 203-205 in *Bulletin d'Information du Comité Euro-International du Béton,* 1991.

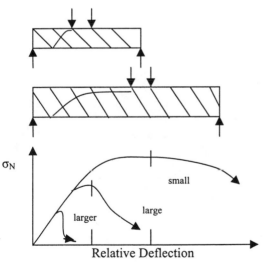

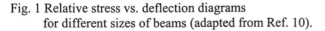

Fig. 1 Relative stress vs. deflection diagrams
for different sizes of beams (adapted from Ref. 10).

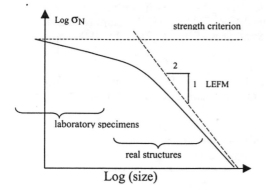

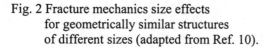

Fig. 2 Fracture mechanics size effects
for geometrically similar structures
of different sizes (adapted from Ref. 10).

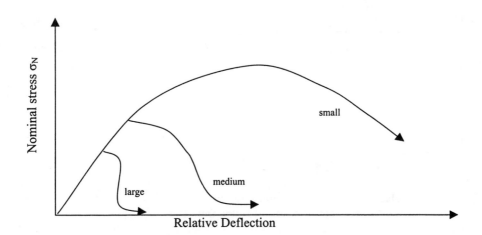

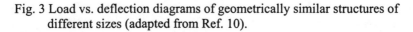

Fig. 3 Load vs. deflection diagrams of geometrically similar structures of
different sizes (adapted from Ref. 10).

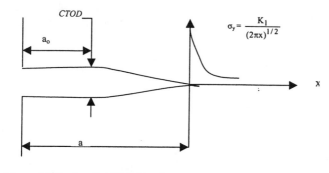

(a)     Effective Griffith Crack

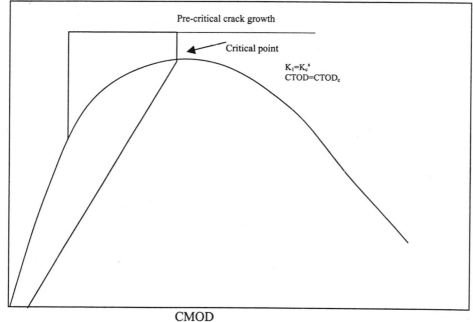

CMOD
(b) Load vs. Crack Mouth Opening Displacement

Fig. 4 Jenq and Shah (22) two-parameter fracture model for concrete.

# Vertex Effect and Confinement of Fracturing Concrete Via Microplane Model M4

## by Z. P. Bažant, F. C. Caner, and J. Červenka

**Synopsis:** A newly developed powerful version of microplane model, labeled model M4, is exploited to study two basic phenomena in fracturing concrete: (a) The vertex effect, i.e., the tangential stiffness for loading increments to the side of a previous radial loading path in the stress space, and (b) the effect of confinement by a steel tube or a spiral on the suppression of softening response of columns. In the former problem, the microplane model is used to simulate the torsional response of concrete cylinders after uniaxial compression preloading to the peak compression load or to a post-peak softening state. Comparisons with new tests carried out at Northwestern University show the microplane model to predict the initial torsional stiffness very closely, while the classical tensorial models with invariants overpredict this stiffness several times (in plasticity of metals, this phenomenon is called the 'vertex effect' because its tensorial modeling requires the yield surface to have a vertex, or corner, at the current state point of the stress space). In the latter problem, microplane model simulations of the so-called 'tube squash' tests are presented and analyzed. In these tests, recently performed at Northwestern University, steel tubes of different thicknesses filled by concrete are squashed to about half of their initial length and very large strains with shear angles up to about 70 degrees are achieved. The tests and their simulations show that, in order to prevent softening (and thus brittle failure and size effect), the cross section of the tube must be at least 16% of the total cross section area, and the volume of the spiral must be at least 14% of the volume of the column. When these conditions are not met, which comprises the typical contemporary designs, one must expect localization of damage and size effect to take place.

**Keywords:** stiffness; strain; uniaxial stress; vertex effect

[1]Reprinted with permission, in a slightly revised form, from the Proceedings of the 4th International Conference on Fracture Mechanics of Concrete and Concrete Structures (FraMCoS-4, Cachan, France), R. de Borst et al., eds., A.A. Balkema, Publishers, pp. 497–504.

**Z.P. Bažant**, F. ACI, M. NAE, is a W.P. Murphy Professor of Civil Engineering and Materials Science at Northwestern University.

**F.C. Caner**, a former Graduate Research Assistant at Northwestern University, is an Assistant Professor of Civil Engineering at MKU Muhendislik ve Mimarlik Fakultesi, 31230 Iskenderun, Turkey.

**J. Červenka**, a former Visiting Scholar at Northwestern University, is a Research Engineer and Partner, Cervenka Consulting, Prague, Czech Republic.

## NATURE OF VERTEX EFFECT AND ITS THEORETICAL IMPLICATIONS

As a convenient way of avoiding material instability, the incremental theory of plasticity is based on Drucker's postulate. This postulate requires the inelastic strain increment vector to be normal to the yield surface in the superposed nine-dimensional spaces of stress and strain components. Adherence to the normality rule (expounded, e.g., in Bažant and Cedolin 1991, ch. 10) is the basic feature of all the practical constitutive relations for plasticity of metals. The normality rule is also adopted for the fracturing and plastic-fracturing constitutive models for concrete with loading surfaces in the strain space (Lin et al. 1997, Červenka et al. 1998).

The normality rule implies that, for load increments that are parallel to the current yield surface (or loading potential surface), called the 'loading to the side', the response is purely elastic. Such behavior was hinted by tests of plastic buckling of plates by Bleich (1952) and was shown patently untrue for metals by Gerard and Becker (1952, 1957) and Phillips and Gray (1961). Gerard and Becker tested axially compressed thin-walled cruciform steel columns that buckle in the plastic range by torsion. The critical load of such columns is proportional to the tangential inelastic stiffness for loading to the side, and it was found that it can be much smaller than the critical load for the elastic stiffness, as small as one half of the elastic critical load (Bažant and Cedolin 1991, Sec. 8.1; in detail Brocca and Bažant 2000).

The existence of inelastic strain increments for loading to the side implies that there must be a corner, or vertex, on the yield surface (or loading potential surface) at the current state point, traveling during loading with the state point as the material hardens or softens. Therefore, the phenomenon is called the 'vertex effect'. The vertex effect is the strongest for abrupt changes of loading direction in the stress space or strain space, but arises for all loading paths (even smooth paths) significantly deviating from proportional loading (radial loading in the stress or strain space).

## MEASUREMENTS OF VERTEX EFFECT IN CONCRETE

To examine the vertex effect in concrete, one may use a loading path in which a uniaxial compressive strain is initially applied. No shear strains develop during this loading. A sudden imposition of incremental shear strain, introduced by torsion, represents 'loading to the side' (relative to the current loading surface). The experiments were designed for cylindrical specimens so that they can be tested in an axial-torsional testing machine. The load path is prescribed in the space of axial displacement versus rotation, and the response in the post-peak regime is of main interest. The loading path includes a sharp corner. A rounded corner could be avoided thanks to the availability of a state-of-the-art testing machine. The specimen was compressed under uniaxial stress (at zero rotation) to strain $\epsilon = 0.2\%$ (which is the axial strain at maximum uniaxial stress $\sigma = f'_c$ for a typical concrete). This was followed by concentric rotation at constant axial strain $\epsilon = 0.2\%$ until the peak torque is reached. Cylindrical specimens of concrete with diameter $D$=4 in (101.6 mm) and height $H$=8 in (203.2 mm) were used.

When a sudden switch from uniaxial compressive loading to torsional loading was made at prepeak stresses up to about 80% of peak, the incremental response was nearly elastic and nearly path-independent. Therefore the vertex effect was tested only in peak and postpeak regions. Concrete was compressed axially at zero rotation to average strain 0.45%, which was followed by concentric rotation until the peak torque was reached while keeping the average axial strain constant.

The specimen was embedded into steel platens at the top and bottom ends to a depth of 12.7 mm. Concrete was glued to the platens by using a high-modulus epoxy (SIKADUR 32 HI–MOD). This means that the ends of the specimen were confined and so the failure processes take place not at the ends but within the gauge region. Two linear variable differential transformers (LVDT's) for axial loading and two LVDT's for rotational loading were used. These LVDT's were fixed to the two steel rings attached directly to the specimen at the top and bottom of the gauge volume, each by four screws. The LVDT's were used for measuring the axial displacements over a gauge length of $L = 114.3$ mm. The distance between the rotational LVDT's was 95.25 mm. For detailed descriptions and figures portraying the instrumentation and the tests, see Caner et al. (2000b).

The tests were performed using an MTS 220 kip = 976.8 kN axial–torsional testing machine operated by digital closed loop control. The stiffness of the machine and the feedback sufficed to keep the present test stable. The fact that the real columns in structures are not loaded under displacement control does not detract from practical relevance of the results. The displacement control which stabilizes the test is merely a way to measure the tangential stiffness in the postpeak softening regime, and the same stiffness then governs the behavior or real columns which is of course dynamic in the postpeak regime, e.g., the response in earthquake cycles reaching into that regime.

Fig. 1 shows a plot of the initial incremental torsional stiffness when the torsional loading started, divided by the elastic torsional stiffness. The results are compared to the prediction of the microplane model M4 (whose parameters were not adjusted to optimize the fit). To get realistic simulations, three-dimensional finite element analysis of the test cylinder was carried out using microplane model M4 (Bažant et al. 2000a) (Fig. 2). For the sake of comparison, the finite element analysis with one of the sophisticated advanced damage-plasticity models for concrete formulated in the classical way, in terms of stress and strain tensors and their invariants (Červenka et al. 1998). A mesh of 2484 elements and 3000 nodes was used, covering only the top half of the specimen since symmetric behavior of the bottom half may be assumed. The boundary conditions at the top of the cylinder were prescribed as the axial displacements and horizontal displacements due to rotation of the platens considered as a rigid body. In each loading step, the boundary displacement and rotation were adjusted so as to match the relative displacements and rotations recorded during the tests at the attached steel rings carrying the LVDT's.

## OBSERVATIONS AND CONCLUSIONS FROM VERTEX STUDY

1. By using a state-of-the art testing machine capable of a sudden switch from compression to torsion and on-specimen gauges with a fast feedback, the vertex effect in the response of concrete to nonproportional loading paths in the stress space has been documented experimentally.

2. At peak compressive load, the vertex effect is strong, and in post-peak very strong. Compared to the elastic torsional stiffness, the initial torsional stiffness after a sudden switch from compression to torsion is reduced to 65% when the torsion begins at compression load peak, and to 23% when the torsion begins at a postpeak state at which the axial load has decreased to 70% of the peak load.

3. The experimental data obtained are modeled using two state-of-the-art but conceptually completely different models. One is microplane model M4, and the other is a fracture-plastic model, a state-of-art tensorial model based on invariants.

4. The initial torsional stiffness after a sudden switch from compression to torsion is predicted by microplane model M4 quite accurately, and without any adjustment of the material parameters previously calibrated by other tests. This demonstrates the microplane model M4 can predict the vertex effect, and does so correctly.

5. The capability of predicting the vertex effect is due to fact that the model implies many simultaneous, independently activated (strain-dependent) yield surfaces on the microplanes and has also independent yield surfaces for volumetric, deviatoric and shear response on each microplane. It is the interaction of these surfaces that produces the vertex effect.

6. The classical invariant-based tensorial models employing only a few yield (or loading potential) surfaces, represented in the present simulation by an advanced model, the fracture-plastic model, are inherently incapable of simulating the vertex effect, and more generally the response to highly nonproportional loading paths. This is documented by the fact that they incorrectly predict the initial torsional stiffness after the switch to be the elastic torsional stiffness.

7. The microplane model prediction of response to torsional loading after a sudden switch from compression to torsion is less accurate but better than the fit with the plastic-fracture model. It must be emphasized that the former is a true prediction, with no adjustment of material parameters previously calibrated by other tests, while the latter is a fit obtained after adjusting some material parameters in the fracture-plastic model.

## NATURE OF CONFINEMENT EFFECT ON DUCTILITY AND THEORETICAL APPROACH

As generally accepted, safe design of concrete columns, beams, piles and other structures requires ensuring a ductile response. This is especially important for columns of tall buildings or bridges. It is well known that from the viewpoint of ductility, circular steel spirals are far preferable to rectangular ties, and that achievement of high ductility calls for confining concrete in steel tubes, as recently used in some very tall concrete buildings. Confinement by circular winding with steel wires or strong fibers, or with bonded fiber-polymer composites, has been shown to be effective for repairing concrete columns or beams damaged by earthquake or other catastrophic events.

Ductility is understood as a plastic behavior, or absence of brittleness. In mechanics terms, the response of a structure is brittle when the tangential stiffness $K_t$ becomes negative, in other words, when the structure undergoes softening (decrease of load at increasing deflection). More precisely, the ductility of a structure is lost when the tangential stiffness matrix $\boldsymbol{K}_t$ ceases being positive definite.

The material at a point of the structure looses ductility (plasticity) and becomes brittle (or quasi-brittle) locally when strain softening begins, i.e., when the tangential moduli matrix $\boldsymbol{E}_t$ ceases being positive definite. It is well established that when the material is softening at some point of the structure, the inelastic deformation localizes and plastic limit analysis is inapplicable because the material strength is not mobilized simultaneously at various points of the structure; rather, localized damage zone propagates during loading, which eventually leads to fracture.

More seriously, when strain-softening develops and damage propagates, a deterministic (energetic) size effect is always present. Thus the failure loads seen in reduced-scale laboratory tests do not scale up for full-size real structures according to material failure criteria expressed in terms of the stress and strain tensors and a fracture mechanics type energy criterion of failure must be used. The larger the structure, the steeper its post-peak softening and the more prone the structure is to explosive dynamic failure driven by a sudden release of its stored energy. The post-buckling response of columns then becomes dynamic and much more sensitive to imperfections, which calls for using higher safety factors for larger structures.

Although the understanding of concrete ductility has advanced enormously, thanks to extensive testing in the aftermath of recent catastrophic earthquakes, a more accurate assessment of concrete ductility consistent with the current knowledge of damage and fracture mechanics is needed. To fill this gap, a recent experimental and theoretical investigation at Northwestern University (Caner et al. 2000b) attempts to clarify the minimum confinement necessary to completely prevent softening, and thus localization into distinct fractures and size effect.

A good historical account of investigation of mechanical behavior of concrete–filled tubular columns is given in Schneider (1998). Roeder et al. (1999) studied the bonding between concrete and the steel tube in a concrete–filled tubular column where the concrete core is loaded to slip against the confining tube. They found out that, average bond strength varied between 0.5 and 3.25 MPa for diameter to wall thickness ratios $D/t < 50$. Schneider (1998) studied the effect of the steel tube shape and the wall thickness on the yield strength of the columns and the confinement of concrete core using the orthotropic material model for concrete ABAQUS uses. None of these studies, however, were able to address the issue of confinement of concrete core in these columns rigorously because they did not use a realistic enough constitutive model.

Toward this goal, the recently developed tube-squash tests (Bažant Kim and Brocca 1999) have been conducted on concrete-filled tubes of various thicknesses and analyzed with a large-strain finite element code. A realistic finite element analysis has been made possible by employing microplane constitutive model M4 (Bažant et al. 2000a,b; Caner and Bažant 2000), which was verified by numerous test data and was originally developed for simulating the penetration of missiles into concrete walls and ground shock effects on buried structures (Bažant et al. 2000c). In this study, after verification by the tube-squash tests, the computational model is used to predict the behavior for various ratios of the cross section areas of steel tube and concrete filling. The predictions are further run for much longer tubes, and also for concrete columns confined by steel spirals.

The steel in the tube is simulated by a microplane model as well. This model (Brocca and Bažant 2000a,b) is equivalent to the classical hardening $J_2$ plasticity for the case of proportional (radial) loading paths. However, unlike that classical theory, this model can correctly reproduce the vertex effects for nonproportional loading paths, which was brought to light long ago by the tests of Gerard and Becker (1957).

The large (finite) strains that occur in the steel tube are handled in step-by-step loading by the updated Lagrangian approach (Crisfield 1997, Zienkiewicz and Taylor 1991). A special finite strain formulation of the microplane constitutive law, combining non-conjugate Green's Lagrangian strain and back-rotated Cauchy stress, is used for concrete (see Bažant et al. 2000c, where certain features permitting non-conjugacy and guaranteeing fulfillment of the energy dissipation inequality are explained in detail).

## TUBE-SQUASH TESTS REVEALING THE EFFECT OF CONFINEMENT

Tubes of thicknesses $t = 3/16$ in $= 4.7625$ mm ($\rho = A_s/A = 36.0\%$; specimen type no.1) and $t = 1/16$ in $= 1.5875$ mm ($\rho = A_s/A = 14.8\%$; specimen type no.2) made of a highly ductile steel alloy ASTM No.1020 with Young's modulus $E = 6800$ ksi $= 46852$ MPa and Poisson's ratio $\nu = 0.25$ were filled with concrete and cured in a fog room for 28 days (Caner et al. 2000b). All the tubes had the same inner diameter $D = 1.5$ in. $= 38.1$ mm and the same length $L = 3.5$ in. $= 88.9$ mm. Normal strength concrete was cast into the tubes; it had a maximum aggregate size of 0.375 in. $= 9.52$ mm, uniaxial compressive strength $f'_c = 6$ ksi $= 41.37$ MPa, and Young's elastic modulus $E = 3500$ ksi $= 24115$ MPa (Poisson's ratio was taken as $\nu = 0.18$). Larger size aggregates used in concrete composed of dolomite, granite and basalt with traces of schist. The sand and cement used in concrete was river sand (regular no. 2) and Portland cement Type I respectively.

The concrete-filled tubes were compressed axially under displacement control in a servo-controlled closed-loop MTS testing machine until the steel tube fractured, which happened when the tube length was reduced to about one half. The test machine used had a custom–built extremely stiff frame with a load capacity of one million lbf. (4.448 N). The loading took place at a constant axial displacement rate of 0.001 in/s (0.0254 mm/s). Because of confinement by the steel tube, shear angles over 70° and axial compressive strains of the order of 50% were achieved in concrete (Bažant et al. 1999).

The axial forces and displacements along with the maximum lateral expansions of the steel tube are recorded during the test. On the two sides of the specimen, LVDT gauges were mounted. After the experiment, some of the specimens were cut into two halves axially and inspected visually for damage distribution in the concrete core. Based on visual inspection of these specimens, it can be noted that for specimens with the thickest tube there was no visible damage except for a barely discernible shear band initiating from the loading platen. On the other hand, for the specimens with the thinner tubes, shear bands developed the concrete around the midsection of the specimen was severely damaged.

## FINITE ELEMENT ANALYSIS WITH
## MICROPLANE MODEL M4 FOR CONCRETE

The finite element analysis of the axisymmetric problem was performed using an explicit dynamic finite element driver which was originally developed by Brocca and Bažant (2000). Since the steel was subjected to very large strains, the finite element driver for steel tube was coded using the updated Lagrangian formulation.

The nonlinear triaxial behavior of concrete was characterized by microplane model M4. The finite strain formulation suitable for the model M4 was originally developed in Bažant et al. (1998). In that formulation, the back–rotated Cauchy stress $s = R^T \sigma R$ and Green's Lagrangian strain $\epsilon = (F^T F - I)/2$, which are *not* work-conjugate, are introduced as the stress and strain measures, respectively ($F = RU$ = deformation gradient, $R$ and $U$ are the material rotation and the right stretch tensor, respectively, and $I$ = unit tensor). The back-rotated Cauchy stress was chosen as the stress measure because it is the only stress measure referred to the initial configuration of the material that allows a physical interpretation of its components on the microplane, so that internal friction, yield limit and pressure sensitivity can have their proper physical meanings. The reason for choosing Green's Lagrangian strain as the strain measure was that it is the only strain measure whose components on a microplane characterize the finite shear angle and normal stretch on that microplane, independently of the components on other microplanes. Even though these stress and strain measures are not work-conjugate, non–negative energy dissipation was ensured by using a stress drop at constant strain in the numerical step-by-step algorithm. The elastic part of strain tensor does not cause negative energy dissipation because the elastic part of strain in concrete is always small.

Since the microplane model for concrete is a local continuum damage model, one must, in general, either use it in the sense of the crack band model, with the proper element size determined as a material property, or introduce a non-local generalization (worked out by Bažant and Ožbolt in the early 1990s). For the sake of simplicity, the former was chosen.

For a short period of time at the beginning of the test, strain softening always develops because the steel, due to its higher Poisson ratio, expands more than concrete. But it does so only until inelastic volume expansion of the concrete core begins. The tube prevents the strain softening zone from developing into a localized band running across the specimen, which would inevitably give rise to size effect. For most of the test, strain softening is suppressed by high confining pressure, except if the tube is very thin.

Because of highly nonproportional loading and the vertex effect, the steel tube was modeled using a microplane model for steel equivalent for proportional loading to $J_2$ plasticity. Since the bond between the steel tube and the concrete core is not perfect and separation or sliding can occur, a very thin layer of transversely isotropic elastic elements was introduced at the interface (with a radial dimension 1% of concrete core radius).

The purpose of encasing concrete columns in tubes is to ensure ductility. Strictly speaking, ductility should be understood as the absence of strain softening damage or fracture. There exists a certain critical (or minimum) reinforcement ratio (ratio of the cross section areas of the steel and concrete) which is necessary to ensure ductility in this sense. Such a condition of ductility, however, can be taken literally only for spiral columns, but not for tubular columns.

In tubular columns, in which the confining tube is under axial compression, the concrete core will always undergo limited temporary strain-softening right at the start of inelastic lateral expansion of concrete, at loads much smaller than the load capacity. The reasons is that the higher Poisson ratio of steel will cause the elastic lateral expansion to be initially higher in the tube than in the concrete core. Therefore, it appears more appropriate not to insists on absence of strain softening anywhere in the specimen, and to determine the critical reinforcement from the condition that the load deflection diagram would exhibit no softening.

Fitting of the experimental results indicate that the finite element model simulated very well the axial load-displacement response. Fig. (3) shows that the deformed shapes predicted by the finite element calculations were quite accurate. These figures also show the contours of equal $E_t \leq 0$ normalized by Young's modulus $E$ of concrete. These contours indicate the concrete to reharden after the initial temporary strain softening caused by a higher elastic Poisson effect in steel.

To ensure perfect ductility of concrete-filled tubular columns, the thickness of the steel tube must exceed a certain critical (or minimum) value. It was assumed that this value should be such that the effective tangential stiffness $K_t$ along the loading path of the column would always remain non-negative.

Fig. 4 shows the minimum values of $K_t$ determined by tests as a function of steel tube thickness $t$. The plot also includes a point with negative $K_t$ which corresponds to the standard (unconfined) compression test of a cylinder ($t = 0$). It may be mentioned that the experimentally determined critical thickness of the tube matches the value predicted numerically using the microplane model M4, without any adjustment in the prediction model.

## OBSERVATIONS AND CONCLUSIONS FROM CONFINEMENT INVESTIGATIONS

1. By performing a series of tube-squash tests on concrete-filled tubes of different wall thicknesses, it was found that, for the concrete used, a fully ductile inelastic response can be ensured only if the ratio of cross section area of steel to the whole cross section area exceeds the critical value of

16%

2. Verification and calibration of state-of-art material models for steel and concrete by the tube squash test makes it possible to predict the inelastic behavior of tubular and spiral columns with higher confidence.

3. A large-strain finite element model previously developed for the analysis of the tube-squash test has been extended to handle uniformly expanding tubular columns and spiral-reinforced columns. Computations indicated that, for the concrete used, the critical steel ratio for tubular columns is $\rho_{cr} = 16.1\%$.

4. The concrete core of tubular columns always softens prior to large inelastic lateral expansion. This is explained by Poisson effect, causing that the expansion of the compressed tube to be initially larger than that of concrete core. The concrete core in this type of column does not harden until the axial strains exceeds 25%.

5. For spirally reinforced columns, the critical steel ratio is found to be

$$\rho_{cr} = 14.2\%$$

which is about the same as that for the tube squash test. Furthermore, it is shown that for $\rho \geq \rho_{cr}$, the concrete core confined by the spiral reinforcement *never* softens, except locally.

6. The aforementioned minimum steel ratio found necessary to completely prevent softening response, i.e., to achieve plastic behavior, is rather high—significantly higher than the steel ratios currently used in design. The implication is that plastic limit analysis is not an adequate design concept for the currently used column dimensions. Therefore, if the current values of steel ratios are not increased, one must pay attention to the localization of softening damage and accept the size effect engendered by it. Because of the brittleness of failure and size effect caused by every strain-softening behavior, very large columns are of particular concern. The safety advantages of moving toward columns with stronger steel confinement should be explored.


## CONSEQUENCES FOR SIMULATIONS OF
## MISSILE IMPACT AND PENETRATIONS


Numerical simulations of missile impact and penetration represent a problem in which it is very important to describe realistically both the plastic-ductile response and the fracturing response of concrete. Under the nose of a penetrating missile, concrete is subjected to an enormous triaxial confining pressure which causes it to behave plastically, dissipating a large amount of energy, thus decelerating the missile. Capturing this dissipation accurately is important for correct predictions of the depth of penetration into a concrete wall or the exit velocity of the a missile getting through. Here the data from tube squash test help to formulate the correct constitutive equations.

Farther away from the penetrating missile, the waves from impact cause fracturing of concrete, which also dissipates much energy, and produce the entry crater and the exit crater. During the impact event, the concrete is subjected to a highly nonproportional loading path, for which the vertex effect must be embodied in the constitutive law in order to ensure a correct tangential stiffness and dissipation.

Large-scale impact simulations have been conducted by M.D. Adley and S.A. Akers at Waterways Experiment Station in Vicksburg, using in super-computer simulations the finite-strain microplane model M4 developed at Northwestern University, which is capable of representing all these phenomena. The details are presented in Bažant et al. (2000b).

## ACKNOWLEDGMENT

Thanks are due to the U.S. Army Engineer Waterways Experiment Station (WES), Vicksburg, Mississippi, for partial funding of Bažant's work under Contracts DACA39-94-C-0025 and DACA42-00-C0012 with Northwestern University. F.C. Caner and J. Červenka thank the U.S. National Science Foundation for funding their research at Northwestern University under Grant CMS-9713944 (directed by Bažant). J. Červenka further thanks the Czech Grant Agency for additional funding of his research in Prague under Contract 103/99/0755.

## REFERENCES

Bažant, Z.P., Caner, F.C., Carol, I., Adley, M.D., and Akers, S.A. (2000). "Microplane model M4 for concrete: I. Formulation with work-conjugate deviatoric stress." *J. of Engrg. Mechanics ASCE* 126 (9), 944–953.

Bažant, Z.P., Adley, M.D., Carol, I., Jirásek, M., Akers, S.A., Rohani, B., Cargile, J.D., and Caner, F.C. (2000). "Large-strain generalization of microplane model for concrete and application." *J. of Engrg. Mechanics ASCE* 126 (9), 971–980.

Bažant, Z.P., and Cedolin, L. (1991). "Stability of structures: Elastic, inelastic, fracture and damage theories." Oxford University Press, New York.

Bleich (1952). *Buckling strength of metal structures*, McGraw Hill, New York.

Budianski, B., Dow, N.F., Peters, R.W., Ghephard, R.P. (1951). "Experimental studies of polyaxial stress-strain laws of plasticity." *Proc., First US Nat. Congr. of Appl. Mech.*, ASME New York, 503–512.

Budianski, B. (1959). "A reassessment of deformation theories of plasticity." *J. of Appl. Mech., Trans. ASME* 26 (259–264).

Brocca, M. and Bažant, Z.P. (2000a). "Microplane constitutive model and metal plasticity." *Applied Mechanics Reviews, ASME* 53 (10), 265–281.

Brocca, M. and Bažant, Z.P. (2000b). "Microplane Finite Element Analysis of Tube–Squash Test of Concrete with Shear Angles up to 70°."*Int. J. of Num. Meth. in Engrg.*, in press.

Caner, F.C., and Bažant, Z.P. (2000a). "Microplane model M4 for concrete: II. Algorithm and Calibration." *J. of Engrg. Mechanics ASCE* 126 (9), 954–961.

Caner, Z.P., Bažant, Z.P., and Červenka, J. (2000b). " Vertex Effect in Strain-Softening Concrete." Report, Northwestern University; submitted to *ASCE J. of Engrg. Mech.*

Caner, F.C., and Bažant, Z.P. (2000c). "Minimum Confinement Ensuring Concrete Column Ductility Via Tube-Squash Tests." Report, Northwestern University; submitted to *ASCE J. of Engrg. Mech.*

Červenka J., Červenka, V., and Eligehausen, R. (1998). "Fracture-Plastic Material Model for Concrete, Application to Analysis of Powder Actuated Anchors", *Proceedings of FramCos 3*, Vol. 2, 1107-1117.

Gerard, G., and Becker, H. (1951). "Column behavior under conditions of impact." *J. Aero. Sci.* 19, 58–65.

Gerard, G., and Becker, H. (1957). "Handbook of structural stability: Part I, Buckling of flat plates." *NACA Techn. Note* No. 3781.

Lin, F.-B., Bažant, Z.P., Chern, J.-C., and Marchertas, A.H. (1987). "Concrete model with normality and sequential identification." *Computers and Structures* 26 (6), 1011–1025.

Phillips, A., and G. A. Gray (1961). "Experimental investigation of corners in the yield surface." *J. of Basic Engineering, Transactions of the ASME*, 83, Series D, 275–289.

Roeder, C.W, B. Cameron, and C.B. Brown (1999). "Composite action in concrete–filled tubes." *J. of Structural Engineering, ASCE*, 125(5),477–484.

Schneider, S.P. (1998). "Axially loaded concrete–filled steel tubes." *J. of Structural Engineering, ASCE*, 124(10),1125–1138.

van Mier, J.G.M. (1986). "Multiaxial strain-softening of concrete; Part I: Fracture, Part II: Load Histories." *Materials and Structures*, 111, (19), 179–200.

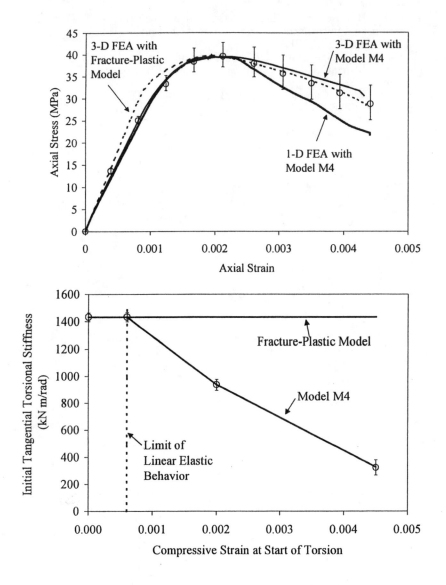

Figure 1: Plot of initial incremental torsional stiffness when the torsional loading starts (bottom), as a function of the axial compressive strain under uniaxial stress (top) at the start of torsion, showing the test data points, the predictions of microplane model M4, and the predictions of damage-fracturing model based on tensorial invariants and loading potentials.

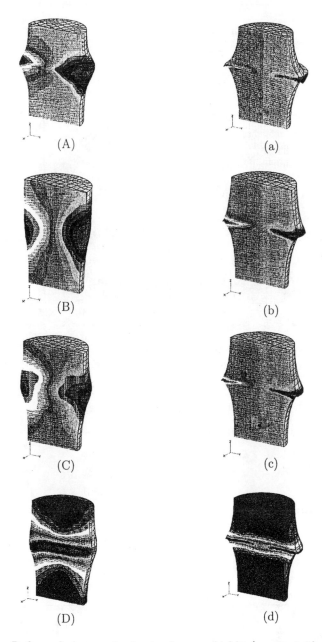

Figure 2: Deformed shapes obtained using model M4 (on the left) and fracture-plastic model (on the right). The loadings paths are proportional (A, a); vertex at peak (B, b) and vertex in the postpeak (at $\epsilon_{zz} = 0.45\%$) (C,D,c,d). The contours of variously shaded zones represent lines of constant strain $\gamma_{xz}$ (A,B,C,a,b,c) and maximum principal strain $\epsilon_I$ (D, d) (for details, see Caner et al. 2000b).

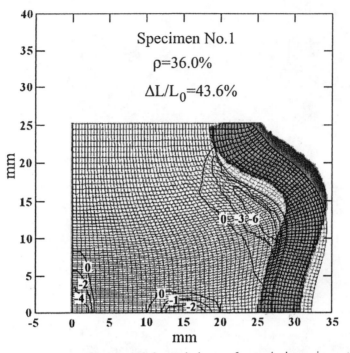

Figure 3: The experimental deformed shape of a typical specimen type no.1 and its prediction by the finite element analysis. Also shown are the contours of equal $K_t \leq 0$ normalized by Young's modulus $E$ of concrete.

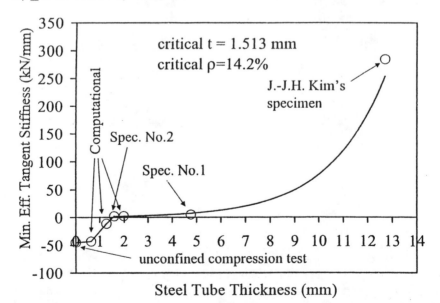

Figure 4: Determination of critical wall thickness of steel tube using the experimental data points in the min. effective tangential stiffness in axial direction vs axial displacement space.

# Concrete and Sustainable Development

## by C. Meyer

Synopsis: The United States is a country known for its wasteful use of natural resources. Efforts to correct the results of past transgressions as well as to balance economic development against legitimate concerns of conservation are pervading almost all aspects of life, including the construction industry. Concrete, being the most widely used material worldwide, is a natural target for conservation of natural resources. The cement industry is a major producer of greenhouse gases and energy user. Recent research has led to the point where numerous by-products of industrial processes with pozzolanic properties can be substituted partially for cement, such as fly ash and ground granulated blast furnace slag. Also other recycled materials are finding increased application in concrete production. For example, recycled concrete has been used successfully in numerous projects, and crushed waste glass is now available as a valuable source of aggregate, since the problem of alkali-silicate reaction has been solved. The key to commercial success is beneficiation, i.e. the targeted utilization of specific properties of the recycled material, which adds value to the end product.

Keywords: concrete; concrete materials science; concrete technology; environmentally friendly construction; fly ash; green buildings; recycling; supplementary cementing materials; sustainable development; waste glass; waste materials

**Christian Meyer** is a professor in the Department of Civil Engineering and Engineering Mechanics at Columbia University in New York. He has worked in engineering practice for eight years before joining the faculty of Columbia University. His main interests are related to analysis and design of concrete structures and earthquake engineering. In recent years he has worked primarily in the field of concrete materials science and technology, in particular searching for ways of using waste materials to develop concrete products.

## INTRODUCTION

The United States is a country blessed with enormous natural resources. This partially explains their wasteful exploitation in the past. To those who grew up in countries less blessed with such natural riches or with scarcities produced by wars or natural disasters, experiencing the wasteful use of those resources in the United States can come as something of a cultural shock.

Recycling and the reuse of natural materials have been traditionally of low priority and often nonexistent in the U.S. This is no longer the case. A dramatically different attitude can be felt today throughout the country. This change came relatively suddenly, gaining significant momentum in the early 1970s. A key event was the celebration of Earth Day in 1970, when a large part of the American public became aware of the limits of the nation's resources and grew concerned about the deteriorating environment, whether soil, water, or air. If any single picture could symbolize this awakening, it was the famous photo taken by our astronauts of "Spaceship Earth", which dramatically illustrated the fact that our planet is indeed finite in size and in resources, and that we had better learn how to live within our means. This applies particularly to Americans, who use those resources vastly out of proportion with their share of the world population.

One other reason why the American public was so slow in realizing the finiteness of its resources was the size of the country. For example, unlike in many countries in Europe, there was plenty of space available to dump its refuse and waste material. Or at least this seemed to be the case and it clearly is no longer true. Not only did the physical space available for landfills become sparse, also, legislation on the federal, state, and local levels imposed severe environmental restrictions on them. As a result, many existing landfills had to be closed, with costly cleanup measures needed for some, and it is now becoming increasingly difficult to open up new landfills.

These developments are probably nowhere as dramatic as in New York City. This largest American metropolis, with 8 million people in the five boroughs of the City proper alone, probably generates more solid waste than any other city in the world, including those with much larger populations. The Freshkill Landfill on Staten Island is the world's largest – an achievement New

Yorkers have no particular cause to be proud of. Moreover, already filled well beyond its original design capacity, it had to be closed in 2001, making a bad situation even worse.

These and various related concerns led to the concept of sustainable development, which can be summarized as follows:

1. Remedy the mistakes of the past by cleaning up our contaminated water and soil.
2. Avoid the pollution of our air, water and soil, including the release of greenhouse gases into the atmosphere that are known to contribute to global warming.
3. Utilize natural resources, whether material or energy, at a rate no greater than at which they can be regenerated.
4. Find a proper balance between economic development and preservation of our environment, i.e. improve the living standard and quality of life without adversely affecting our environment.

These goals describe an ideal state and are obviously difficult to achieve. Yet, we do not have much of a choice, lest the liveability of our planet take a rapid turn for the worse. As the World Earth Summits in Rio de Janeiro (1990) and Kyoto (1997) demonstrated very clearly, this worldwide problem can be solved only through concerted international efforts. The industrialized countries are called upon to reduce the emission of greenhouse gases and the wasteful use of natural resources, and the developing countries need to avoid the mistakes made by the industrialized world in the past and develop their economies using technologies that make optimal use of energy and natural materials, without polluting the environment.

## THE ROLE OF THE CONCRETE INDUSTRY

The construction industry is not exempt from the above-mentioned requirements. The concrete industry in particular is called upon to improve its record, because it is both a major contributor to air pollution and consumer of vast quantities of natural materials (1). For each ton of cement produced, one ton of $CO_2$, a greenhouse gas, is released into the atmosphere. Worldwide, the cement industry produced about 1.4 billion tons in 1995, which caused the emission of as much $CO_2$ gas as 300 million automobiles – accounting for almost 7% of the total world production of $CO_2$ (2). Concrete is the most widely used material worldwide. Our industry has a responsibility and societal duty to make a contribution towards sustainable development that is commensurate with its size. There is still a large difference between simply declaring concrete to be a "green" (in the sense of "environmentally friendly") material and actually taking the steps necessary to achieve sustainable development.

There are two major opportunities to achieve such a goal that shall be addressed here. As portland cement production is known to require large amounts of energy and is responsible for the release of greenhouse gases, any effort to reduce the cement content in concrete will be beneficial. For this reason, researchers who develop cementitious materials that require comparably small amounts of energy to produce or that are waste products from industrial or combustion processes are making a major contribution to achieve this goal. The other possibility is to substitute recycled materials for aggregate or reinforcement. This includes the recycling of concrete itself. By one estimate, the concrete industry is currently consuming 8 billion tons of natural material each year (3). Any efforts to reduce such dependence of virgin materials will therefore be a contribution towards sustainable development. In the following, some of the recent advances in both areas shall be discussed to point out directions for future research to make concrete a more environmentally friendly material.

A separate approach towards conserving natural resources is the targeted increase in durability, because more durable structures need to be replaced less frequently. Such durability increase can be achieved by choosing appropriate mix designs and selecting suitable aggregates and admixtures. This issue has been discussed in detail in (1) and shall not be dealt with here any further.

## CEMENT SUBSTITUTES

Cement is the key component of concrete that binds the other components together and gives the composite its strength. A considerable amount of work has been reported in the literature on how to use waste products of combustion or industrial processes as supplementary cementitious materials (3,4,5). Because of their cementitious or pozzolanic properties these can serve as partial cement replacement. Ideally, the development of such materials serves three separate purposes simultaneously. On the one hand, waste byproducts have an inherent negative value, as they require disposal, typically in landfills, subject to tipping fees that can be substantial. When used in concrete, the material's value increases considerably. The increase in value is referred to as "beneficiation". As this supplementary cementitious material (SCM) replaces a certain fraction of the cement, its market value may approach that of cement. A second benefit is the reduction of environmental costs of cement production in terms of energy use, depletion of natural resources, and air pollution. Also, the tangible as well as intangible costs associated with landfilling the original waste materials are eliminated.

Finally, such materials may offer intriguing additional benefits. Most concrete mixes can be engineered such that the SCM will give the mix certain properties (mechanical strength, workability, or durability) which it would not have without it. It is the challenge for the concrete technologist when developing a mix design, to combine these three different goals in an optimal way such that the economic benefits become transparent. The key task is to turn waste material with a large inherent negative value into a potentially valuable product. The increase in value should be both real, in terms of converting a liability into a commodity with an increased market value, as well as intangible in terms of reduced environmental costs. The fundamental challenge for the researcher is to identify waste materials with inherent properties that lend themselves to such beneficiation. Below, a few examples shall be mentioned.

*Fly ash* is the byproduct of coal burning power plants and is known to have excellent pozzolanic properties (4,5). Its use in the concrete industry has a long and successful tradition. However, in terms of the ratio of fly ash utilized to fly ash produced, there remains considerable room for improvement. For example, of the 60 million tons of ash produced in 1995 in the U.S., only 8.1 million tons were utilized (2). India beneficiated only 2 million of the 57 million tons produced there in the same year. This latter figure is worrisome, because India, like many other developing countries, is expected to increase considerably its coal-based power generation and cement production capacities in the years to come. Without a major concurrent effort to make productive use of the ash, the environmental pollution can be expected to worsen at a comparable rate (2,3).

The use of fly ash as partial cement replacement is not without its challenges. There are limits as to how much of the cement may be replaced. 20% is an often mentioned and easily achieved goal. Malhotra has shown that as much of 60% cement replacement by ASTM Class F fly ash is feasible (6). Recent research has shown that it is possible to replace 100% of the cement with chemically self-activated fly ash (7). However, the activators proposed so far either need to be added in unreasonable amounts or are relatively expensive. A major point of concern is the generally slow strength development of fly ash concretes. However, in construction practice, high early strength is important only for some projects. In many others, such as those involving mass concrete, slow strength development may even be an advantage, as it generates lower heat of hydration rates. Another potential problem is quality control, because the exact properties of the fly ash may change from batch to batch, depending on the source material. Therefore, concrete containing large amounts of fly is less likely to be suitable for architectural or other applications with specific esthetic requirements. On the positive side, fly ash is known to suppress alkali-silica reaction to some extent.

*Ground granulated blast furnace slag* is another industrial waste product with beneficial properties that have been well documented in the literature. The glassy granular material is formed when molten blast-furnace slag is rapidly chilled in water. The material is then ground to specification. Whereas in the past it was typically deposited in landfills, because of its excellent cementitious properties, it is now being used as a partial cement substitute. If it is of the proper quality and used in properly designed mixes, it can increase the strength, durability and other properties of the end product (8).

Possibly the greatest success story in this regard is that of *condensed silica fume*. A byproduct of the semiconductor industry, this siliceous material is known to improve both strength and durability of concrete to such an extent that modern high-performance concrete mix designs as a rule call for the addition of silica fume. Even though the material is difficult to handle because of its extreme fineness, its benefits are so obvious that its market value is considerable. This makes silica fume a classic example of waste material beneficiation.

A final example is *solid waste incinerator ash*. To be specific, solid waste incinerators generate both bottom ash and fly ash. Because the fly ash is typically contaminated with various hazardous substances, its disposal creates additional problems and also complicates its beneficiation. Some technologies that have been proposed render the toxic elements harmless by vitrification, which requires large amounts of energy (9). If less effective detoxification methods are employed, extensive leaching tests need to be performed to assure that the harmful substances cannot leach out under most service exposure conditions. Protocols for such tests can simulate actual service conditions only inadequately, and therefore environmental protection agencies prescribe a variety of different tests (10).

## RECYCLED AGGREGATE

Aggregate constitutes approximately 70% of concrete volume. Worldwide, this amounts to billions of tons of crushed stone, gravel, and sand that need to be mined, processed, and transported. In some parts of the country, suitable gravel pits and sources of construction grade sand are depleting, while the opening of new sources requires time-consuming environmental impact statements and the procuring of the necessary permits, aside from the need to overcome public opposition. For this reason, in certain geographic regions such as Long Island of New York State, the search for alternate sources of aggregate is increasingly becoming a necessity.

The substitute material that comes to mind first is *recycled concrete*. Construction debris and demolition waste constitute 23% to 33% of municipal solid waste, and demolished concrete contributes the largest share of this waste material (11). A sizeable amount of literature is available on the use of recycled concrete (12), and in North America, several projects have been completed successfully (11,13). The use of recycled concrete poses many interesting research problems. The fines and dust produced during demolition and crushing, together with the pore structure of old concrete, increase the water absorption, which has to be considered in the mix design. An additional challenge is posed by the quality control, due to the wide range of material properties. Compared with the properties of virgin material, those of recycled concrete are not as easily controlled. The material may be contaminated from various sources. As a result, it is typically of lower quality compared to virgin material. But unless high-strength or high-performance concrete is called for, the performance specifications of the end product may be achievable just as well with reprocessed concrete or a blend of virgin and reprocessed material.

The economics of recycled concrete depends on a number of factors. Since virgin material is generally very inexpensive, it is not easy to offer reprocessed concrete at a comparable price. But as transportation constitutes a major cost component of aggregate, the location of a suitable source of virgin material and the distance to the nearest landfill for the demolished concrete are likely to be decisive factors in determining whether the recycling of concrete is economical or not. Even with the avoidance of tipping fees, the costs of crushing and processing the old concrete are likely to reduce profit margins to such an extent, that additional economic incentives may be needed for a producer to use recycled concrete. For example, tax credits for recycled material content can offer this additional incentive. Thus, the real challenge to the researcher is again the search for properties of recycled concrete that virgin material does not have and that lend themselves to beneficiation.

Another source of aggregate is *waste glass*. Unlike in many European countries, separation of post-consumer glass by color in major U.S. metropolitan areas is the exception. As it is difficult for the glass industry to use mixed-color cullet, most of such cullet at this time finds its way into landfills. There are no significant markets for such glass, which is more or less contaminated with paper, bottle caps, and organic material from residuals of the original contents. The use of glass as an aggregate for concrete has been contemplated some time ago, but those early efforts have been unsuccessful because of the alkali-silica reaction (ASR) problem. A major research program at Columbia University was undertaken to study this problem in great detail, and various solutions of the problem are now available (14-17).

A considerable effort has been made to investigate the economic aspects of using waste glass as aggregate. The key to success lies again in the extent to which the special properties of the glass are exploited. For this purpose, we may distinguish between "commodity" and "value-added" products. The primary objective of commodity products is to utilize as much waste glass as possible. An example is the concrete masonry block, which is typically mass-produced in highly automated facilities with good quality control. It is possible to replace part of the fine aggregate by glass, because it was found that finely ground glass particles passing US standard sieve #100 cause only negligible ASR-induced expansions in the mortar bar test according to ASTM C 1260. The naked eye cannot distinguish glass particles of such size from regular sand. The economics of such substitution is questionable, because ordinary sand is relatively inexpensive, whereas the glass needs to be crushed and washed to remove the harmful sugars and other contaminants.

Very finely ground glass powder passing mesh #400 has been shown to have pozzolanic properties, so that it may be used as partial cement substitute. Since such fine powder is generated as dust during crushing anyway, the economics are improved if both part of the sand and cement are replaced at the same time.

The economic picture changes dramatically, when the special properties of the glass are exploited in "value-added" products. Aside from its pozzolanic property, glass has indeed some properties that make it quite special (17):

- Because it has basically zero water absorption, it is one of the most durable materials known to man. With the current emphasis on durability of high-performance concrete, it is only natural to rely on extremely durable ingredients.
- The excellent hardness of glass gives the concrete an abrasion resistance that can be reached only with few natural stone aggregates.
- For a number of reasons, glass aggregate improves the flow properties of fresh concrete, so that very high strengths can be obtained even without the use of superplasticizers.
- The esthetic potential of color-sorted post-consumer glass, not to mention specialty glass, has barely been explored at all and offers numerous novel possibilities for design professionals.

One example of a value-added product, in which all of these special properties are exploited, are decorative terrazzo tiles. Such tiles typically use relatively expensive aggregate that often is imported. Crushed waste glass, even if color sorted and washed, can easily compete with such aggregate. As a result, glass concrete tiles are already being mass-produced commercially. Other examples, still under development, are wall panels, building façade elements, table-top counters, benches, planters, etc. Architects, designers, and artists are intrigued by the potential of this new material. Once realized in such high-end products, the glass becomes a valuable resource, and the goal of beneficiation has been achieved completely.

*Dredged material* is a further important example that is being evaluated for use in concrete (18). One of the most pressing problems confronting most major seaports of the world is the need for dredging in order to keep the shipping lanes open. Until recently, the dredged material was simply disposed of in the open ocean. But since it may be highly contaminated, national legislation and international agreements are now prohibiting such practice. The Port Authority of New York and New Jersey, for example, is now facing the task of properly disposing of up to four million cubic yards of dredged material each year. The material consists mostly of clays and silts, much of it highly contaminated with oils, heavy metals, PCB's and other toxic substances. A major research project is currently underway at Columbia University to search for a beneficiation technology that renders the toxic components harmless. It is contemplated to use the treated material as aggregate or filler in concrete, aside from other applications. The research challenge is formidable, because the bulk of the material does not readily lend itself to utilization in concrete.

The economics of this example is unlike that of the others, because of the very large negative value inherent in the material. If the processing cost can be kept well below that of disposal in specially designed facilities, the retail value of the end product becomes less significant. Used as an ingredient in concrete, it would have again achieved the goal of adding value. A solution of this pressing problem is not only of concern to environmentalists. The economic well-being of entire regions may depend on it.

*Waste wood*, such as sawdust and shavings, has also been used to produce specialty concrete products (19). Also in this case, potentially adverse chemical reactions between the sugars and other organic substances in the wood and the cement present a challenge that needs to be addressed. The primary advantage of waste wood aggregate is the low weight and high thermal insulation value of the material. The economics depends on several factors, including the geographical area where the material originates. In the U.S., much of the material is utilized in manufacturing of particle boards. But if products can be developed that utilize the special properties of the wood in concrete, commercialization should be possible.

## CONCLUSIONS

The concrete industry is a major contributor to air pollution and user of natural resources. As such it bears a special responsibility to make a contribution towards sustainable development that is commensurate with its size. It can do so by pursuing three goals:

1. Searching for cement production technologies that are less energy-intensive and cause less air pollution. Since such technologies will not be available in the foreseeable future, the more realistic approach is to reduce the need for portland cement, primarily by increased use of supplementary cementitious materials, especially waste materials.
2. Replacing concrete ingredients by recycled materials, such as recycled concrete or waste glass.
3. Through careful concrete mix design and prudent choice of admixtures, improve the durability of structures such that they need to be replaced less frequently.

The development of practical technologies requires the solution of technical and economic problems. It appears that often the technical problems are easier to solve, because the required research is based on sound scientific principles and methods. The commercial aspects of product development are subject to numerous economic influences and sometimes to political and psychological ones as well. The economic feasibility is determined primarily by the forces of supply and demand, neither of which can be controlled easily. These relationships have to be well understood to assure commercial success.

There are a few success stories for illustration. They all have in common that certain inherent properties are exploited in an optimum way to create value. By utilizing the pozzolanic properties of fly ash, for example, a waste material of the coal-burning power industry with inherent negative value is converted into a value-added material with a market value close to that of the cement it replaces. The benefits of silica fume for high-performance concrete are so obvious that this by-product of the semiconductor industry achieves a market value that is even higher. Color sorted waste glass gives rise to new decorative and architectural concrete applications and thereby achieves a market value comparable to that of the high-end aggregates used for specialty products.

In each case, the market value will be determined partially by the competing materials on strictly economic terms and often by political considerations as well. For example, if the political process determines that increased recycling and reuse of resources is desirable, legislation can promote this with tax incentives or outright requirements. For example, the Federal Government can require that a certain percent of recycled material be used on all federal construction projects. Such a requirement changes the economics of reprocessing technologies. But the better the inherent properties of the materials are utilized, the less dependence on legislative or regulatory initiatives is needed for such technologies to be commercially successful.

As a final point, it should be stressed that our educational system can make a useful contribution towards achieving the goals stated above. Our colleges and universities have already been called upon to increase their students' awareness of societal issues. As our profession reaches a more and more mature state, the educational emphasis is bound to shift away from some of the traditional areas of study anyway, as these will be covered more and more by computer-aided technologists. By stressing fundamental materials science and engineering and basic principles of sustainable development, educators can help prepare the next generation of engineers for a changing world, which will be much less tolerant of the old wasteful ways of using our natural resources.

## REFERENCES

1. Gjorv, O.E. and Sakai, K., eds., *Concrete Technology for a Sustainable Development in the 21st Century*, E&FN Spon, London, 2000.

2. Malhotra, V.M., "Role of Supplementary Cementing Materials in Reducing Greenhouse Gas Emissions", in *Concrete Technology for a Sustainable Development in the 21st Century*, O.E. Gjorv and K. Sakai, eds., E&FN Spon, London, 2000.

3. Mehta, P.H., "Concrete Technology for Sustainable Development – An Overview of Essential Elements", in *Concrete Technology for a Sustainable Development in the 21st Century*, O.E. Gjorv and K. Sakai, eds., E&FN Spon, London, 2000.

4. *Fly Ash, Slag, Silica Fume and Other Natural Pozzolans*, Proceedings, 6th International Conference, Special Publication 178, American Concrete Institute, Farmington Hills, MI, 1998.

5. *Fly Ash, Slag, Silica Fume and Other Natural Pozzolans*, Proceedings, 5th International Conference, Special Publication 153, American Concrete Institute, Farmington Hills, MI, 1995.

6. Malhotra, V.M. and Ramezanianpour, A.R., *Fly Ash in Concrete*, 2nd Ed., CANMET, Energy, Mines and Resources Canada, Ottawa, Canada, 1994.

7. Samadi, A. "Treatment of Fly Ash to Increase its Cementitious Characteristics", PhD Dissertation, Drexel University, Philadelphia, PA. 1996.

8. Sehgal, J.P., "Environmentally Friendly Concrete", The Concrete Industry Board Bulletin, New York, March 2001.

9. *ASME/Bureau of Mines Investigative Program on Vitrification of Residue from Municipal Waste Combustion Systems*, American Society of Mechanical Engineers Report CRTD-24, 1993.

10. Hohberg, I., de Groot, G.J., van der Veer, A.M.H., and Wassing, W., "Development of a Leaching Protocol for Concrete", Waste Management 20 (2000) 177-184.

11. "Recycling Concrete Saves Resources, Eliminates Dumping", Environmental Council of Concrete Organizations, Skokie, IL, 1997.

12. Hansen, T.C., ed., *Recycling of Demolished Concrete and Masonry*, RILEM Report 6, E&FN Spon, London, 1992.

13. "Recycling Concrete Pavements", *Concrete Paving Technology*, TB-014P, American Concrete Pavement Association, Skokie, IL, 1996.

14. Meyer, C. and Baxter, S., "Use of Recycled Glass for Concrete Masonry Blocks", Final Report to New York State Energy Research and Development Authority, Albany, NY, Report 97-15, Nov. 1997.

15. Meyer, C. and Baxter, S., "Use of Recycled Glass and Fly Ash for Precast Concrete", Final Report to New York State Energy Research and Development Authority, Albany, NY, Report 98-18, Oct. 1998.

16. Jin, W., Meyer, C., and Baxter, S., "Glascrete – Concrete with Glass Aggregate", ACI Materials Journal, March-April 2000.

17. Meyer, C., "Recycled Glass – From Waste Material to Valuable Resource", *Recycling and Reuse of Glass Cullet*, R.K. Dhir et al, eds. Thomas Telford, London, 2001.

18. Millrath, K., Kozlova, S., Shimanovich, S., and Meyer, C., "Beneficial Use of Dredge Material", Progress Report prepared for Echo Environmental, Inc., Columbia University, New York, Feb. 2001.

19. Gliniorz, K.-U. and Natterer, J., "Structural Elements of Wood Lightweight Concrete" (in German), Report IBOIS 00:03, Ecole Polytechnique Federale de Lausanne, March 2000.

SP 206–32

# Concrete for Freshmen

## by E. N. Landis and W. P. Manion

**Synopsis:** An introductory construction materials course was developed for first year civil and environmental engineering students at the University of Maine. Because it is typically the first engineering course in which most civil engineering students enroll, the course also serves as an introduction to engineering. In addition to the title subject ("Materials") we introduce general principles of engineering analysis and design. We have adopted a materials science approach in our coverage of construction materials. That is, we emphasize how material properties are a function of their microstructure, and that we control microstructure through processing. We have found concrete to be an ideal material with which to illustrate this concept. Students are easily able to observe the processing-microstructure-properties links through laboratory and homework exercises in concrete mix design and testing. In the broader context we have found materials in general, and concrete in particular, to be excellent model topics for introducing the general topics of engineering analysis and design.

**Keywords:** concrete education; freshman engineering; teaching laboratories

513

ACI member Eric N. Landis is an Associate Professor of Civil Engineering at the University of Maine. He received his BS (1985) from the University of Wisconsin, and his PhD (1993) from Northwestern University. His teaching and research interests include but are not limited to microstructure-property relationships for cement-based and wood-based composite materials.

William P. Manion is an Instructor in the Department of Civil & Environmental Engineering at the University of Maine. He received is BS (1989) from the SUNY College of Environmental Science and Forestry, and his MS (1992) from the University of Maine. His teaching and research interests include use of recycled materials in construction and instructional technologies.

## INTRODUCTION

There is a recent interest introducing engineering analysis and design at a much earlier point in the curriculum (e.g. 1). The traditional approach is to spend the first year or two in the curriculum loading up on the basic math and science skills that are required for engineering analysis and design. However, a recent paradigm shift is to introduce design problems early on in the curriculum. One motivation for this is that when the students see an array of engineering problems, they will have a better appreciation for the basic skills that are necessary for rational solutions.

In the spirit of this movement, civil and environmental engineering students at the University of Maine get an introduction to engineering through a construction materials course they take during their first semester at the university. The course uses issues in materials engineering as a model for general problems in engineering design and analysis, and thus covers a fairly wide range of subjects. The course is part of an overall curriculum modification intended in part to improve retention among civil engineering students during their first year or two by putting at least one required civil engineering course in each semester in the suggested curriculum.

The basic theme or thesis of this paper is that through our experience teaching this introductory course, the subject of construction materials in general, and concrete technology in particular, are excellent vehicles for introducing broad concepts of engineering analysis and design to first year engineering students.

## COURSE GOALS & OBJECTIVES

The course being described here is typical of undergraduate construction materials courses. A three credit-hour lecture course is taken concurrently with a one credit hour laboratory session. The topics covered include structure, properties and testing of metals, portland cement concrete, wood, bituminous concrete mixtures, and composite materials. The ABET course Goals are as follows:

- The students will learn basic physical, mechanical and chemical properties of different construction materials.
- The students will be introduced to various tools of engineering problem solving including statistical and experimental analysis.

The ABET Course Outcomes are:

- The student will demonstrate the ability to perform a simple statistical analysis of experimental data, including fitting data to standard models, and making predictions based on those models.
- The student will demonstrate understanding of stress, strain, strength, toughness, durability and fatigue, and the student will be able to perform calculations that involve those quantities.
- The student will demonstrate an understanding of the role of microstructure in material properties.
- The student will demonstrate an understanding of production and properties of steel, concrete, wood and wood composites, and FRP composites.
- The student will be able to identify the critical design issues in material selection.

While these goals and outcomes are typical, the challenge in this course is to cover the topics in sufficient depth without the aid of traditional prerequisites such as chemistry, physics, and/or calculus.

In addition, because it is the first engineering course in which most students enroll, we try to cover as broad an array of general engineering topics as possible. These topics include the engineering design process, marginal economic analysis, probability-of-failure and safety factors. Additionally, we like to cover as many "professional engineering" issues as is practical. In the laboratory sessions we emphasize broad concepts of statistical distributions of properties, precision and accuracy. Design problems include design of concrete mixtures and sizing of structural members. Here the students must take into account technical considerations uncertainty, safety, economy, and intangibles.

## LECTURE SESSIONS

As mentioned above, the topics covered are fairly representative of undergraduate construction materials courses. The only difference is the amount of background material covered, and the depth of coverage of new material. The course outline is summarized as follows:

- Introduction
- Experimentation and Laboratory Analysis
    - Role of experimentation and testing in engineering analysis and design
    - Variability and statistical analysis
    - Sampling and testing
- Properties of Materials
    - Mechanical
    - Chemical
    - Thermal
    - Role of microstructure
- Materials of Construction
    - Steel, aluminum
    - Portland cement concrete
    - Bituminous concrete
    - Wood
    - Composites
    - others
- Other Topics
    - Design process
    - Codes
    - Various professional issues

The time devoted to each subject in the list varies considerably. For example we spend nearly a third of the course on Portland cement concrete. There are three primary reasons for this. First, it is a favorite of the instructor. Second, it is a material for which civil engineers not only specify properties, but are also responsible for producing those properties. Third (and most important pedagogically) it represents a model material for which microstructure, processing and properties can all be influenced and related in a basic teaching laboratory. This third issue is expanded upon below.

Because the students do not have much basic math and science background, the availability of a suitable textbook is problematic. In lieu of a general textbook we use a large number of handouts developed by the instructor. We do require the students purchase a copy of the PCA design manual (2) as a guide for our coverage of concrete.

The unifying framework for the discussion of all materials is the general relationship between microstructure and properties. No student passes the course without recognizing microstructure as the key to modifying properties. An underlying theme is the relationship between microstructural order and performance predictability. This is done through discussions of both microstructural features, and statistical distributions of different properties.

## LABORATORY SESSIONS

The laboratory sessions, as with most good lab courses, are used to highlight and expand upon the principles discussed in lecture sessions. A schedule of the lab sessions in the 2001 fall semester is shown in Table 1 (3). The general approach is simply to use specific problems to introduce and reinforce broader concepts. (For example, we measure the strength and stiffness of wood specimens of different moisture contents along different material axes to illustrate anisotropic behavior.) Each experiment session leads to a comprehensive report describing methods, data development, and interpretation of results. In addition, as part of the report the student must solve a particular design problem that requires information obtained from their experiments.

One concept we work very hard to reinforce throughout the semester is the inherent variability and uncertainty associated both with material properties and measurements. The very first lab session has the students measuring the compressive strength along the grain of 30 small samples of wood. This exercise serves a large number of functions. First, it is a way for them to get their "hands dirty" right from the start. We go over general lab operations and safety procedures. Second, they are introduced to some of the laboratory instruments they use all semester, such as the Instron load frame and its associated control software. Third and perhaps most significant, they get a feel for the variations inherent in any experimental process. As a part of the laboratory analysis the students must do a statistical analysis of their data, fit the data to different probability distributions, and use the distributions to come up with performance predictions and safety factors. It should be noted that the statistical analysis is done the following week as a part of their introduction to spreadsheets. We've found the students seem to have more incentive to learn the software when there is real data to provide the motivation.

## SIGNIFICANCE OF CEMENT & CONCRETE COVERAGE

As has previously been mentioned, there is significant lecture and laboratory time devoted to portland cement concrete. Roughly a third of lecture time, and almost half of the laboratory sessions are devoted to concrete and related materials. In addition to the relative importance of concrete as a most-used construction material, we feel there are a number of educational reasons for the extensive coverage of concrete. These reasons range from illustration of general materials issues to a general introduction to engineering design.

### Materials Science Issues

The concept of relating microstructure and processing to properties can be quite conveniently illustrated with cement and concrete. By spending a few lecture periods on cement hydration and microstructure, we are easily able to tie all the parameters for which we design back to the microstructure. Strength, durability, unit weight, toughness and other properties all are shown to have a microstructural basis. Through this relationship we are able to illustrate how changes in processing will affect microstructure and therefore properties. While conceptually this coverage is no different from steel or any other material discussed in the course, what makes concrete different, is that we are able to illustrate this in the laboratory through fabrication and testing of a number of different mix designs. By covering the microstructural basis for properties, the standard design procedures (e.g. 2) can be shown to have a rational foundation.

## Engineering Design Issues

The design of concrete mixes tailored for particular applications is a nice way to illustrate general concepts of engineering design. In this course we introduce design in general as a compromise of competing parameters. In concrete mix design we can present an application that has certain functional standards to meet. From these functional standards we can produce a list of specific design objectives. We are then able to look at different design routes to meet those objectives. The design process quickly shows itself to be a cyclic optimization process. Although economic analysis is not formally covered, we do discuss the time value of money, and how that might affect life-cycle costs.

## OTHER EDUCATIONAL ISSUES

Clearly, a strength of the course is it hands-on nature. The students seem to have great appreciation for the "real" problems on which they work. (Since most of the students are concurrently taking calculus and chemistry, they seem particularly happy to do something "concrete!") The course fits in well with the department's scheme of putting a required engineering course into each semester in the four-year curriculum. By doing this, we are able to reduce the number of students who become disenchanted with engineering early on, because they are not granted the opportunity to do engineering while loading up on basic math and science courses.

A battle we continually fight with these first year students is weaning them of the "right answer" mentality developed through high school. Many of these students are very uncomfortable dealing with problems that have multiple solutions. In addition, students tend not to be used to thoroughly documenting the process they went through to arrive at the particular solution. Our approach in combating these problems has been to give assignments early in the semester that require solutions to open-ended problems, but apply simple concepts in which all students are familiar. For example a typical problem might require an estimate of how many miles one has to walk in order to cut the grass on a one acre lawn given a certain type of lawn mower. The solution obviously requires a number of assumptions, and depending on the assumptions, the solution changes. Clearly we are most interested in their developing a logical approach to obtaining solutions.

## COURSE ASSESSMENT

In terms of our ABET 2000 accreditation, we have used the course to develop student proficiency in the following areas:

- Proficiency in construction materials.
- Application of mathematical and physical principles.
- Ability to perform civil and environmental engineering design.
- Understanding of professional practice issues.
- Ability to conduct laboratory experiments and to critically analyze and interpret data in the area of materials.
- Ability to communicate effectively both orally and in writing.

This is a large number of outcomes, and in fact it has more outcomes than any other course in the four-year curriculum. We believe we have been reasonably successful in meeting these outcomes based on both instructor assessments and student assessments.

Meeting all these outcomes, however, does not come without some cost to the basic subject matter of materials. First, as the students do not necessarily have much background in basic math or sciences, our coverage, particularly of materials science, can not be as deep as would be found in a junior or even sophomore level course. There is some sacrifice of chemistry and chemical properties of materials in this course. Second, as the students have not had basic courses in mechanics (statics and strength of materials) we must spend time introducing basic concepts of stress and strain, at the expense of more materials coverage. However, as the broad scope of the course extends beyond the title subject, we accept the assumption that there is an overall net gain in the undergraduate's engineering experience.

## CONCLUSIONS

We have found the topic of materials in general, and concrete in particular, as taught in a first year, first semester civil engineering materials course to be excellent vehicles to teach a broad array of engineering issues to students just starting their careers. Through simple analysis and design problems we can introduce elementary engineering concepts, and develop a variety of analytical tools that the students will apply throughout their college and engineering careers.

## REFERENCES

1. National Research Council, *Engineering Education: Designing an Adaptive System*, National Academy Press, Washington, D.C., 1995.
2. S. H. Kosmatka and W. C. Panarese, *Design and Control of Concrete Mixtures, 13th Edition*, Portland Cement Association, Skokie, 1988.
3. Course web site: http://www.umeciv.maine.edu/cie111.

**Table 1. Laboratory Sessions in First Year Materials Course**

| Session | Topic | Theme |
|---|---|---|
| 1 | Material Variability | Role of experimentation |
| 2 | MS Excel | Introduction to spreadsheets for engineering data analysis |
| 3 | Plastics | Temperature-dependence, viscoelasticity |
| 4 | Design Lab | Design problem with multiple constraints |
| 5 | Steel and Aluminum Tension | Elastic-plastic behavior |
| 6 | Aggregates | Size gradation, specific gravity, moisture content, unit weight |
| 7 | Concrete Mix Design | Group design problem |
| 8 | Concrete Mix | Concrete placement, Q/C testing, batch yield |
| 9 | 7-Day Concrete Strength | compression testing, NDE |
| 10 | Wood | anisotropy, moisture effects |
| 11 | Ready-mix Plant Tour | How it's done in "real world" |

# The Education Connection: From Research to the Web

## by S. H. Kosmatka

Synopsis:  This paper reviews the opportunities that the Portland Cement Association has taken to address part of the education needs of the cement and concrete industries. Addressed are current educational efforts and a review of how research at universities addresses both the educational and technical needs of the industry. A list of concrete related web sites is included. Through education the concrete industry can meet the need for informed professionals who are necessary to sustain concrete as the building material of choice for this century.

Keywords:  concrete research; education; web sites

Steven H. Kosmatka is the Managing Director of
Research and Technical Services at the Portland
Cement Association. He is a member of ACI committees
123—Research, 225—Hydraulic Cement, 232—Fly Ash, TAC
High-Performance Concrete, Concrete Research Council,
and the Strategic Development Council.

## INTRODUCTION

The concrete industry, including cement companies,
concrete producers, construction companies,
engineering firms, and vendors to the concrete
industry, experienced difficulty finding people
educated and skilled in concrete technology to hire
for positions vacated by retirees or positions
created by the economic boom of the 1990s. Educating
new people to the industry and expanding the skills
of those already in the industry is key to filling
staff vacancies and to maintaining the economic
health of the concrete industry. This paper reviews
some of the efforts of the Portland Cement
Association (PCA) in education and how research, the
internet, and other resources can be combined to meet
current education needs.

## EDUCATIONAL NEED AND VISION

The ultimate result of education is skilled workers
ranging from concrete laborers and design engineers
to executives and instructors. Pre- and Post-high
school education about the concrete industry is
critical to attract new workers. Continuing education
and training of existing workers is essential to keep
pace with industry innovations. New technologies and
practices require adjustment to new work
environments.  Only education and training can bridge
this gap.

One of the top four goals of the Portland Cement
Association's current strategic plan is education.
Other goals include market expansion, advocacy, and
technology and standards. In order for most of these
goals to be successful, education must be properly

implemented. The PCA education goal states that industry personnel, specifiers, constructors, and educators will benefit from programs that provide knowledge and understanding of cement-based products and their use. Objectives for the education goal include:

- Create industry alliances to establish and strengthen educational programs for the cement and concrete industry, including the workforce
- Increase the number of schools teaching and the number of students studying concrete-related courses
- Increase knowledge about concrete design and construction among relevant officials and decision-makers
- Provide industry-specific education for cement industry personnel.

Education is one of the concrete industries goals to be addressed over the next 3 decades. ACI (2001) states that "By 2030, the concrete industry will be seen as a source of safe, well paying, and challenging careers resulting in the creation of a committed, diverse, and skilled workforce. The successful future of the U. S. concrete industry depends greatly upon the industry's ability to attract high-quality, well-trained personnel." But where will the education come from to create "high-quality, well-trained personnel" and in what form?

## RESEARCH

Research and development is critical to the growth of the concrete industry. Research has provided major advances ranging from frost resistant air-entrained concrete of the late 1930s to self-compacting high-performance concrete and 800 MPa reactive powder concrete of today. Without these advances and educational efforts transferring new innovative technologies to the field, concrete would not enjoy the competitive advantages and extensive use currently experienced.

Research not only advances the state of knowledge, it also provides an opportunity to educate young, future professionals for the workforce. Part of the Portland

Cement Association's efforts to educate young professionals in the university setting is shown by PCA funded research at universities performed by professors and their students. The Portland Cement Association is currently funding $1.3 million in research in the following areas at universities:

Soil Cement and Roller-Compacted Concrete
- Performance of Soil Cement, Texas A&M
- Freeze-Thaw Resistance of Roller-Compacted Concrete, Laval University

Engineered Structures and Bridges
- Structural Design Load and Resistance Factors, University of Michigan
- Effect of High Performance Concrete on Corrosion, University of Waterloo
- Effect of Bridge Deck Flexibility on Durability, University of Illinois
- Corrosion of Post-Tensioned Tendons, Penn State
- Curing of High Performance Concrete Bridge Decks, University of Toronto

Cement and Concrete Technology
- Optimum Sulfate Content of Portland Cement, Aberdeen University
- Delayed Ettringite Formation, Purdue University
- Test Methods for Delayed Ettringite Formation, University of Toronto
- Masonry Flexural Strength, Clemson
- Concrete Laboratory Manual, University of Illinois
- Chemical Path of Ettringite, Northwestern University
- Detached Plume in Cement Manufacture, Northwestern University

The Portland Cement Association research at the above universities provides hands on learning about cement and concrete for individuals in a variety of disciplines including materials science, civil engineering, structural engineering, environmental engineering, chemical engineering, industrial engineering, chemistry, geology, ceramics, and physics.

## PCA RESEARCH FELLOWSHIP PROGRAM

The Portland Cement Association had its first research fellowship with the National Bureau of Standards (now National Institute of Standards and Technology) from 1924 to 1965. The purpose was to study the constitution and properties of portland cement. 77 research reports evolved from the program that created a foundation for cement technology in the United States. Many of the results of that program, such as Bogue calculations for estimating compounds in cement, are still used today.

Since 1994 the Portland Cement Association has sponsored a fellowship program at universities to advance cement and concrete technology. Each year, two $20,000 one-year grants are awarded to provide students with a financial resource to study and advance the science and technology of cement and concrete. The research award competition is open to any student completing studies toward a masters or doctoral degree from an institution of higher education accredited by a regional or national agency. The applicant must have been accepted for graduate study in an engineering, science, material science, or architectural program.

Recipients of the Portland Cement Association Research Fellowship and the respective topics include the following (1994 to 2001):

Colorado State University
- Development of a Microwave Inspection Technique for Determining Water-to-Cement Ratio of Fresh Cement Based Materials

Cornell University
- Frost and Scaling Resistance of High Strength Concrete as a Function of Mixture Proportions and Time of Finishing Operations

Georgia Institute of Technology
- Seismic Resistance of Steel Confined High Strength Reinforced Concrete Columns

Michigan State University

- Optimizing the Efficiency of Joints and Cracks in Roller-Compacted Concrete Pavements

Princeton University
- Frost Protection Using Porous Ceramic Shells

Purdue University
- Volume Stability of Concrete Relative to Delayed Ettringite Formation (DEF) and Optimum SO3 Content in Portland Cement

University of Illinois
- Masonry Mortars for Controlled Curing and Performance
- Examination of Strut-and-Tie Methodologies and their Implications on Practice Through the Use of a Computer-Based Design and Analysis Tool

University of Missouri-Columbia
- Optimizing Tertiary Blends for High Performance Concrete in Bridge Applications

University of Texas at Arlington
- Evaluation of Sulfate Resistant Cement for Effective Stabilization of Natural Sulfate Rich Subgrades

University of Texas at El Paso
- Long-Term Performance of Cement-Treated Base Materials

University of Toronto
- Mechanisms of Optimum Sulfate Content of Cement
- Effect of High-Temperature Curing on Durability
- Effects of Mix Proportions and Curing on Concrete Permeability and its Resistance to Ingress of Aggressive Fluids

## EDUCATION AND TRAINING

The Portland Cement Association provides distance learning, cement industry programs, paving programs, and concrete industry programs to anyone interested in expanding their knowledge about cement and concrete. Classes and workshops address topics such as: logistics; mill grinding; kiln process; microscopy of cement and clinker; masonry; concrete pavement construction; principles of concrete; mix design; dispatching; petrography; repair materials and methods; and troubleshooting concrete. Many concrete-related organizations, such as the American Concrete Institute, the American Concrete Pavement Association, and the National Ready Mixed Concrete Association, provide educational classes, workshops, and seminars. Consult association web sites for more information.

## PCA EDUCATION FOUNDATION

In 2001, the Portland Cement Association established the PCA Education Foundation. The objective of the foundation is to fund a variety of educational activities that will increase public knowledge regarding appropriate uses of cement and concrete. It will address technological advancements and standards and try to advance the scientific understanding of cement and concrete. The Foundation will also provide scholarships and grants for the study of engineering and the physical sciences relating to the production and use of cement and concrete.

## CEMENT 101

The Manufacturing Technical Committee of the Portland Cement Association works with twenty universities in Canada and the United States to develop awareness of careers in cement manufacturing. In an effort to reach more people, the Committee worked directly with *Cement Americas* magazine to create a supplement titled *Cement 101: A Guide to Careers in the Cement Industry*. *Cement 101* is designed to create an awareness of and interest in the cement industry for students. It shows students the many different disciplines needed in the cement industry and the kinds of research performed by the industry. *Cement 101* provides an introduction to the production of cementitious products; a look at careers and career paths in the cement industry; profiles of both cement companies and equipment and service suppliers making a difference in their communities and for their employees; and a resource and reference guide that will direct professors and candidates to companies and organizations they may want to contact regarding employment opportunities in the cement industry. More information can be obtained at www.cementcareers.org and www.cementamericas.com.

### CENTER FOR ADVANCED CEMENT BASED MATERIALS

The Portland Cement Association participates in the Center for Advanced Cement Based Materials (ACBM) which consists of facilities and resources at Northwestern University, University of Illinois, University of Michigan, Purdue University, and the National Institute of Standards and Technology.

ACBM research currently addresses: Development of CKD-Slag Blended Cements for Durable Concrete; Interactions of Chemical and Mineral Admixtures in Fresh Cement Paste; Early Age Crack Resistance of Concrete; Determining Early Stiffening and Strength Gain by Non-Destructive Techniques for Early Serviceability of Structures; Chloride Transport in Blended Cements; Blended Fiber-Reinforcement for Improving Cracking Resistance of Concrete Pavements, Slabs, Bridge Decks, and Industrial Floors.

ACBM was originally created as a National Science Foundation Center in 1989 and then became a solely industry supported organization in 2000. Individuals from numerous disciplines and numerous countries have been involved with ACBM. 91 graduating students and 40 professors have prepared over 200 research reports on the science of cement and concrete through the ACBM program.

### FACULTY WORKSHOPS AND SEMINARS

Faculty workshops have recently become a popular way to educate instructors. They are offered by many organizations. ACBM and PCA both cosponsor Faculty Enhancement Workshops each year. The program focuses on educating the instructor on the principles of cement and concrete technology. Many colleges no longer provide materials courses addressing concrete. This program is an attempt to turn that tide by giving instructors an opportunity to feel comfortable with concrete technology and it provides tools that can be used in a classroom setting.

A Professors Seminar on Buildings is sponsored by PCA, the Concrete Reinforcing Steel Institute, Precast/Prestressed Concrete Institute (PCI), and the National Ready Mixed Concrete Association. A Professors Seminar on Bridges is sponsored by PCA and PCI. Hundreds of professors and instructors from throughout North America have participated in these valuable programs.

## VIRTUAL CEMENT AND CONCRETE

Computer modeling provides an excellent opportunity for individuals to "experiment" with cement and concrete. The Virtual Cement and Concrete Testing Laboratory (VCCTL) developed by the Building Materials Division of the National Institute of Standards and Technology is an excellent example of how computer technology can be used as a self-teaching tool about cement systems. The Web-based virtual laboratory can be used for evaluating and optimizing cement-based materials. Substantial savings in time, materials, labor, and money can be achieved by reducing the number of physical concrete tests performed. The core of the virtual lab is a computer model for the hydration and microstructure development of cement-based systems that is based on 11 years of research at NIST. The virtual lab can be accessed at http://ciks.cbt.nist.gov/vcctl/.

## WEB SITE RESOURCES

The internet provides a great opportunity for individuals to learn about the concrete industry. Distance learning via computer provides specific education for individuals anywhere in the world. The internet can certainly be a first start in the search for knowledge about cement and concrete. The author is biased toward the following web sites: www.aci-int.org, www.portcement.org, and www.cementcareers.org. Following is a partial list of cement and concrete oriented web sites:

Cement Associations

British Cement Association--www.bca.org.uk/
Bundesverband der Deutschen Zementindustrie (BDZ)--www.bdzement.de/
Cement Association of Canada--www.cpca.ca/cpca/cpca.nsf
Cembureau (European Cement Association)--www.cembureau.be/
Cement and Concrete Association of Australia--aqua.civag.unimelb.edu.au/
Cement and Concrete Institute of South Africa--www.cnci.org.za
Cement Manufacturers' Association (India)--www.cementindia.com/
Federacion Interamericana del Cemento--www.ficem.org
Instituto Mexicano del Cemento y del Concreto A.C.--www.imcyc.com
Portland Cement Association--www.portcement.org/
South African Cement and Concrete Institute--www.cnci.org.za
Verein Deutscher Zementwerke (VDZ)--www.vdz-online.de/

Concrete Associations/Organizations

American Association of State Highway Officials--www.aashto.org
American Ceramic Society--www.acers.org
American Coal Ash Association--www.ACAA-USA.org
American Concrete Institute (ACI)--www.aci-int.org
American Concrete Pavement Association (ACPA)--www.pavement.com
American Concrete Pipe Association--www.concrete-pipe.org
American Concrete Pressure Pipe Association--www.acppa.org
American Concrete Pumping Association--concretepumping.com/acpa/
American Shotcrete Association--www.shotcrete.org
American Society of Concrete Contractors--www.ascconc.org

American Society of Engineering Education (ASEE)--
www.asee.org/
American Society for Testing and Materials--
www.astm.org
American Society of Civil Engineers--www.asce.org
American Underground Construction Association--
www.auca.org
Architectural Precast Association--
www.archprecast.org
Building Officials and Code Administrators
International--www.bocai.org
Building Research Establishment (U.K.)--
www.bre.co.uk/
Canadian Engineering Network (CEN)--
www.transenco.com/
Canadian Society for Civil Engineering--www.csce.ca
Cast Stone Institute--www.caststone.org
Cement Careers--www.cementcareers.org
Center for Transportation Research and Education--
www.ctre.iastate.edu/
Civil Engineering Research Foundation--www.cerf.org/
Concrete Foundations Association--
www.concreteworld.com/cfa
Concrete Homes--www.concretehomes.com/
Concrete Reinforcing Steel Institute--www.crsi.org
Concrete Sawing and Drilling Association--
www.csda.org
Construction Innovation Forum--www.cif.org
Construction Specifications Institute--www.csinet.org
Council on Tall Buildings and Urban Habitat--
www.lehigh.edu/~inctbuh/inctbuh.html
Council for Masonry Research--www.masonryresearch.org
Environmental Council of Concrete Organizations--
www.ecco.org
European Concrete--www.europeanconcrete.com
Expanded Shale, Clay & Slate Institute--www.escsi.org
High-Performance Concrete Network of Centres of
Excellence (Canada)--www.usherb.ca/concrete
HITEC (Highway Innovative Technology Evaluation
Center)--www.cenet.org/hitec
Institute for Research in Construction (Canada)--
www.nrc.ca/irc
Institute of Electrical and Electronics Engineers
(IEEE)--www.ieee.org/
Insulating Concrete Forms Association--www.forms.org
Intelligent Transportation Systems (tour of DOT's)--
www.itsonline.com/dot_onl.html

Interlocking Concrete Pavement Institute--
www.icpi.org/ICPI
International Cement Microscopy Association--
www.cemmicro.org
International Center for Aggregates Research--
www.ce.utexas.edu/org/icar/
International Code Council--www.intlcode.org/
International Concrete Repair Institute--www.icri.org
International Ferrocement Information Center (IFIC)--
www.ait.ac.th/
International Ferrocement Society--
www.ferrocement.org
International Masonry Institute--
www.imiweb.org/imihome.htm
International Standards Organization (ISO)--
www.iso.ch/
Japan Concrete Institute--www.jci-
net.or.jp/index_e.html
Masonry Advisory Council Online--www.maconline.org
Masonry Heating Home Page--mha-net.org
The Masonry Society--www.masonrysociety.org
Materials Research Society--www.mrs.org/
Midwest Concrete Consortium--
www.ctre.iastate.edu/mcc/
National Stone, Sand, and Gravel Association--
www.nssga.org/index.shtml
National Association of Corrosion Engineers--
www.nace.org
National Association of Home Builders--www.nahb.com
National Concrete Masonry Association--www.ncma.org/
National Mining Association--www.nma.org
National Precast Concrete Association--
www.precast.org
National Ready Mixed Concrete Association--
www.nrmca.org
National Slag Association--www/nationalslagassoc.org
National Spa & Pool Institute--www.nspi.org
National Terrazzo & Mosaic Association Inc.--
www.ntma.com
Perlite Institute Inc.--www.perlite.org
Post-Tensioning Institute--www.pti-usa.org
Precast/Prestressed Concrete Institute (PCI)--
www.pci.org
Silica Fume Association--www.silicafume.org
Tilt-Up Concrete Association--www.tilt-up.org/
U.S. Committee on Large Dams (USCOLD)--
www2.private1.com/~uscold

Valtion Teknillinen Tutkimuskeskus (VTT)--www.vtt.fi/
Vermiculite Association--www.vermiculite.org
Virginia Transportation Research Council--
www.vdot.state.va.us/vtrc/
Wire Reinforcement Institute--www.bright.net/~wwri/

Government Sites

Construction Metrication Council--
www.nibs.org/cmcnews.htm
High Performance Construction Materials and Systems--
titan.cbt.nist.gov/
Highway TechNet--www.ota.fhwa.dot.gov
Library of Congress--www.loc.gov
Long Term Pavement Performance --
www.ltppdatabase.com/main.htm
National Institute of Building Sciences--www.nibs.org
National Institute of Standards and Technology--
www.nist.gov
National Technical Information Service (NTIS)--
www.ntis.gov
National Transportation Agency of Canada--
www.ncf.carleton.ca/
Ontario Ministry of Transportation--www.mto.gov.on.ca
State Transportation Web Sites--
www.fhwa.dot.gov/webstate.htm
SUCCEED Engineering Visual Database (NSF)--
succeed.ee.vt.edu/index.html
Technology Transfer Information Center (NIST)--
www.nal.usda.gov/ttic/
Transport Canada--www.governmentsource.com
Transportation Research Board--www.nas.edu/trb/
     State DOT's--
     www.nas.edu/trb/directory/states.html
     Research in Progress--
     http://nationalacademies.org/trb/trip
Transportation Research Information Service Database-
-tris.amti.com/search.cfm
Turner Fairbank Highway Research Center--
www.tfhrc.gov
U.S. Army Corps of Engineers Information--
www.usace.army.mil/
U.S. Army Engineer Waterways Experiment Station--
www.wes.army.mil
REMR Database--www.wes.army.mil/REMR/remr.html
Handbook for Concrete and Cement--
www.ws.army.mil/SL/MTC/handbook/handbook.htm

U.S. Bureau of Mines--www.usbm.gov
U.S. Bureau of Reclamation Materials Engineering
Research--www.usbr.gov/merl
U.S. Patent and Trademark Office--
www.uspto.gov/patft/index.html
Virtual Cement and Concrete Testing Laboratory--
http://ciks.cbt.nist.gov/vcctl/

## Journals

Advances in Cement Research-- www.ice.org.uk/journals
Aggregates Manager--
www.aggman.com/Pages/archivesearch.html
CENews--www.cenews.com/edconcmain.html
Cement and Concrete Composites-- www.elsevier.nl
Cement and Concrete Research-- www.elsevier.nl
Concrete Construction--www.worldofconcrete.com
Concrete Producer--www.worldofconcrete.com
Concrete Network--www.concretenetwork.com
Magazine of Concrete Research--
www.ice.org.uk/journals
Nordic Concrete Research--www.itn.is/ncr
Roads and Bridges Online-- www.roadsbridges.com

## University Sites

Advanced Cement-Based Materials (at Northwestern U)--
www.civil.nwu.edu/ACBM
Center for By-Products Utilization (U of Milwaukee)--
www.uwm.edu/Dept/CBU/
Infrastructure Technology Institute (at Northwestern
University)--iti.acns.nwu.edu
Northwestern Univ. Transportation Sources--
www.library.nwu.edu/transportation/
Princeton University Transportation Resources--
dragon.princeton.edu/~dhb/
Texas Transportation Institute (Texas A&M
University)--tti.tamu.edu
University of California Berkeley--
www.lib.berkeley.edu/ITSL/transpub.html
University of Illinois College of Engineering--
www.engr.uiuc.edu/

Thousands of additional concrete-related web sites
are available on the internet. They can be found by
searching on one of many search providers such as:

```
AltaVista--www.altavista.com
Ask Jeeves--www.askjeeves.com
Dogpile--www.dogpile.com
Excite--www.excite.com
FAST Search--www.alltheweb.com
Google--www.google.com
HotBot--www.hotbot.com
Mining Co--www.miningco.com
Northern Light--www.northernlight.com
Yahoo--www.yahoo.com
GOVBOT (Government)--ciir2.cs.umass.edu/Govbot
DOTBOT (Transportation)--search.bts.gov/ntl/
```

**SUMMARY**

The demand for education is great in the concrete
industry. This paper provides a glimpse of resources
and opportunities available. By connecting
educational needs with research needs and other
resources, the concrete industry can meet the need to
have informed professionals who are necessary to
sustain concrete as the choice building material of
this century.

**REFERENCES**

ACI, *Vision 2030: The U.S. Concrete Industry*,
American Concrete Institute, Farmington Hills,
Michigan, January 2001, 32 pages.

SP 206—34

# Improving Effectiveness of Research in the Classroom Through Teamwork

## by D. C. Jansen

Synopsis: In most curricula, students have numerous opportunities to work as part of a team, but they are seldom instructed on how to function as part of a team, a valuable skill in the corporate environment. As part of a larger initiative to develop five fundamental skills, Tufts University's School of Engineering has implemented a program to introduce teamwork skills to all engineering students. This program is designed to develop good habits for functioning as part of a team. In context of the work being presented, the team working skills were introduced into a sophomore level civil engineering materials course (including concrete, of course!). Students were given two lectures followed by laboratory exercises to emphasize the teamwork concept of defining and working towards a common goal and not being so driven by the task. The students then practiced their skills throughout the course by functioning as teams in all their laboratory exercises and report writing. Through proper functioning teams, cooperative learning is promoted, and students learn the material better and more efficiently. Team performances were periodically assessed. Overviews of the School of Engineering's five fundamental skills program and teamwork initiative are presented. Outcomes from the teamwork program incorporated into the civil engineering materials course were assessed at several stages during the class and are reported.

Keywords: cooperative learning; education; teaching; teamwork

**Daniel C. Jansen** is currently an assistant professor in the Department of Civil and Environmental Engineering at Tufts University, Medford, MA. He teaches courses in civil engineering materials, concrete properties, and reinforced concrete design. His areas of research include compressive behavior of concrete, fracture properties of concrete, quality control testing, high performance concrete, and use of reclaimed materials in concrete. He was a co-recipient of the 2001 ACI Wason Medal for Materials Research and is a member of ACI Committee 363 High Strength Concrete.

## INTRODUCTION

Working as part of a team, whether there are two members or 20, is something nearly everyone experiences during their lifetime. Teams can be two people joined in marriage for eternity or a large team assigned to produce a new product design in a day, but if the teams do not practice proper teamwork, they are apt to function less efficiently. While some people naturally develop good teamwork skills, many people do not, and nearly everyone can benefit from a combination of training and practice.

To provide necessary practice, students are often required to work in teams on projects or homeworks with the assumption that they will naturally develop teamwork skills on their own. However, good teamwork skills are not developed efficiently, or in some cases not at all, without being taught like any other skill.[1] As an added bonus, working in teams, when instituted properly, promotes cooperative learning, CL (1-3). Under the paradigm of CL, students work together on homework or projects and learn from one another.

As part of a larger initiative to develop five primary skills for engineering students at Tufts University, teamwork skill training was incorporated into a sophomore level civil engineering materials course for the first time in the Spring semester of 2001. These are skills that the students will continue to use and develop throughout their academic careers and will hopefully bring with them into their everyday lives and professional careers.

## OVERVIEW OF FIVE SKILL WORKSHOPS

The school of Engineering integrated the professional skill workshops into its undergraduate curriculum in response to a survey of engineering alumni and industry professionals. Five professional skill sets were identified that would greatly benefit graduating engineers as they transition from engineering school and adapt to the workplace. Young engineers historically have learned these basic skills on the job with varying success. This program gives students specific learning experiences to develop the skills through workshops, lectures and exercises.

The workshops are integrated into selected regular engineering courses. Freshmen have Presentation Skills Workshops in the Fall semester learning

PowerPoint and how to give oral presentations; each student makes timed presentations that are videotaped and critiqued. In the Spring semester freshmen have Prototype Building Workshops with instructions on tool and prototype building to develop and practice three-dimensional building skills. Sophomores receive Team Skill Workshops with lectures, exercises and team assignments. Juniors have Informational Retrieval Workshops to learn research skills using traditional and electronic research techniques followed by a research assignment. Seniors receive Leadership Skill Workshops with lectures and seminar discussions on leadership in the workplace and in life; these workshops are integrated into their Senior Capstone courses that involve team projects and already have similar workshops incorporated into them.

To ensure that students from all engineering disciplines receive similar training, the Dean of Engineering coordinated the efforts with each department. All first year students take the same two introduction to engineering courses, so the first two skill sets were incorporated into those courses. Since sophomores, juniors, and seniors have no courses which overlap all disciplines, each department selected one course that all students from each discipline take, and the workshops were incorporated into those courses. Creation of the workshops were coordinated and assisted by the Dean of Engineering's office.

Although the students have demanding courses, many view the professional workshops as a welcome alternative to the typical engineering lecture format. Many seniors who work in project teams with specific goals and short deadlines are stressed and are quite eager to learn team management and leadership techniques to help them with their task at hand.

Team Skill Workshops for Sophomore Engineering Students

A single sophomore course was identified in each of the School of Engineering's Departments to integrate the team skill workshops. The goal was to reach all sophomore engineering students and to encourage the instructors to use team-based learning in the course.

The School of Engineering retains the services of an experienced professional corporate trainer and consultant who designs and leads the Team Skill and Leadership workshops. Part of his work is to help Tufts faculty incorporate team learning in their classes so they will eventually be able to conduct the workshops themselves. The corporate trainer is assisted by a Tufts alumnus from a local company, the Director of the Engineering Project Development Center (EPDC) from the Dean of Engineering's office at Tufts University, as well as the course instructor.

The model is to have two or three team skill workshop sessions in place of or in addition to course lectures, depending on the course instructor's preference. The workshops are tailored to each course and its use of student teams as well as what stage the teams are in. Each consists of an interactive discussion on teamwork followed by individual and team group exercises to demonstrate the efficiency of teams and the skills needed to be effective team members.  Most courses incorporate teamwork for labs, quizzes, written assignments, or projects.  The workshops were incorporated into *"Thermodynamics"* for Mechanical Engineering, *"Introduction to Digital Logical Circuits"* for Electrical Engineering and Computer Science, *"Thermodynamics and Process Calculations"* for Chemical Engineering, and *"Civil Engineering Materials and Measurements"* for Civil and Environmental Engineering.  An in-depth discussion of teamwork in *"Civil Engineering Materials and Measurements"* follows.

## TEAM SKILLS IN CIVIL ENGINEERING MATERIALS COURSE

Teamwork skills was introduced into *CEE-2 Civil Engineering Materials and Measurements* at Tufts University for the first time in the Spring semester of 2001.  The author was the instructor, and has taught the course for five consecutive years (Spring semesters 1997 – 2001).

Course Overview

All sophomores in the Department of Civil and Environmental Engineering at Tufts University take *CEE-2 Civil Engineering Materials and Measurements*.  It is a one credit course with three hours of lecture and a 2 ½ hour lab each week.  Students do not receive additional credit for the laboratory, so typically the course has a heavy workload in comparison to other one credit courses.  The course covers:
- Introduction to Civil Engineering & materials
- Purpose for testing materials for Civil Engineering applications
- Variability of materials (introduction to probability and statistics)
- Metals
- Aggregates
- Portland Cement
- Concrete
- Timber
- Time dependent strains (creep, shrinkage, thermal)
- Failure criteria

The students in the class (typically 35 to 40) all attend the same lectures, but are divided into four laboratory sections with no more than 12 students in a section.

Laboratory Sessions - - The laboratory sessions, given in Table 1, are designed to emphasize concepts (for example, Experiment 1 demonstrates the variability of the tensile strength of wire), to give students the opportunity to perform standard tests (for example, in Experiment 6 while mixing concrete, students run air content, slump, and unit weight tests on fresh concrete), and to develop engineering report writing skills.  In the Spring 2001, labs were held every week except weeks 1, 6, 10, 14, and 16 only because not all sections met those weeks due to the semester schedule (weeks 1 and 16) or holidays (weeks 6, 10, and 14). The students have always worked in groups of two to four while running tests and gathering data, however, each student was responsible for writing the laboratory reports from 1997 to 2000; however, in 2001, in conjunction with the teamwork concept, each team submitted a single report. Typically there have been eight written reports, six of which were one to two pages of text with two to five figures and tables, and two full reports which would be about eight pages of text and 10 or more tables and figures. The longer reports would typically be from the later labs where they have more time to work on the reports since there are weeks without labs.

Philosophy of Learning Through Writing - - The author (instructor) believes the best way for students to develop and improve writing skills is through practice. Writing reports helps the students to learn to express numbers in meaningful ways by having them present results in tables, graphical format, or contained within written sentences. Writing also forces the students to actually think about what they did in lab (by making them write the procedures in their own words) and to think about the numbers which they obtained (by having them explain the results in words rather than just have them calculating a number and putting a box around it). In general, engineering students do not like writing, because they find it difficult, but this is the very reason to force them to get practice. So the author feels report writing is fundamental to the understanding of the course material and developing as a better engineer. It is for these reasons that in previous years with CEE-2, each student had to write his/her own lab report. In the past, the reports usually have a large variation in quality due to the amount of work put forth by the student; students who put more time and effort into the reports turned in ones of higher quality. The better reports would result in better grades for the course, so the students would see direct results of their hard work.

Incorporation of Team Skills

The teamwork skills were incorporated into CEE-2 for the first time in the Spring 2001 semester. The professional corporate trainer gave two lectures, on the Mondays of weeks three and 11: 1) introduction to teamwork and 2) team assessment, feedback, and conflict resolution. Lab sessions those weeks had exercises to help the students practice, develop and understand the teamwork concepts. There was an additional 15 minute oral review and evaluation of teamwork during the last lab session in week 15 performed by the Director of the EPDC. Descriptions of the two lectures and accompanying labs are given below.

Introduction to Teamwork Lecture and Lab Session - - This lecture began by defining what makes a group a team. The corporate trainer, who gave the lecture, defined it as:

*People working together in a committed way to achieve a common goal or mission. A common goal or purpose means people share the responsibility for delivery of the same output whether it be winning the game, creating the vision, solving the problem, or achieving whatever task they have in common. Teams have a common goal or purpose that is significant enough to warrant the members' commitment and there is an interdependence. Being on a team requires a person to operate as a team member which may be different from his/her behavior as an individual contributor. A high performance team will have agreements, and effective group process, skills and practices which facilitate its performance.*

Furthermore, the lecture described why and how teamwork can be more efficient than working alone, if the team functions correctly. The corporate trainer also emphasized that teams must examine the process before starting in on the task, apparently a significant problem with engineers since they are problem solvers and want to get to solving the problem right away. Examples and in class demonstrations emphasized these points. The Tufts Alumnus, whose company has hired the same corporate trainer to help his teams function more effectively, interjected with experiences of managing and working with teams at his own company.

The laboratory session consisted of several exercises to drive home the importance of teamwork, how properly functioning teams work better than individuals, how to set up teams and delegate responsibilities, and the students created written ground rules for their expectations of one another. The ground rules included whether they would expect one another to show up to meetings on time, how they would contact one another (e-mail or by phone), and what their expectations were for the course (did they just want to get through the course or were they shooting for an 'A'). Effective listening was taught by having one student on a team talking for two minutes on why they wanted to be Civil Engineers and the team members listening would use rephrasing or clarifying techniques to help their listening and to let the speaker know that they are paying attention. To practice organization, listening, and establishing process instead of jumping into the task, the entire lab section (approximately 10 students), the students were required to determine how long a fictitious ancient culture took to build an obelisk. Each student was given three cards, two with a vital piece of information and one with a useless fact, and they had to verbally share their information with their teammates and gather it to determine how long the culture took to build the obelisk. So each student had some piece of knowledge to contribute and the others had to listen while they added their information.

Team Assessment, Feedback, and Conflict Resolution - - The second lecture, a little more than half way through the semester, focused on assessing how their team was functioning, how to provide constructive feedback, and how to resolve conflicts within their teams. The laboratory session included reviewing their written ground rules to see how well they maintained expectations and followed the rules they established for themselves. They also had to state what each team member's biggest strength was and what each member could most improve; this was an exercise in how to provide feedback to team members. As an exercise for the entire lab section, each student was given a list of 15 items which they had available as their boat was sinking 2000 miles from the nearest point of land. Each student individually prioritized the list, then they worked together as a team to construct a single list. As compared to the obelisk exercise, the lost at sea exercise did not require everyone's input, so structuring the team and listening to each member's opinion was more difficult although just as important. It also re-emphasized the team's need to focus on process before jumping into the task.

Reflection - - In the very last lab, each team met with the Director of the EPDC and discussed how they liked working in teams and how they felt the lectures and exercises helped them accomplish their lab experiments and reports. These discussions served two purposes: 1) for us to see what worked and what might need improvement in following years, and 2) so the students would reflect back on the importance of good teamwork skills and how the skills helped them function as a better team.

Grading and Reorganization

Since all members of a team received identical grades on the lab reports, worth 30% of their total grade, it would be unfair if one member of a team did not contribute equally to the reports; this was the primary reason the author had every student hand in individual reports in past years. In addition, students perform better when there is some sense of individual accountability when working in a cooperative learning environment (3).

To account for individual contribution and ability to function in a team, 10% of the overall course grade was based on contribution to their team. To determine each individual's contribution, each student was asked to evaluate him or herself and his/her team members on several key points: amount of work, quality of work, keeping to agreed upon ground rules, and overall contribution. Finally they were each assigned to distribute $100 among their team (including themselves). These evaluations took place mid-way through the semester and at the end of the semester. If a student did poorly in their reviews and received less than $100 total, then they would get less than the full 10%. A very poor student would get 0 out of the 10% which would effectively lower their course grade by a full letter grade. On the other hand, students who received over $100 total because they did much more than their fair share, then they would still only receive the complete 10%, effectively penalizing them for not working well as a team and taking additional responsibilities instead of trying to get the other students to contribute evenly. Others describe more effective tools for peer rating which do not rely as heavily on students assigning numbers to their teammates, but instead use more descriptive assessments (3,4). The instructor can then translate the descriptive ratings into points to adjust student grades.

Midway through the semester (week 9), the teams were re-organized. This to give the students opportunities to learn to work with more individuals, to prevent a few students in one team from being penalized by being stuck with one or two weak members throughout the semester, and so the evaluations could be more fairly assessed.

Assessment

Instructor's Perspective - - From the instructor's perspective, there were many advantages to having students work in teams and the advantages were strengthened by good team working skills. The reports that the students were turning in were of much higher quality than ones which had been done in the past by individual students. When students were working in teams of four, it reduced the number of reports which needed to be graded from 40 down to 10. This was a considerable time saver, and it also allowed the grader to spend more time carefully examining each report and to provide considerably more constructive feedback. The students also benefit from producing thorough reports of higher quality instead of sloppy and hurried reports. In fact, the reports had much more in depth analysis and the students put more thought into their analyses. The interaction of working on reports as a team provided important discussion which helped them develop thoughts and ideas which an individual would not be able to develop on one's own. Also, as Vygotsky stated, "a child exhibits increased and demonstrable mental capacity and ability when assisted by more skilled peers," which means the weaker students were able to learn more from the stronger students. Lastly, poorer students were able to see the quality of work and the amount of effort of strong students.

Student's Perspective - - From the students' written evaluations of teamwork and from the Director of the EPDC's discussion with the teams in the last lab, a number of students mentioned that when they could not remember how to perform a calculation, they could freely ask a team member for help in cases where they would ordinarily be reluctant to seek help from the instructor. If the entire team could not figure out the problem, then they would feel justified seeking help from the instructor. The students also stated that they developed close relationships with their team members. Finally, many mentioned they enjoyed working together more than working alone, making the entire course more enjoyable.

Student Evaluations of Course and Instructor - - The major surprise came with the student evaluations of the course and instructor. Table 2 shows the average student ratings on four questions over the last five years the author has been the instructor for the course. The student ratings are on a scale of 1 to 5 where 1 is poor, 2 is below average, 3 is average, 4 is above average, and 5 is excellent. In the four years he has taught the course without teamwork, the average instructor evaluation (question 8) was a 3.7. With teamwork this jumped remarkably to a 4.6. In comparison to equivalently sized courses in the Department of Civil and Environmental Engineering at Tufts University over the same five year period, a score of 3.7 places the instructor in the bottom 26% while the 4.6 is in the top 13%. Likewise the course evaluations (question 15) improved from an average of 3.6 (1997 – 2000) to a 4.4.  A 3.6 corresponds to the bottom $34^{th}$ percentile while the 4.4 is in the top $15^{th}$ percentile.  Some of this remarkable improvement could be attributed to the fact that the students perceived the workload to be considerably less in 2001 than in previous years, as can be seen from their responses to question 19 as given in Table 2; however, in 1997 when the students response to amount of work was similar to 2001, there was not a corresponding increase in the instructor or course evaluations.  It should be pointed out that they perceived the amount they learned to be slightly higher than in all previous years, although the amount is statistically insignificant.

A few of the choice comments that students provided in the course evaluations included some of these wonderful compliments:

"I really enjoyed this course, and I didn't expect to."

"This was a great first exposure to civil engineering.  It made me realize that I had chosen the right major."

And the words which can bring tears to the eyes of almost any Civil Engineer:

"Concrete is the <u>coolest</u>."

## CONCLUSIONS

Introducing teamwork skills can benefit laboratory or project based courses on many levels.  Through cooperative learning, students actually learn more information, they learn better report writing skills from one another, and they enjoy the work more than working alone.  The fact that they develop skills to function as part of a team is a bonus which will help them throughout their academic and professional careers.

## REFERENCES

1.  Seat, E., and Lord, S. M., "Enabling Effective Engineering Teams: A Program for Teaching Interaction Skills," *Journal of Engineering Education*, Vol. 88, No. 4, Oct. 1999, pp. 385-390.
2.  Felder, R. M., and Brent, R., "Cooperative Learning in Technical Courses: Procedures, Pitfalls, and Payoffs," ERIC Document Reproduction Service http://www.ncsu.edu/effective_teaching/Papers/Coopreport.html, Report ED 377038, 1994, accessed June 11, 2001.
3.  Kaufman, D. B.; Felder, R. M., and Fuller, H., "Accounting for Individual Effort in Cooperative Learning Teams," *Journal of Engineering Education*, Vol. 89, No. 2, April 2000, pp. 133-140.
4.  Felder, R. M., and Brent, R., "Effective Strategies for Cooperative Learning," *Journal of Cooperation and Collaboration in College Teaching*, Vol. 10, No. 2, Spring 2001, pp. 69-75.

**Table 1. Schedule of Laboratory Sessions (2001)**

| Week | Laboratory |
|------|-----------|
| 1 | No Lab – Classes begin Wednesday |
| 2 | Experiment 1: Statistical Analysis of Wire in Tension |
| 3 | **Teamwork Exercises** |
| 4 | Experiment 2: Elastic Properties of Aluminum and Steel |
| 5 | Experiment 3: Metals (3 Types of Steel, Aluminum, and Cast Iron); Complete Tensile Stress-Strain Curve, Brinell Hardness, and Charpy Impact |
| 6 | No Lab – Holiday on Monday (President's Day) |
| 7 | Experiment 4: Aggregates (Fine and Coarse Aggregates); Gradation, Bulk Specific Gravity, and Absorption Tests |
| 8 | Experiment 5: Evaluation of Portland Cements (Type I and III); Cube Strength and Fineness |
| 9 | Experiment 6: Casting Concrete and Determination of Fresh Properties of Concrete<br>Experiment 5 Continued |
| 10 | No Lab – Spring Break |
| 11 | **Teamwork Exercises**<br>Experiment 5 Continued |
| 12 | Experiment 5 Completed |
| 13 | Experiment 7 (Continuation of Experiment 6): Testing of Hardened Properties of Concrete; Compressive Strength, Modulus of Elasticity, Modulus of Rupture, and Splitting Tensile Strength |
| 14 | No Lab – Holiday on Monday (Patriot's Day) |
| 15 | Experiment 8: Properties of Timber (Pine and Redwood); Compression, Tension, and Flexure |
| 16 | No Lab – Classes finish on Monday (rest of week is reading period and Exams) |

**Table 2. Summary of CEE-2 Materials and Measurements Course Evaluations**

| Year | Instructor Evaluation (Question 8) | Course Evaluation (Question 15) | Amount Learned (Question 18) | Amount of Work (Question 19) |
|------|------|------|------|------|
| 1997 | 3.8 | 3.4 | 3.6 | 4.1 |
| 1998 | 3.4 | 3.5 | 3.6 | 3.8 |
| 1999 | 3.9 | 3.7 | 3.6 | 3.3 |
| 2000 | 3.8 | 3.9 | 3.8 | 3.8 |
| Average 1997-2000 | 3.7 ± 0.2 | 3.6 ± 0.2 | 3.65 ± 0.1 | 3.75 ± 0.3 |
| **2001*** | **4.6** | **4.4** | **3.9** | **3.2** |

*Teamwork skills introduced

SP 206-35

# Engineering Mechanics and Structural Materials Research in the Twenty-First Century

## by K. P. Chong

**Synopsis:** Mechanics and materials are essential elements in all of the transcendent technologies in the twenty first century and in the New Economy. These transcendent technologies include nanotechnology, microelectronics, information technology and biotechnology. Research opportunities and challenges in theoretical and applied mechanics as well as engineering materials, including cement-based materials, in the exciting information age are presented and discussed.

Keywords: condition asssessment; designer materials; durability; information technology; multi-scales; nanotechnolgy; research needs

**KEN P. CHONG**, F.ASCE, Director of Mechanics & Materials at NSF where he formulates and administers the U.S. policy and research, educational programs in solid mechanics and engineering materials. In FY 2002 he is on sabbatical leave as an Embassy Fellow at the US Embassy in Bern, Switzerland, and a Visiting Scientist at NIST. He has published over 150 refereed technical papers and several books on mechanics and structures; edits an Elsevier journal and a book series; recipient of the ASCE Edmund Friedman Award.

## INTRODUCTION

The National Science Foundation (NSF) has supported basic research in engineering and the sciences in the United States for a half century and it is expected to continue this mandate through the next century. As a consequence the United States is likely to continue to dominate vital markets because diligent funding of basic research confers a preferential economic advantage (1). Concurrently over this past half century, technologies have been the major drivers of the U. S. economy, and as well, NSF has been a major supporter of these technological developments. According to the former NSF Director for Engineering, Eugene Wong, there are three *transcendental* technologies:

- Microelectronics – Moore's Law: doubling the capabilities every two years for the last 30 years; unlimited scalability; nanotechnology is essential to continue the miniaturization process.
- Information Technology [IT] – NSF and DARPA started the Internet revolution about three decades ago; confluence of computing and communications.
- Biotechnology – molecular secrets of life with advanced computational tools as well as advances in biological engineering, biology, chemistry, physics, engineering including mechanics and materials.

By promoting research and development at critical points where these technological areas intersect, NSF can foster major developments in engineering. The solid mechanics and materials engineering (M&M) communities will be well served if some specific linkages or alignments are made toward these technologies. Some thoughtful examples for the M&M communities are:

| | |
|---|---|
| • Bio-mechanics/materials | • Simulations/modeling |
| • Thin-film mechanics/materials | • MEMS |
| • Wave Propagation; NDT | • Smart materials/structures |
| • Nano-mechanics/materials | • Designer materials |
| • Scale effects | • Virtual modeling and testing |

Designer materials, Virtual modeling and testing , NDT, smart materials, smart structures and others have important applications in cement and concrete. Considerable NSF resources and funding will be available to support basic research related to these technologies. These opportunities will be available for the individual investigator, teams, small groups and larger interdisciplinary groups of investigators. Nevertheless, most of the funding at NSF will continue to support unsolicited individual investigator proposals on innovative "blue sky" ideas.

## NANOTECHNOLOGY

Initiated by the author, with the organization and help of researchers from Brown [K. S. Kim, et al], Stanford, Princeton and other universities, a NSF Workshop on Nano- and Micro-Mechanics of Solids for Emerging Science and Technology was held at Stanford in October 1999. The following is extracted from the Workshop Executive Summary. Recent developments in science have advanced capabilities to fabricate and control material systems on the scale of nanometers, bringing problems of material behavior on the nanometer scale into the domain of engineering. Immediate applications of nanostructures and nano-devices include quantum electronic devices, bio-surgical instruments, micro-electrical sensors, functionally graded materials [including civil and structural materials], and many others with great promise for commercialization. The branch of mechanics research in this emerging field can be termed nano- and micro-mechanics of materials. A particularly challenging aspect of fostering research in the nano- and micro-mechanics of materials is its highly cross-disciplinary character. Important studies of relevance to the area have been initiated in many different branches of science and engineering. A subset of these, which is both scientifically rich and technologically significant, has mechanics of solids as a distinct and unifying theme. It was also revealed, however, that the study of complex behavior of materials on the nanometer scale is in its infancy. More basic research, which is well coordinated and which capitalizes on progress being made in other disciplines, is needed if this potential for impact is to be realized. In addition to the expected benefit to the target areas, such research invariably advances other technologies, conventional or emerging, through a spill-over effect; this serendipitous benefit can also be anticipated from focused mechanics research in nano technology.

Recognizing that this area of nanotechnology is in its infancy, substantial basic research must be done to establish an engineering science base; this link between the discoveries of basic science and the design of commercial devices must be completed to realize the potential of this area. Such a commitment to nano-and micro-mechanics will lead to a strong foundation of understanding and confidence underlying this technology based on capabilities in modeling and experiment embodying a high degree of rigor. The potential of various concepts

in nanotechnology will be enhanced, in particular, by exploring the nano- and micro-mechanics of coupled phenomena and of multi-scale phenomena. Examples of coupled phenomena discussed in this workshop include modification of quantum states of materials caused by mechanical strains, ferroelectric transformations induced by electric field and mechanical stresses, chemical reaction processes biased by mechanical stresses, and change of bio-molecular conformality of proteins caused by environmental mechanical strain rates. Multi-scale phenomena arise in situations where properties of materials to be exploited in applications at a certain size scale are controlled by physical processes occurring on a size scale which are orders of magnitude smaller. Important problems of this kind arise, for example, in thermo-mechanical behavior of thin-film nano-structures, evolution of surface as well as bulk nano-structures caused by various material defects, nano-indentation, nano-tribological response of solids, and failure processes of MEMS structures. Details of this workshop report can be found in:
[ http://en732c.engin.brown.edu/nsfreport.html ].

Coordinated by M. Roco, NSF recently announced a program [NSF 00-119; see: www.nsf.gov] on collaborative research in the area of nanoscale science and engineering. The goal of this program is to catalyze synergistic science and engineering research in emerging areas of nanoscale science and technology, including: biosystems at nanoscale; nanoscale structures, novel phenomena, and quantum control; device and system architecture; design tools and nanosystems specific software; nanoscale processes in the environment; multi-scale, multi-phenomena modeling and simulation at the nanoscale; and studies on societal implications of nanoscale science and engineering. This program or initiative provides support for: Nanoscale Interdisciplinary Research Teams (NIRT – one has been awarded to the UC-Riverside, Stanford and U. of Illinois –Urbana team on nanomechanics), Nanoscale Science and Engineering Centers (NSEC), and Nanoscale Exploratory Research (NER). Key research areas have been identified in advanced materials, nanobiotechnology (e.g. nano-photosynthesis), nanoelectronics, advanced healthcare, environmental improvement, efficient energy conversion and storage, space exploration, economical transportation, and bionanosensors. The National Nanotechnology Initiative (NNI - published on February 7, 2000 and is available on www.nano.gov) will ensure that investments in this area are made in a coordinated and timely manner (including participating federal agencies – NSF, DOD, DOE, EPA and others), and will accelerate the pace of revolutionary discoveries now occurring . In addition, individual investigator research in nanoscale science and engineering will continue to be supported in the relevant Programs and Divisions outside of this initiative.

## CEMENT-BASED AND OTHER MATERIALS

The National Science Foundation, through its engineering programs in Civil and Mechanical Systems Division, in addition to the Engineering Research Centers (e.g. at Lehigh University) and the Science and Technology Centers (NSF Center for Advanced Cement-Based Materials at Northwestern University), supported research in concrete materials and structures. Other strongholds include the NIST Building and Fire Research Laboratory in which the Inorganic Building Materials Group is operating the Virtual Cement and Concrete Testing Laboratory [VCCTL]. Virtual testing can be performed with variables such as cement particle sizes and composition, mineral admixtures, temperature and moisture in curing, and different aggregates (2).

Some of the NSF supported projects are listed below:

* Durability of Concrete
* Low temperature Behavior
* Mathematical Modeling of Concrete Creep and Shrinkage
* Self-healing Concrete; Smart Concrete
* Research Needs for Concrete Masonry
* Rehabilitation, Renovation and Reconstruction of Bldgs
* New Rib Geometries for Re-bars
* High Strength and/or High Performance Concrete
* Fiber Reinforced Concrete
* Fracture Toughness and Behavior
* Size effects
* Micro-mechanics of Concrete
* Shear in Reinforced Concrete
* Physics and Chemistry of Cement-based Materials
* Behavior of Concrete in Cold Climate
* Fiber Optics (sensors) in Concrete
* Continuous Lightly Reinforced Concrete Joist Systems
* Highway Bridges
* Construction Methods
* Automation/Robotics
* Non-Destructive Evaluation
* Seismic Precast Structures
* Structural Controls
* International Cooperative Initiatives
* Full-Scale Seismic Pseudo Dynamic Testing of Reinforced Concrete

## HIGH PERFORMANCE CONCRETE

More than half of the NSF supported research projects are related to high performance concrete (HPC). One way to define HPC is (3) "Concrete having desired properties and uniformity which cannot be obtained routinely using only conventional constituents and normal mixing, placing, and curing practices." These properties include: high strength, high toughness, durability, enhanced workability, freeze-thaw durability, etc. National programs on HPC exist in US, China, Japan, Canada, Norway, and France. Generally chemical admixtures or fibers can be added in concrete to achieve certain mechanical properties. Concrete admixtures (4) include: high-range water reducers (superplasticizers), retarders, accelerators and air-entraining agents. In the United States, W.R. Grace and Master Builders are the two largest suppliers of the concrete admixtures, accounting for about two-thirds of the U.S. admixture market. A development (4) was on extended-set retarders and activators, which indefinitely keeps fresh concrete from setting (until it is mixed with an activator).

Besides the basic concrete research sponsored by the NSF, other federal programs, such as the Strategic Highway Research Program (SHRP), Ideas Deserving Exploratory Analysis (IDEA) and NIST supports HPC and concrete research (4,5).

## NSF PROJECTS

The following is a list of examples of current NSF research projects in concrete. Highlights of some of these projects will be presented.

Reza Zoughi, U of Missouri Rolla ; Detection and Profiling of Accelerated Chloride Penetration in Concrete Using Near-field Microwave Techniques.

Suru Shah, Northwestern University; Development of Non-Clinker Based Cement for Hazard Reduction.

Kimberly E. Kurtis, Georgia Tech ; POWRE: Examination of the Mechanisms of ASR Gel Expansion Control by Lithium Additive in Concrete.

James K. Morton, CSC Palatine, IL;  SBIR Phase II: Noncorroding Steel Reinforced Concrete.

Sookie S. Bang , Venkataswamy . Ramakrishnan, S. Dakota School of Mines; Application of a Microbial Immobilization Technique in Remediation of Concrete Cracks.

George W. Scherer, Princeton University ; Novel Method for Measuring Permeability of Concrete.

Ronald R. Berliner and  Mark S. Conradi, U of Missouri Columbia;  The Durability of Concrete: The Crystal Chemistry of the Calcium Aluminosulfate Hydrates and Related Compounds.

Hamlin M. Jennings, Northwestern University ; The Effect of Composition on the Colloid Structure of Calcium Silicate Hydrate: Implications for Concrete Durability and a Basis for a Fundamental Understanding of Cement.

Lois Schwarz , U of Arkansas ; CAREER: Roles of Rheology, Chemical & Mechanical Filtration, and Applied Gradient on Injectability of Cement Grouts, Morphology of Grouted Soil, and Improvement in Soil Properties.

The following NSF projects are mostly completed.

* Bazant (Northwestern) "Nonlinear and Probabilistic Theory for Concrete Creep."

* Brock (Kentucky) "Transient Studies of Dislocation-based Micromechanical Effects in Fracture."

* Buyukozturk (MIT) "A Study of Fracture Behavior in High Strength Concrete."

* Darwin/McCabe (Kansas) "Improving Development Characteristic of Reinforcing Bars" (Co-sponsored by Civil Engineering Research Foundation).

* Gopalaratnam (Missouri), Shah (Northwestern), Batson (Clarkson), Criswell (Colorado St.), Ramakrishnam (SDSM&T) and Wecharatana (NJIT) "Fracture Toughness of Fiber Reinforced Concrete (FRC)"

   - study specimen size, loading configuration/rate.
   - evaluate currently used toughness and specifications.
   - provide basis for the use of strength/toughness data.
   - recommend guidelines in specifications.

* Hawkins (Illinois) "Mixed Mode Fracture of Concrete."
   - innovative methods to investigate mixed mode concrete fracture

* Hinze/Holt (Washington) "Expansive Cements"

- mix with water in bore holes, etc.
- use in lieu of explosives
- measure expansive stresses
- geomechanics

* Hover (Cornell) - durable concrete.

  - 60 ksi automatic mercury porosimeters were built,
    have been used in a microsilica-based concrete
    parking garage.

* Hsu (Houston) "Shear in Reinforced Concrete."

  - facilities include giant steel frame with 40 jacks,
    100 ton each

  - test panels up to 55" square, 16" thick.
  - limited in HPC tests.
  - material laws: softening, constitutive
    For "Reinforced Concrete Framed Shearwalls in
    Earthquake." See Reference (6).

* Kangari (Georgia Tech) "Advanced Technologies and
  Materials in Concrete Slip-forming."

* Lee and Chung (SUNY-Buffalo) "Effect of Freezing
  Cycles on the Mechanical Behavior of Concrete."

* Li (Michigan) "Fracture Testing Technology for Advanced
  Ceramic and Cementitious Matrix Based Composites."

* Monteiro (Berkeley) - Micromechanics in Concrete.

* Naaman (Michigan) "High Performance Fiber Reinforced
  Cement Composites:  International Workshops."  (with Reinhardt and
  German DFG).

* Nanni (Penn State), "Aramid Fiber Reinforced Concrete."

* Nawy/Maher (Rutgers) "Fiber Optic Sensors for Strength
  Evaluation and Early Warning of Impending Structural
  Failure."

* Ross (New Mexico) "Dynamic Fracture in Quasi-Brittle
  Materials:  An Experimental Study."

* Sansalone/Pratt (Cornell) "Neural Network in Non-Destructive Testing."
  - impact-echo develop by Carino and Sansalone.
  - automating and simplifying impact-echo signal analysis/ reduction.
  - applicable to HPC.

* Englekirk (Englekirk & Hart) "Concept Developments for Precast Concrete Structural Systems."

* Stanton (Washington) "Connection Classification and Modelling for Precast Seismic Structural Systems."

* Keiser (Illinois) "An Experimental Setup to Investigate Dynamic Torsional-Translational Response of Reinforced Concrete."

* Azizinamini (Nebraska) "Design and Detailing of Transverse Reinforcements for High Strength Reinforced Concrete Columns Subjected to Seismic Loading."

* Wood (Illinois) "Experimental Investigation of the Strength Stiffness and Deformation Capacity of Slender Reinforced Concrete Walls."

* Wight (Illinois) "Earthquake Type Loading on R/C Beam-Column Connections:  Special Cases of Wide Beams and Eccentric Beams."

* Moehle (Berkeley) "Seismic Resistance and Retrofit of Post-Tensioned Flat-Plate Floors."

* Eberhard/Sozen (Illinois) "Experiments and Analysis to Study the Seismic Response of Reinforced Concrete Frame-wall Structures with Yielding Columns."

NSF CENTER FOR ADVANCED CEMENT-BASED MATERIALS [ACBM]: started by Principal Investigator (Suru Shah) as a STC in 1989, currently with strong support from industry and other sources. It consists of four universities and NIST:

| | |
|---|---|
| Northwestern (Shah) | Michigan |
| Illinois | NIST |
| Purdue | |

<u>Original Areas of Investigation of ACBM</u>
* Processing procedures (control of initial/final structure).
* Microstructure (porosity, damage, interfaces)
* Bulk Properties (fracture mech., durability, FRC)
* Modeling (generic, interfacing, simulation, database)
* Design of New Materials (improved properties through material science)

## RELATED INITIATIVES

The initiative *Engineering Sciences for Modeling and Simulation-Based Life-Cycle Engineering* (Program Announcement NSF 99-56) is a three-year collaborative research program by NSF and the Sandia National Laboratories (Sandia) focusing on advancing the fundamental knowledge needed to support advanced computer simulations. This collaborative initiative capitalizes on the missions of both organizations. NSF's mission is to advance the fundamental science and engineering base of the United States. Sandia has the responsibility to provide solutions to a wide range of engineering problems pertinent to national security and other national issues. It is moving toward engineering processes in which decisions are based heavily on computational simulations [see e.g. Ref. 3]; thus, capitalizing on the available high performance computing platforms. This initiative has sought modeling and simulation advances in key engineering focus areas such as thermal sciences, mechanics and design. About 24 awards were made totaling $7m.

The NSF Civil and Mechanical Systems (CMS) Division developed an initiative on *Model-based Simulation* (MBS), see (NSF 00-26). Model-based simulation is a process that integrates physical test equipment with system simulation software in a virtual test environment aimed at dramatically reducing product development time and cost. This initiative will impact many civil/mechanical areas: "structural, geotechnicial, materials, mechanics, surface science, and natural hazards (e.g., earthquake, wind, tsunami, flooding and land-slides)." MBS would involve "combining numerical methods such as finite element and finite difference methods, together with statistical methods and reliability, heuristics, stochastic processes, etc., all combined using super-computer systems to enable simulations, visualizations, and virtual testing." Expected results could be fewer physical testing, or at best, better strategically planned physical testing in the conduct of R&D. Examples of the use of MBS in research, design and development exist in the atmospheric sciences, biological sciences, and the aerospace, automotive and defense industries. The manufacturing of the prototype Boeing 777 aircraft, for example, was based on computer-aided design and simulation.

In the future one should expect the continued introduction of bold innovative research initiatives related to important national agenda such as the nano-technology, environment, civil and mechanical infrastructure.

## CHALLENGES

The challenge to the mechanics and materials research communities is: How can we contribute to these broad-base and diverse research agenda? Although the mainstay of research funding will support the traditional programs for the foreseeable future, considerable research funding will be directed toward addressing these research initiatives of national focuses. At the request of the author, a NSF research workshop has been organized by F. Moon of Cornell University to look into the research needs and challenges facing the mechanics communities.

Mechanics and materials engineering are really two sides of a coin, closely integrated and related. For the last decade this cooperative effort of the M&M Program has resulted in better understanding and design of materials and structures across all physical scales, even though the seamless and realistic modeling of different scales from nano-level to system integration-level (Fig. 1) is not yet attainable. In the past, engineers and material scientists have been involved extensively with the characterization of given materials. With the availability of advanced computing, and new developments in material sciences, researchers can now characterize processes, design and manufacture materials with desirable performance and properties in cement-based and other materials. One of the challenges is to model short-term micro-scale material behavior, through meso-scale and macro-scale behavior into long term structural systems performance. Accelerated tests to simulate various environmental forces and impacts are needed [NSF 98-42]. Supercomputers and/or workstations used in parallel are useful tools to solve this scaling problem by taking into account the large number of variables and unknowns to project micro-behavior into infrastructure systems performance, and to model or extrapolate short term test results into long term life-cycle behavior (8,9). Using elaborate synergistic combination of databases, models and computational tools (2) virtual testing is a viable research tool in designer materials with optimized desirable properties.

| MATERIALS | | STRUCTURES | INFRASTRUCTURE | |
|---|---|---|---|---|
| nano-level $(10^{-9})$ | micro-level $(10^{-6})$ | meso-level $(10^{-3})$ | macro-level $(10^{+0})$ | systems-level $(10^{+3})$ m |
| *Molecular Scale* | *Microns* | | *Meters* | *Up to Km Scale* |
| *nanomehanics *self-assembly *nanofabrication | *micromechanics *microstructures * smart materials | *meso-mechanics *interfacial-structures *composites | *beams *columns *plates | *bridges * lifelines *airports |

*Fig. 1. Physical scales in materials and structural systems* (10).

## ACKNOWLEDGMENTS AND FEEDBACK

The author would like to thank his colleagues and many members of the research communities for their comments and inputs during the writing of this opinion paper. He would appreciate further feedback from the research communities at large. Information on NSF initiatives, announcements and awards can be found in the NSF website:[www.nsf.gov].

## DISCLOSURE AND COPYRIGHT POLICY

## REFERENCES

1. Wong, E. "An Economic Case for Basic Research", *Nature*, Vol. 381, pp.187-188, May. (1996).

2. Garboczi, E. J.; Bentz, D. P.;and Frohnsdorff, G. F. "Knowledge-Based Systems and Computational Tools for Concrete," *Concrete International*, V. 22, No. 12, Dec. 2000, pp. 24-27.

3. N.J. Carino and J.R. Clifton, "Outline of a National Plan on High-Performance Concrete:  Report on the NIST/ACI Workshop, May 16-18, 1990," *NISTIR 4465*, National Institute of Standards and Technology, USA, Dec., 1990.

4. D. B. Rosenbaum, "Concrete Admixtures," *Engineering News Record*, pp. 35-40, Jan. 21, 1991.

5. 5. "Nationally Coordinated Program of Highway Research, Development, and Technology-Annual Progress Report  FY 1990," *FHWA-RD-90-109*, Federal Highway Administration, USA.

6. 6.  T. Hsu and S.T. Mau, "International Workshop on Concrete Shear in Earthquake," University of Houston, USA,  January 1991.

7. Boresi, A. P., Chong, K. P. and Saigal, S., *Approximate Solution Methods in Engineering Mechanics*, John Wiley, New York (2001).

8. NSF "Long Term Durability of Materials and Structures: Modeling and Accelerated Techniques", *NSF 98-42*, National Science Foundation, Arlington, VA. (1998)

9. Chong, K. P. "Smart Structures Research in the U.S.", *Keynote paper, Proc. NATO Adv. Res. Workshop on Smart Structures*, held in Pultusk, Poland, 6/98, *Smart Structures,* Kluwer Publ. (1999) pp. 37-44.

10. Boresi, A. P. and Chong, K. P. *Elasticity in Engineering Mechanics*, John Wiley, New York (2000).

APPENDIX A

## Current and Former Masters and Doctorate Students, Post Doctoral Fellows, and Visiting Researchers

Juergen Adolphs, POROTEC GmbH, Germany

Shuaib Ahmad, American Concrete Institute International, Michigan, USA

Yilmaz Akkaya, Istanbul Technical University, Turkey

Corina Aldea, Bayex, Canada

Manuel Alvarado

A. Anandereja

Farhad Ansari, University of Illinois at Chicago, Illinois, USA

Matthew Aquino

Peter Babaian, Simpson Gumpertz & Heger Inc., Massachusetts, USA

Perumalsamy Balaguru, Rutgers University, New Jersey, USA

Robert Ballarini, Case Western Reserve University, Ohio, USA

Emilie Becq-Giraudon, Chicago Dept. of Transportation - Bridges, Illinois, USA

Lanny Betterman, Consulting Engineer, Minnesota, USA

Joseph Biernacki, Tennessee Technological University, Tennessee, USA

Jose Umberto Arnaud Borges, University of Sao Paulo, Brazil

Van Bui, Northwestern University, Illinois, USA

Penelope Burke, Christopher B. Burke Engineering Ltd., Illinois, USA

Alberto Castro-Montero, Consultant, Texas, USA

Sushil Chandra, Bechtel, California, USA

Li Liang Chang

Ralph Chapman, John Brown E&C, USA

Hung-Liang Chen, West Virginia University, West Virginia, USA

Yit-Jin Chen

Cheng Jang Chen, British Columbia, Canada

Karthik Chermakani, Microsoft, Washington, USA

Sokhwan Choi, Kookmin University, Korea

Lan Chung, Dankook University, Korea

Richard Claxton, Marquette University, Wisconsin, USA

Kathryn A. Cox

Michele Cyr, Northwestern University, Illinois, USA

Avraham Dancygier, Technion-Israel Inst. of Technology, Israel

J.K. Dattatreya, Structural Engineering Research Center, India

Dale DeFord, James Hardie, California, USA

M. Divikar, Consultant, California, USA

Katja Dombrowski, Bauhaus-University Weimar, Germany

Herbert Duda, GermanyGill Roy Eapen, Deloitte & Touche Consulting, USA

Thomas Easley, General Electric, Ohio, USA

Frank Ensslen, Ruhr-Universität Bochum, Germany

APPENDIX A

## Current and Former Masters and Doctorate Students, Post Doctoral Fellows, and Visiting Researchers

Apostolis Fafitis, Arizona State University, Arizona, USA
S. Frondistou, Consultant, Greece
Sanjiv Garg
Nicole Gersomke, Germany
Ravindra Gettu, Universitat Politecnica de Catalunya, Spain
Masoud Ghandehari, Polytechnic University of New York, New York, USA
Odd Gjorv, The Norwegian Inst. of Technology NTH, Norway
Ulker Gokoz, Enka Construction Company, Turkey
Vellore Gopalaratnam, University of Missouri-Columbia, Missouri, USA
Roman Gromotka, Krupp Uhde GmbH, Germany
Miroslaw Grzybowski, Royal Institute of Technology, Sweden
Antonio Guerra, A.J. Guerra & Associates, Dominican Republic
Kadir Güler, Technical University of Istanbul, Turkey
Jan Hamm, Swiss School of Engineering for the Wood Industry, Germany
Sanjay Jaiswal, Consultant, Illinois, USA
Daniel Jansen, Tufts University, Massachusetts, USA
Yeou-Shang Jenq, Columbus Engineering Consulting, Inc., Ohio, USA
Zongliang Jia
Da-Hua Jiang, China (deceased)
Reji John, US Air Force Research Laboratory, Ohio, USA
Mustafa Karaguler, I.T.U. Mimarlik Fakultesi, Turkey
William Key
Hungare Kim
Katharina Klemt, Technical University of Darmstadt, Germany
George Koliarakis
Maria Konsta-Gdoutos, Democritus University of Thrace, Greece
Osamu Kontani, Kajima Corporation, Japan
Krishna Kulkarni, Caterpillar, Illinois, USA
Joseph Labuz, University of Minnesota, Minnesota, USA
Eric Landis, University of Maine, Maine, USA
David Lange, University of Illinois at Urbana-Champaign, Illinois, USA
John Lawler, Wiss, Janney, Elstner Associates, Illinois, USA
Byeong-Cheol Lho, Sangji University, Korea
Shuh-Huei Li
Zongjin Li, The Hong Kong University of Science and Technology, China
Walter Libardi, Universidade Federal de Sao Carlos, Brazil
Daryl Lee Logan
Dharmawan Ludirdja, National Ready Mixed Concrete Association, Maryland, USA

APPENDIX A
## Current and Former Masters and Doctorate Students, Post Doctoral Fellows, and Visiting Researchers

Arup Maji, University of New Mexico, New Mexico, USA
Tim Malonn, Technical University of Braunschweig, Germany
Shashi Marikunte, Southern Illinois University, Illinois, USA
Ilena Maria Metzner, Switzerland
Richard Miller, The University of Cincinnati, Ohio, USA
Kenro Mitsui, Takenaka Corporation, Japan
Barzin Mobasher, Arizona State University, Arizona, USA
Jill Kathleen Morrison
Monsango Moukwa, Master Builders, Ohio, USA
Bin Mu, Northwestern University, Illinois, USA
Munwam Muzo, Construction Technology Laboratories, Illinois, USA
Antoine Naaman, University of Michigan, Michigan, USA
Moncef Nehdi, University of Western Ontario, Canada
Shigeru Niiseki, Tohoku University, Japan
Eiji Nouchi, Nihon University, Japan
Chengsheng Ouyang, Iowa Department of Transportation, Iowa, USA
Turgay Öztürk, Darmstadt University of Technology, Germany
Antonia Pacios-Alvarez, Universidad Politécnica de Madrid, Spain
Ranga Palaniswamy, Bechtel Savannah River Inc., Georgia, USA
Alva Peled, Ben-Gurion University, Israel
Pariya Phuaksuk, Structural Engineer, Thailand
Jeffrey Picka, University of Maryland, Maryland, USA
Publio Pintado, University of Sevilla, Spain
Michael Pistilli, Prairie Group, Illinois, USA
L. Pogula
John S. Popovics, The University of Illinois at Urbana-Champaign, Illinois, USA
Qing Qin
Vijay Rangan, Curtin University of Technology, Australia
Julie Rapoport, Northwestern University, Illinois, USA
Pierre Rossi, Laboratoire Central des Ponts et Chaussees, France
Thomas Rowe, Wiss, Janney, Elstner, & Associates, Illinois, USA
Carlo V. Ramos Royo
Wan Rui, China
Aaron Saak, General Electric, Ohio, USA
Ramamurthy Sankar, Madras, India
Manasit Sarigaphuti, Siam Cement Co., Ltd., Thailand
Yoshiaki Sato, Oita University, Japan
Angelika Schiessl, Techical University of Munich, Germany

APPENDIX A
Current and Former Masters and Doctorate Students,
Post Doctoral Fellows, and Visiting Researchers

Scott Selleck, Pratt & Whitney, Connecticut, USA
Yixin Shao, McGill University, Canada
H.F. Shaw
Stanley Shitote, MOI University, Kenya
Russell Smith
Shantharama Somayaji, California State University at San Luis Obispo,
      California, USA
Ramesh Srinivasan, Tek Systems, Illinois, USA
Henrik Stang, Danish Technical University, Denmark
Andrea Steinwedel, Germany
Patricia Styer
Wimal Suaris, University of Miami, Florida, USA
Kolluru Subramaniam, New York City College, New York, USA
Joseph Gee Sun
Junji Takagi, Shimiju, Japan
Tianxi Tang, J. Ray McDermott, Texas, USA
Mehemet Tasdemir, I.T.U. Insaat Facultesi, Istanbul, Turkey
Karl-Christian Thienel, Liapor GmbH & Co. KG, Germany
Ilker Bekir Topcu, Osmangazi University, Turkey
Gilberto Jesus Velazco
Tom E. Virding
Karine Vittuari, LCPC, France
Thomas Voigt, Northwestern University, Illinois, USA
Kejin Wang, Iowa State University, Iowa, USA
Ming L. Wang, University of Illinois at Chicago, Illinois, USA
Pao-Tsan Wang
Methi Wecharatana, New Jersey Institute of Technology, New Jersey, USA
Jason Weiss, Purdue University, Indiana, USA
Karl Wiegrink, Technical University of Munich, Germany
Katie Wierman, Northwestern University, Illinois, USA
Tina Wilhelm, Darmstadt University of Technology, Germany
Ekkehard Wollrab, Germany
Hung-fu Xin
Pan Xuewan, China
Chung-Chia Yang, National Taiwan Ocean University, Taiwan
Wei Yang, Simmcode, Co., Massachusetts, USA
Sang Chun Yoon, Housing Research Institute of KNHC, Korea

APPENDIX A
Current and Former Masters and Doctorate Students,
Post Doctoral Fellows, and Visiting Researchers

Dong-Jin Yoon, Korea Research Institute of Standards and Science (KRISS), Korea
Orwin Paul Youngquist, Illinois, USA
Yingshu Yuan, CUMT, China
Davide Zampini, MBT (Schweiz) AG, Switzerland
Steve Zimmerman, Northwestern University, Illinois, USA
Michele Zulli, Università dell'Aquila, Italy

# SURENDRA P. SHAH

| | |
|---|---|
| **Education** | B.E., B.V.M. College, Bombay |
| | M.S., Lehigh University |
| | Ph.D., Cornell University |

## Academic Experience

Present Position:   Walter P. Murphy Professor of Civil Engineering, Northwestern University

Prior Positions:   Professor of Civil Engineering, Department of Materials Engineering, University of Illinois at Chicago

Visiting Associate Professor, Department of Civil Engineering, Massachusetts Institute of Technology

## Administrative Experience

Present Position:   Director, Center for Advanced Cement-Based Materials

Prior Positions:   Director, Center for Concrete and Geomaterials, Northwestern University

Coordinator, Graduate Program Structural Engineering, Department of Civil Engineering, Northwestern University

Director, Graduate Program, Department of Materials Engineering, University of Illinois, Chicago

## Industrial Experience

Present:   Short-term consultant for several industrial companies in U.S. and abroad

Prior Positions:   Research Consultant, Elborg Technology Company, Norway
Research Consultant, AKZO-ARMAK Laboratories, Netherlands
Research Consultant, Amoco Chemical Corporation
Research Consultant, The Science Museum of Virginia, Richmond, Virginia
Research Consultant, U. S. G., Des Plaines, Illinois
Research Consultant, Wiss, Janney, and Elstner, Northbrook, Illinois
Research Consultant, Holderbank Management, Ltd., Switzerland
Research Consultant, Corning Glass Works, Corning, New York
Research Engineer, Portland Cement Association, Skokie, Illinois
Design Engineer, Modjeski and Masters, Harrisburg, Pennsylvania

## HONORS AND OTHER PROFESSIONAL ACTIVITIES

The Richard J. Carroll Memorial Lectureship, Johns Hopkins University, 2001
American Concrete Institute, Arthur R. Anderson Award to ACBM Center
Civil Engineering Research Foundation Charles Pankow Award for Innovation (Collaborative Work with W.R. Grace and ARCO)
Engineering-News Record (ENR) Newsmaker Award
Swedish Concrete Award
Anderson Award, American Concrete Institute

ASTM Sanford E. Thompson Award
RILEM Gold Medal Award
Fellow, American Concrete Institute
Fellow, RILEM
Teaching Excellence Award, CE students
Distinguished Visiting Professor, National University of Singapore
Alexander von Humboldt Senior Visiting Scientist Award
NATO Visiting Senior Scientist to Turkey
Guest Professor, Denmark Technical University
Guest Professor, Delft University of Technology, Delft, The Netherlands
Consultant to NATO Science for Stability Program
Visiting Professor, University of Sydney
NATO Visiting Senior Scientist to France
Member of the Evaluation Team of Danish Research Groups in the field of Concrete
UNIDO Consultant to People's Republic of China
UNESCO Expert to India
Member, Editorial Board, ASCE Journal of Civil Engineering Materials
Member, Editorial Board, Journal of Ferrocement
Member, Editorial Board, RILEM Journal of Materials and Structures
Editor-in-Chief, *Concrete Science and Engineering*

**Keynote Speaker**
Advances in Civil Engineering, IIT, Khagpine, India, January 2002
Italian Fracture Workshop, Brescia, Italy, November 2001
Ferrocement 7, Singapore, July 2001
Concrete Under Severe Conditions, CONSEC IV, Vancouver, British Columbia, June 2001
American Society of Civil Engineers, Baltimore, MD, March 2001
"Early-Age Cracking of High Performance Concrete," University of Houston, April 2001
University of Chile, Santiago, Chile, April 2001
"Advances in Concrete Through Interdisciplinary Research," Gent University, Belgium, May 2001
High Performance Concrete, Hong Kong, December 2000
Cement and Concrete Technology in 2000, Istanbul, Turkey, September 2000
Conference on Nondestructive Testing, Torres, Brazil, November 1999
Euromat Conference, "Fibre Reinforced Concrete," Munich, Germany, September 1999
Creating with Concrete Conference, University of Dundee, Scotland, September 1999
12 International Conference on Composite Materials, "Advanced Cement-Based Composites," Paris, France, July 1999.
8$^{th}$ DBMC, Vancouver, Canada, May-June 1999
First International Symposium on SnFRC, Orlando, Florida, February 1998
Sixth International Symposium on Ferrocement, University of Michigan, June 1998
Fifth International Conference on Failure, Durability & Retrofiltering, Singapore, November 1997
U.S. - Japan Workshop on Civil Infrastructure, Honolulu, August 1997
Engineering Foundation Conference on High Performance Concrete, Hawaii, July 1997
TCDC Workshop on "Advances in High Performance Concrete Technology and its Applications," Madras, India, April 1997
U.S. - India Workshop on NDT of Concrete, Roorkie, India, December 1996
IUPAC/Chemrawn IX , Seoul, South Korea, September 1996
Concrete in the Service of Mankind, Dundee, Scotland, June 1996.
Fourth International Symposium on Utilization of High-Strength/High Performance Concrete, Paris, France, May 1996
Concrete Under Severe Conditions (CONSEC '95), Sapporo, Japan, August 1995
Fourth Congress on Mechanics, Xanthi, Greece 1995.
XIth European Ready Mixed Concrete Congress & Exhibition, Istanbul, Turkey, June 1995
Tenth ASCE Engineering Mechanics Specialty Conference, Boulder, Colorado, May 1995.
ASCE Mechanics Conferences, Boulder, Colorado, May 1995

Industry-University Conference on Fiber Reinforced Concrete, Toronto, March 1995
ASCE Materials Conference, San Diego, November 1994
RILEM Workshop on Technology transfer, Barcelona, Spain, November 1994
IUTAM Conference, Torino, Italy, September 1994
Lectureship Program, National Science Council, Republic of China, December 1993
Symposium on Utilization of High-Strength Concrete, Lillehammer, Norway, June 1993
Conference on Micromechanics of Concrete and Cementitious Composites, Lausanne,
    Switzerland, March 1993
ASME Conference, Anaheim, California, November 1992
RILEM Conference on Interface in Cementitious Composites, Toulouse, October 1992
Concrete Technology of the Future, RILEM Workshop, Finland, September 1992
FRAMCO Conference, Colorado, June 1992
Spanish-French Meeting on Fracture Mechanics, Costa-Brava, Spain, April 1992
50th Anniversary of Swedish Cement and Concrete Institute, Stockholm, June 1992
Advances in Cement-Based Composites for Construction Industry, SAMPE Symposium, Tokyo,
    Japan, December 1991
Symposium on Fatigue in Steel and Concrete Structures, Madras, India, December 1991
Noordwijk Conference on Disordered Materials, Holland, June 1991
First Canadian University - Industry Workshop on Fibre Reinforced Concrete, Quebec, October
    1991
Damage and Diagnosis of Materials and Structures, Torino, Italy, September 1991
Fiber Reinforced Cementitious Materials, Materials Research Society Symposium, November
    1990
Workshop on Inorganic Bonded Wood Composites, Idaho, October 1990
International Conference on Micromechanics of Failure of Quasi-brittle Materials, Albuquerque,
    June 1990
Materials Research Society Symposium on Fiber Reinforced Cementitious Materials, December
    1990
International Conference on Fracture of Concrete and Rock, Cardiff, 1989.
Workshop on Fracture Toughness and Fracture Energy, Tohoku University, Sendai, Japan,
    October 1988
International Conference on Fracture and Damage of Concrete and Rocks, Vienna, July 1988
Materials, Research Society Symposium on Bonding in Cementitious Composites, Boston,
    December 1987
International Symposium on Fiber Reinforced Concrete, Madras, India, December 1987
40th Anniversary, RILEM Conference on Materials Science and Materials Engineering
    Versailles, France, September 1987
RILEM Conference on Fracture of Concrete and Rock, Houston, June 1987
International Conference on Urban Shelter in Developing Countries, London, September 1987
Council on Tall Buildings, Chicago, January 1986
International Conference on Natural Fiber Reinforced Concrete, Iraq, October 1986
PCI Conference on Durability of Glass Fiber Reinforced Concrete Panels, Chicago, November
    1985
Materials Research Society, Annual Meeting, Boston, December 1985
International Conference on Ferrocement, Bangkok, January 1985
Sixth International Conference on Fracture, New Delhi, December 1984
RILEM Conference on Multiaxial Loading of Concrete, Toulouse, May 1984
7th International Conference on Structural Mechanics in Reactor Technology, SMiRT-7,
    Chicago, August 1983
RILEM-CEB-IABSE-IASS Symposium on Concrete Structures under Impact and Impulsive
    Loading, Berlin, June 1982
International Conference on Constitutive Laws for Engineering Materials, Tucson, Arizona,
    January 1982
UN Conference on Modernization of Concrete Construction, Madras, India, January 1982
Annual Conference of the Materials Research Society, Boston 1980
NATO Advanced Research Institute, Paris, November 1980
Caribbean Conference on Concrete Structures, Santo Domingo, July 1979

Scandinavian Conference on Fiber Reinforced Concrete, Stockholm, June 1977
Pan American Structural Conference, Caracas, Venezuela, 1974

## TECHNICAL COMMITTEES

Member, Bureau, RILEM (2002-)
Chair, Advisory Committee, Engineering Mechanics Division, ASCE (1999-2001)
Chairman, Executive Committee, Engineering Mechanics Division, American Society of Civil
    Engineers (1996-1997)
Chairman, Properties of Concrete, Transportation Research Board (1993-96)
Member, National Initiative on High-Performance Concrete (past)
Member, Materials Research Council, American Concrete Institute (past)
Member, Management Advisory Board, RILEM
Member, Advisory Committee on Cement and Concrete, Strategic Transportation Research
    Study (1985-1986)
Chairman, RILEM Committee on Strain-Softening of Concrete
Chairman, RILEM Committee on Fracture of Concrete (1986-1992)
Chairman, Fiber Reinforced Concrete, American Concrete Institute (past)
Vice Chairman, Fracture of Concrete and Rock, Society of Experimental Mechanics
Member, High Strength Concrete, American Concrete Institute
Member, Ferrocement, American Concrete Institute
Member, Fracture Mechanics, American Concrete Institute
Member, Polymer Concrete, American Concrete Institute
Chairman, Fatigue of Concrete Structures, American Concrete Institute (past)
President, Chicago Chapter, American Concrete Institute (1974-75)
Chairman, Properties of Materials, Engineering Mechanics Division, American Society of Civil
    Engineers (1973-75, 1985-87)
Member, Ad Hoc Committee on Ferrocement for Developing Countries, National Academy of
    Sciences (1971-1972)

## CONFERENCES

Co-chair, ACI-RILEM Symposium in Non-Destructive Evaluation, Dallas, October 2001
Chair, Symposium on High Performance Fiber Reinforced Thin Products, ACI, Chicago, March
    1999
Co-Chairman, Symposium on Materials Science of Concrete, ACI, Chicago, March 1999
Co-Chairman, Engineering Foundation Conference, Canada  July, 1998
Co-Chairman, Materials for Infrastructure, Institute of Mechanics and Materials, UCSCD, April
    1998
Co-Chairman, Symposium on Nondestructive Characterization of Materials in Aging Systems,
    Materials Research Society, Boston, December 1997
Co-Chairman, Symposium on Advanced Cement-Based Materials, McNU '97, Evanston, IL
    June 1997
Co-Chairman, Symposium HH: Structure-Property Relationships in Hardened Cement Paste and
    Composites, Materials Research Society, 1996 Fall Meeting, Boston, December 1996
Co-Chairman, Synthesizing Cement-Based Materials for the 21st Century, American Chemical
    Society, National Meeting, Chicago, 1995
Co-Chairman, International Conference Workshop on Technology Transfer of the New Trends in
    Concrete, Barcelona, Spain, November 1994
Co-Chairman, SEM Conference on Nondestructive Testing of Concrete in the Infrastructure,
    Dearborn, Michigan, June 1993
Co-Chairman, ACI Symposium on Materials Science in Concrete, Boston, March 1991
Co-Chairman, ACI Symposium on Fiber Reinforced Concrete, Dallas, November 1991

Co-Chairman, International Conference on Micromechanics of Failure of Quasi-Brittle Materials, Albuquerque, June 6-8, 1990

Chairman, NATO-ARW on Toughening Mechanism of Quasi-Brittle Materials, Northwestern University, July 1990

Co-Chairman, International Conference on Fracture of Concrete and Rock, Cardiff, September 1989

Co-Chairman, Symposium on Bond in Cement Based Composites, Materials Research Society, Boston, December 1987

Co-Chairman, International Conference on Fracture of Concrete and Rock, Society of Experimental Mechanics, Houston, June 1987

Co-Chairman, Symposium on Strain Rate Effects in Cement-Based Composites, Materials Research Society, Boston, December 1985

Member, Organizing Committee, RILEM Symposium on Fracture of Concrete, Laussane, October 1985.

Co-Chairman, NSF-STU Seminar on Steel Fiber Reinforced Concrete, Stockholm, June 1985

Co-Chairman, International Symposium on Ferrocement, Bangkok, January 1985

Chairman, NATO Advanced Research Workshop on Nonlinear Fracture Mechanics, Northwestern University, Evanston, September 1984

Member, Scientific Committee, RILEM-CEB Conference on Multiaxial Loading, Toulouse, April 1984

Chairman, ACI-RILEM Symposium on Fatigue, Detroit, September 1982

Member, Advisory Panel and Chairman of the Session, International Conference on Bond in Concrete, Scotland, June 1982

Co-Chairman, RILEM Symposium on Ferrocement, Bergamo, Italy, July 1981

Chairman, Symposium on Recent Research on Fatigue of Concrete Structures, ACI, Puerto Rico, Sept. 1980; Dallas, February 1981

Chairman, National Science Foundation Sponsored Workshop on High Strength Concrete, 1979

Member, Steering Committee, Gordon Conference on Building Materials, 1973

Chairman, Conference on New Materials in Concrete Construction, University of Illinois at Chicago Circle, Chicago 1971

## PUBLICATIONS

Dr. Shah has published over four hundred papers in journals, proceedings, and books, and has edited twelve books and has co-authored the book Fiber Reinforced Cement Composites, McGraw-Hill, 1992; High Performance Concretes and Applications, Edward Arnold, 1994.; and Fracture Mechanics of Concrete, John Wiley, 1996. He was the Editor-In-Chief of the Elsevier-ACI journal *Advanced Cement Based Materials* and currently is the Editor-in-Chief of the RILEM *Concrete Science and Engineering Journal*. For a full list of Dr. Shah's article publications see the ACBM Website (www.acbm.info).

## Research Awards

*Ultrasonic Technique for Monitoring the Setting and Hardening of Concrete*, Infrastructure Technology Institute, Northwestern University

*Development of Non-Clink Cement for Environmental Hazard Reduction,* National Science Foundation

*Hybrid Fiber Reinforced Composites,* Center for Advanced Cement-Based Materials

*CKD-Slag Blended Cements,* Center for Advanced Cement-Based Materials

*Extruded Fiber Reinforced Concrete Panels for Residential Construction,* National Science Foundation

*Effect of Pressure on Manufactured Cement Board*, Saint-Gobain

*Concrete Reinforced with Cellulose Fibers,* Weyerhaeuser

*Rheology of Cement Matrix for Self-Compacting Concrete,* Center for Advanced Cement-Based Materials

*Durability of Glass Fiber Reinforced Cement-Based Composites*, Nippon Electric Glass Fibers, America

*Ultrafine Fly Ash*, Boral Materials Technology

*High Performance, Non Corroding Steel Reinforced Concrete*, NSF-SBIR

*Injection System Pilot Study*, Hilti Entwicklungsgesellschaft mbH

*Instrumentation and Laboratory Improvement*, National Science Foundation

*Extruded Fiber Reinforced Cement Composites,* Illinois Clean Coal Institute

*General Wall System Specification*, Butler Mfg. Co. Research Center

*ACBM-Howard Joint Research Collaboration*, National Science Foundation

*Studies of Fracture Processes with Computer Vision and Microtomography*, Air Force Office of Scientific Research

*Constitutive Modeling of Concrete*, FAA Center of Excellence for Airport Pavement, University of Illinois

*Microstructure, Transport Property and Statistical Science*, National Institute of Statistical Sciences

Stonecraft, Incorporated, General Research

*Improved Condition Monitoring for Bridge Management*, Infrastructure Technology Institute

*Concrete Research Needs Symposium*, ITT

*Immobilization of Waste in Grout-Main,* Westinghouse Hanford

*The Faculty Enhancement Program*, National Science Foundation

*Simplified Boiling Water Reactor Project*, Department of Energy

*Strain Softening Response of High Strength Concrete*, National Science Foundation

*Rate of Loading Dependency of Reinforced Concrete*, National Science Foundation - Pennsylvania State

*Characterization of Fracture Using Acoustic Emission*, AFOSR

*Removing Barriers to the Increased Use of High Strength Concrete, State of Illinois* (with E. Rossow, F. Young and R. Burg)

NATO-ARW on *Toughening Mechanism of Quasi-Brittle Materials*

State of Illinois Challenge Grant, Illinois Business Partnership Program

*Innovative Infrastructure*, Department of Education (with C. H. Dowding).

*Reinforcement of Concrete with Cellulose Fibers*, Proctor & Gamble

*A Study of Fracture Processes in Concrete Using Laser Holography*, AFOSR

*Mixed-Mode Fracture of Concrete at High Strain Rate*, AFES

ACBM Industrial Affiliate Program

*Shrinkage Reducing Admixture*, ARCO Chemical

Center for Science and Technology of Advanced Cement-Based Materials (ACBM), National Science Foundation (1989-2000); Industrial Consortia (2000-). The ACBM Center is a consortium of five institutions: Northwestern University, University of Illinois, University of Michigan, Purdue University, and the National Institute of Standards and Technology. Dr. Shah is the Principal Investigator (PI) and the Director of the Center.

*Toughness Data in Specification of Fiber Reinforced Concrete*, NSF.

*Modification of the Physico-Chemical Properties of Cement Paste*, AFOSR (with Barbara-Ann Lewis)

*Influence and Specimen Size Loading Configuration Loading Rate and Fiber Type in the Flexural Behavior of Fiber Reinforced Concrete*, Concrete Materials Research Council (with V. S. Gopalaratnam)

*Symposium on Bonding in Cementitious Materials*, AFOSR (with MRS)

*A System for Microscopic Image Analysis for Studying Fracture Toughness and Cement Composites*, NSF Equipment Grant

*High Rate, Closed-Loop Triaxial Testing System for Concrete*, Rock and Soil, DOD Equipment Grant

*Dynamic Response of Embedded Structures*, AFOSR (with L. M. Keer)

*Long-Term Ductility of Glass Fiber Reinforced Concrete Panels*, NSF

*Expansive Cement Induced Fracture and Activity Detection*, U. S. Army Corps of Engineers, Waterways Experiment Station (with C. H. Dowding).

*Symposium on Strain Rate Effects and Fracture in Cement Composites*, AFOSR (with Material Research Society).

Cooperative Research with Denmark Technical University, NATO

*Microstructure, Crack Initiation, Propagation and Localization in Concrete*, AFOSR

*U. S.-Sweden Joint Seminar on Steel Fiber Reinforced Concrete*, NSF

*NATO Advanced Research Workshop on Application of Fracture Mechanics to Cementitious Composites*, NATO

*Post-Peak Tensile Response of Concrete*, U. S. Bureau of Reclamation

*Effect of Grain Size, Strain Rate and Moisture in Rock*, NSF (with C. H. Dowding).

*Strain Rate Effects for Concrete and Fiber Reinforced Concrete*, ARO

*Cyclic Stress-Strain Curves of Confined Concrete*, NSF

*Fracture Process Zone and R-Curves for Cementitious Composites*, NSF

*Fracture Toughness of Fiber Reinforced Concrete*, AFOSR

*Nondestructive Testing of Concrete*, James Electronics

German Government Visiting Scientist Award (DAAD)

*Instrumented Impact Testing System,* University of Illinois Research Board

*Envelope Curves for Confined Concrete,* NSF

*Workshop on High Strength Concrete,* NSF

*Dynamic Properties of Fiber Reinforced Concrete Subjected to Impact Loading,* ARO (with A. Naaman).

*Fracture and Multiple Cracking of Fiber Reinforced Concrete,* NSF (with A. Naaman).

*Cooperative Research with Delft Technical University,* NATO

*Static and Dynamic Properties of Ferrocement,* NSF (with A. Naaman).

*Development of Joint Research with Indian Institute of Science,* NSF

*Materials for Housing Construction in Developing Countries,* University of Illinois Research Board

*Ferrocement Panels,* University of Illinois Research Board

*Triaxial Behavior of Concrete,* NSF

*Environmental Chamber for Concrete Research,* Title VI Equipment Grant

*Nature of Critical Load and the Effect on the Behavior of Concrete Structures,* NSF

# CONVERSION FACTORS—INCH-POUND TO SI (METRIC)*

| To convert from | to | multiply by |
|---|---|---|
| **Length** | | |
| inch | millimeter (mm) | 25.4E† |
| foot | meter (m) | 0.3048E |
| yard | meter (m) | 0.9144E |
| mile (statute) | kilometer (km) | 1.609 |
| **Area** | | |
| square inch | square centimeter ($cm^2$) | 6.451 |
| square foot | square meter ($m^2$) | 0.0929 |
| square yard | square meter ($m^2$) | 0.8361 |
| **Volume (capacity)** | | |
| ounce | cubic centimeter ($cm^3$) | 29.57 |
| gallon | cubic meter ($m^3$)‡ | 0.003785 |
| cubic inch | cubic centimeter ($cm^3$) | 16.4 |
| cubic foot | cubic meter ($m^3$) | 0.02832 |
| cubic yard | cubic meter ($m^3$)‡ | 0.7646 |
| **Force** | | |
| kilogram-force | newton (N) | 9.807 |
| kip-force | newton (N) | 4448 |
| pound-force | newton (N) | 4.448 |
| **Pressure or stress** **(force per area)** | | |
| kilogram-force/square meter | pascal (Pa) | 9.807 |
| kip-force/square inch (ksi) | megapascal (MPa) | 6.895 |
| newton/square meter ($N/m^2$) | pascal (Pa) | 1.000E |
| pound-force/square foot | pascal (Pa) | 47.88 |
| pound-force/square inch (psi) | kilopascal (kPa) | 6.895 |
| **Bending moment or torque** | | |
| inch-pound-force | newton-meter (Nm) | 0.1130 |
| foot-pound-force | newton-meter (Nm) | 1.356 |
| meter-kilogram-force | newton-meter (Nm) | 9.807 |

| To convert from | to | multiply by |
|---|---|---|
| | **Mass** | |
| ounce-mass (avoirdupois) | gram (g) | 28.34 |
| pound-mass (avoirdupois) | kilogram (kg) | 0.4536 |
| ton (metric) | megagram (Mg) | 1.000E |
| ton (short, 2000 lbm) | megagram (Mg) | 0.9072 |
| | **Mass per volume** | |
| pound-mass/cubic foot | kilogram/cubic meter (kg/m$^3$) | 16.02 |
| pound-mass/cubic yard | kilogram/cubic meter (kg/m$^3$) | 0.5933 |
| pound-mass/gallon | kilogram/cubic meter (kg/m$^3$) | 119.8 |
| | **Temperature§** | |
| deg Fahrenheit (F) | deg Celsius (C) | $t_C = (t_F - 32)/1.8$ |
| deg Celsius (C) | deg Fahrenheit (F) | $t_F = 1.8t_C + 32$ |

\* This selected list gives practical conversion factors of units found in concrete technology. The reference source for information on SI units and more exact conversion factors is "Standard for Metric Practice" ASTM E 380.  Symbols of metric units are given in parentheses.

† E indicates that the factor given is exact.

‡ One liter (cubic decimeter) equals 0.001 m$^3$ or 1000 cm$^3$.

§ These equations convert one temperature reading to another and include the necessary scale corrections. To convert a difference in temperature from Fahrenheit to Celsius degrees, divide by 1.8 only, i.e., a change from 70 to 88 F represents a change of 18 F or 18/1.8 = 10 C.

# Index